THE SCIENCE OF COMPOSTING

THE SCIENCE OF COMPOSTING

Eliot Epstein

President, E&A Environmental Consultants, Inc.
Canton, Massachusetts
Adjunct Professor of Public Health
Boston University School of Public Health

CRC Press
Taylor & Francis Group
Boca Raton London New York

CRC Press is an imprint of the
Taylor & Francis Group, an **informa** business

CRC Press
Taylor & Francis Group
6000 Broken Sound Parkway NW, Suite 300
Boca Raton, FL 33487-2742

First issued in paperback 2019

© 1997 by Taylor & Francis Group, LLC
CRC Press is an imprint of Taylor & Francis Group, an Informa business

No claim to original U.S. Government works

ISBN-13: 978-1-56676-478-0 (hbk)
ISBN-13: 978-0-367-40112-2 (pbk)

Library of Congress Cataloging-in-Publication Data
Main entry under title: The Science of Composting

Visit the CRC Web site at www.crcpress.com

Library of Congress Card Number 96-61583

Visit the Taylor & Francis Web site at
http://www.taylorandfrancis.com

and the CRC Press Web site at
http://www.crcpress.com

To my wife Esther
my children Beth, Jonathan, and Lisa
who have provided me with love, guidance and support

A Sense of Humus

To consider *humus* is to get a hint of the oneness of the universe. All flesh is grass, in more than the figurative sense the prophet intended. During the long history of this planet, weather has disintegrated rock, tiny lichens have made a speck of vegetable mold, countless generations of short-lived weeds have waxed fat for summer, giant forests have flourished for an aeon, and all in turn have died and given back to the earth more goodness than they have taken from it. All have been composted into humus. And the life of insects and of animals and of men which was sustained upon the life of these plants and upon the life of other animals, all these creatures too have enriched the surface of the earth with their excreta and finally with their bodies. All in turn have been composted into humus.

BERTH DAMON

Preface   *xiii*

Chapter 1. Composting: A Prospective **1**
Introduction   1
Composting and Recycling   2
History   3
Philosophical Aspects and the Future of Composting
   in the United States   13
Advantages and Disadvantages of Composting   16
Conclusion   16
References   17

Chapter 2. Basic Concepts **19**
Introduction   19
Oxygen and Aeration   22
Moisture   32
Temperature   36
Nutrients   39
pH   48
Summary   50
References   50

Chapter 3. Microbiology **53**
Introduction   53
Microbial Populations   56
Temperature   56

Moisture   70
Nutrients   70
Inoculants   73
Summary   74
References   74

Chapter 4. Biochemistry 77
Introduction   77
Organic Matter   79
Biochemical Manifestations Occurring
    during Composting   83
Biochemical Manifestations Occurring When Compost
    Is Applied to Soil   98
Humus Formation   101
Summary   104
References   104

Chapter 5. Stability, Maturity, and Phytotoxicity 107
Introduction   107
Stability and Maturity   112
Summary   133
References   134

**Chapter 6. Trace Elements, Heavy Metals,
and Micronutrients** 137
Introduction   137
Essentiality and Toxicity   138
Occurrence in the Environment   146
Environmental Consequences   152
Soil-Plant Interactions   155
Summary   163
References   164

Chapter 7. Organic Compounds 171
Introduction   171
Organic Compounds in Various Compost
    Materials and Feedstocks   173
Fate of Organic Compounds during Composting   194
Reactions and Movement of Toxic Organics in Soil   198
Uptake by Plants and Potential Entry
    into the Food Chain   203
Conclusion   206
References   207

Chapter 8. Pathogens **213**
Introduction   213
Primary Pathogens in Wastes and Compost   215
Worker Health Risks of Solid Waste Composting   223
The Effect of Composting on Pathogen Destruction   225
Survival of Pathogens in Soils and on Plants   235
Conclusions   241
References   242

Chapter 9. Bioaerosols **247**
Introduction   247
Aspergillus Fumigatus   253
Endotoxin and Organic Dusts   283
Conclusion   290
References   291

Chapter 10. Odors and Volatile Organic Compounds **301**
Introduction   301
Odorous Compounds and Odors Emitted
    at Composting Facilities   304
Volatile Organic Compounds   316
Air Dispersion Modeling for Composting Facilities   330
Conclusion   340
References   340

**Chapter 11. Soil Physical and
Chemical Manifestations** **343**
Introduction   343
Effect of Compost Application on Soil
    Physical Properties   347
Effect of Compost Application on
    Soil Chemical Properties   363
Summary   378
References   379

Chapter 12. Utilization of Compost **383**
Introduction   383
Horticulture   389
Agricultural Crops   400
Silviculture   408
Conclusion   411
References   412

Chapter 13. Compost Utilization II **417**
Introduction   417
Plant Pathogen Destruction during Composting   418
Plant Disease Suppression   420
Biofiltration   431
Summary   442
References   442

Chapter 14. Regulations **447**
Introduction   447
Concepts and Approaches to Regulations   448
United States   449
Canada   470
Europe   476
Criteria for Compost Quality and Facility Design   479
Conclusion   480
References   481

Index   *483*

Composting has evolved during the twentieth century from an art to a science. For the past 40 years, I have been involved with various aspects of organic matter decomposition and its effect on soils and crops. In 1956 I wrote my dissertation on "The Effect of Organic Matter on Soil Aeration with Particular Reference to Nutrient Uptake by Plants." My knowledge and experience at Purdue University, University of Maine, United States Department of Agriculture at Beltsville, Maryland, and at E&A Environmental Consultants, Inc., have provided me with an appreciation for the impact compost can have on the environment, waste management, humus formation, plant growth, and conservation of our most important resource, the soil.

It is in these institutions that I met and had the fortunate experience to work with outstanding individuals and scientists. From such individuals as Helmut Kohnke, my Ph.D. advisor, I began to learn of the importance of organic matter in preserving soil. Rufus Chaney, USDA, guided me in applying sound scientific principles when evaluating research, as well as realizing that true scientists can modify their views as scientific data change. Dr. Chaney provided me with much insight and knowledge about heavy metals. For the past 17 years my closest associate, Joel Alpert, supported my efforts, helped me critically examine issues, and helped guide E&A Environmental Consultants, Inc. into becoming the foremost company dealing with composting. There were others in E&A, Todd Williams, Mark Gould, Larry Sasser and their staff, who helped E&A achieve its prominence and stature in the science and technology of composting. Without their help, we would not be where we are today. I appreciate very much Gerry Croteau's review and comments on the chapter on microbiology and Ron Alexander's review of the chapter on compost uses.

I am especially grateful to Nerrisa Wu for editing and offering helpful suggestions on the manuscript. Her assistance was extremely useful in making this book a better text.

Much have I learned from my teachers, but much more have I learned from my students and colleagues.

Composting to many is a simple process. The lack of appreciation for the complexity and involvement of the biological, physical and chemical aspects of composting has resulted in major failings and setbacks. Composting is a biological process that affects and is affected by physical and chemical forces. This volume covers the very complex nature of composting. It is hoped that the reader understands and appreciates the myriad interactions between microbiology, biological and inorganic chemistry, and physics. The reader must evaluate what is presented here and in other treatises with an inquisitive and open mind. *Too often we believe what someone has researched, written or published as the scientific truth.* For example, data obtained from greenhouse studies, small composting apparatuses, and laboratory experiments need to be recognized as often being far from the real world of nature. However, they often provide us with direction and guidance to seek the truth. Greenhouse studies can only provide us with trends and comparisons between treatments. Plants grown in containers are often root-bound, and the chemical and physical conditions greatly impact their behavior. Greenhouse studies are not a substitute for field studies. Even field studies, if not properly designed, can be influenced by surrounding conditions that affect soil moisture, light and interrelating forces impacting plant physiology and plant response. Thus, both greenhouse and field studies should be viewed with caution.

Data obtained from small laboratory composting units distort the microbiological and environmental effects prominent in the real world. Many scientists cling to the work of early researchers who conducted studies in the laboratory under controlled or ideal conditions. Unfortunately, their data have been used as gospel and many of us have failed to recognize that this artificial world did not provide the correct insight into microbial behavior on a large scale.

The microbial environment of composting is not only affected by the heterogeneity and diversity of the population, but also by interactions between organisms, production of microbial products, chemical aspects of feedstocks, temperature, water, aeration and many other conditions discussed in this book.

This book, titled *The Science of Composting,* is the first in a two-volume set. The companion volume is titled *The Technology of Composting.* This first book is divided into three major units: Chapters 1 through 5 discuss the fundamentals, microbiology, biochemistry and physical/chemical aspects of composting; Chapters 6 through 9 deal with environmental and health issues; and Chapters 10 through 14 cover the beneficial impacts of compost on soils, uses, and regulatory aspects.

Composting is the highest form of recycling and the reuse of resources. What greater benefit can be had by mankind to perpetuate the living soil, which provides us with sustenance, food, and life, by returning and utilizing organic matter? What better utilization of organic wastes can be had than composting?

ELIOT EPSTEIN

CHAPTER 1

Composting: A Prospective

INTRODUCTION

The subject of composting can be subdivided into two major areas: the composting process and the compost product. Composting as a process involves the biological decomposition of organic matter. This definition is all encompassing, from simplistic backyard composting to large composting facilities. In this text composting means: The biological decomposition of organic matter under controlled, aerobic conditions into a humus-like stable product. The term *controlled* indicates that the process is managed or optimized to achieve the objectives desired. Some of the major objectives are to

- decompose potentially putrescible organic matter into a stable state and produce a material that may be used for soil improvement or other beneficial uses
- decompose waste into a beneficial product; composting may be economically favorable as compared to alternative disposal costs and may be more environmentally acceptable than more conventional solid waste management methods
- disinfect pathogenically infected organic wastes so that they may be beneficially used in a safe manner
- bioremediate or biodegrade hazardous wastes by means of the composting process

This chapter provides a brief overview of the history of composting, its current status, and philosophical aspects regarding the role of composting in the treatment of wastes and their ultimate beneficial use.

1

The use of several general terms needs to be explained. Biosolids will be used throughout in lieu of sewage sludge. This term has been adopted by The Water Environment Federation and is currently used by the U.S. Environmental Protection Agency (USEPA) in their manual *A Plain English Guide to the EPA Part 503 Biosolids Rule* (USEPA, 1994). Many different sludges exist: some chemical and some biological. Biosolids indicate that the solid is a product of biological activity and is primarily organic. There are several types of biosolids, e.g., pulp and paper mill sludge; a better term would be *municipal biosolids* to distinguish from industrial biosolids.

The past literature often uses the terms garbage or refuse. In this text, the term municipal solid waste (MSW) is used since it is often difficult to distinguish in the literature what was the composition of the waste stream. Source-separated organic wastes or other types of separated wastes will be so designated to distinguish them from mixed MSW.

COMPOSTING AND RECYCLING

In recent years our society has developed a hierarchy of waste management (USEPA, 1989). At the top of the hierarchy of integrated waste management and the most desirable is source reduction. This includes reduction of waste generated through improved packaging and reuse of materials. The second tier is recycling, including composting. Waste combustion and landfilling are placed at the bottom of acceptable technologies.

Composting is the highest form of recycling. An organic, discarded material is converted (recycled) for reuse in a manner that *only can benefit mankind.* The major use for compost is beneficial. Compost can improve soil conditions and plant growth, and reduce the potential for erosion, runoff, and non-source pollution. *Compost is an organic matter resource.* Properly produced compost adds humus to soils.

A considerable portion of the domestic waste stream is amenable to composting. Yard waste and biosolids are organic and are easily composted. In 1989 USEPA in the introduction to the 40CFR503 regulations estimated that the 15,300 publicly owned treatment works generated 7.7 million dry metric tons of biosolids. The volume of biosolids generated is expected to double by the year 2000 due to population growth. Distribution and marketing of biosolids products, primarily composting, but also including heat dried material, accounted for 9.1% of biosolids disposal at that time (USEPA, 1989).

Sixty-eight percent of the MSW waste stream is organic and therefore can be composted (Figure 1.1). Paper constitutes 37.5%. A considerable portion of the paper can be effectively recycled. However there is always a portion

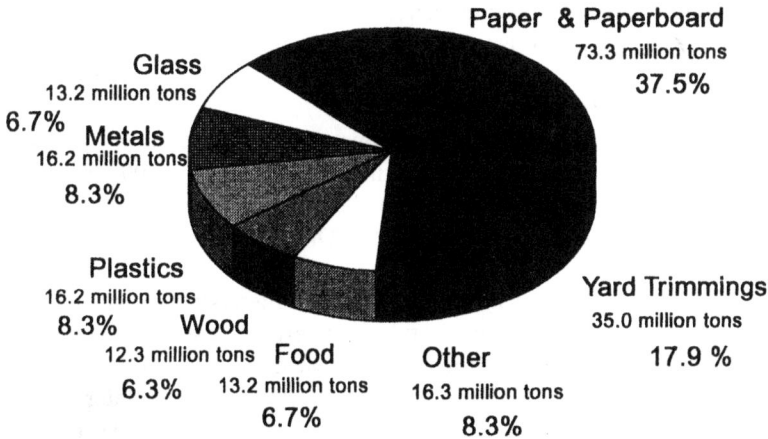

FIGURE 1.1. Materials generated in MSW by weight and percentage, 1990. (From USEPA, 1992.)

of the paper fraction that is soiled and cannot be recycled. Hyatt (1995) estimates that this fraction represents approximately 16%. USEPA estimated that, in 1990, 17% of the MSW waste stream was recycled or composted (USEPA, 1992). In 1990 over 177 million tons of MSW was generated in the United States. This represented approximately 1.95 kg/person/day. This number is projected to increase to 189 million tons or 2.05 kg/person/day by the year 2000.

Legislature by states to discourage landfilling will encourage recycling and composting. The only way communities can meet goals exceeding 35% (except in areas producing very heavy amounts of green material) is to compost a portion of the MSW stream. In addition to residential recovery of organic materials for composting, opportunities exist for large scale composting of food wastes generated in markets and restaurants.

HISTORY

Composting in the broadest sense undoubtedly was practiced in ancient times. The Israelites, Greeks, and Romans used organic wastes directly or composted them (Martin and Gershuny, 1992). The early civilizations of South America, China, Japan and India practiced intensive agriculture and used animal and human wastes as fertilizers (Howard, 1943). Many of these organic wastes were heaped and allowed to rot for long periods of time, producing compost.

Research on composting in the United States appears to have begun in the

1880s. One of the earliest publications on composting in the United States was Bulletin No. 61 by North Carolina Agricultural Experiment Station, published in December 1888, XI. Composts—Formulas, Analyses, and Value (Figure 1.2). Maynard (1994) cites 70 years of research on waste composting and utilization at the Connecticut Agricultural Experiment station. Hyatt (1995) reported that during the period 1971 to 1993 the number of citations, both in the United States and internationally, relating to the subject of compost has grown to 11,353. This indicates strong interest and research effort on the subject.

The concept of large-scale composting in a methodical manner is often attributed to Sir Albert Howard and his Indore process at the Institute of Plant Industry, Indore, Central India, between the years 1924 and 1931 (Howard, 1935). Initially, the process was anaerobic, but was later modified to an aerobic process and renamed the Bangalore process. The basic concept was to utilize vegetable and animal wastes and night soil (human excrement), mixing them with an alkaline material for neutralizing acidity, and managing the mass through turning for aeration and water addition. The process used either shallow pits or piles, that at times contained 909 tonnes (1000 tons). The dimensions of some compost pits were 9 m (30 ft) by 4 m (14 ft) by 0.9 m (3 ft) with sloping sides. It is interesting to note that Sir Howard observed that "air percolates the fermenting mass to a depth of about 18 to 24 inches only, so for a height of 36 inches, extra aeration must be provided." Only in recent years has it been documented that oxygen levels at the bottom of windrows are very limited and that this zone becomes anaerobic (see Chapter 2). Van Vuren (1949) published results of composting of urban wastes in 1939 in South Africa based on Sir Howard's principles (Figure 1.3). He viewed the composting of urban wastes as a method of disinfection while producing organic matter which could restore soil humus. One of the earliest systems was patented by Beccari in 1922 (Beccari, 1922).

The first full-scale, refuse composting facility in Europe was established in the Netherlands in 1932. The process used was the Van Maanen process, that was a modification of the Indore method. The plant was operated by N.V. Vuilafvoer Maatschapii (VAM). Unground refuse was piled in large windrows and turned by overhead cranes (Breidenbach, 1971; Epstein, 1976). Following World War II composting increased in Europe, with practically no composting being conducted in the United States. In the Netherlands two large composting plants were constructed in Mierlo and Wijster using the VAM method. It was estimated that nearly one-third of the Dutch refuse was composted in these two plants. Unit trains from The Hague would bring the waste onto an elevated ramp and dump the waste forming large piles. Overhead or mobile cranes would move and turn the waste for several months. Screening would then produce several different grades of

Dec., 1888, Jan., 1889.

North Carolina

Agricultural Experiment Station.

Bulletin No. 61.

XI. Composts—Formulas, Analyses, and Value.

Publications will be sent to any address upon application.

FIGURE 1.2. An early publication by North Carolina Agricultural Experiment Station, Bulletin No. 61, XI. Composts—Formulas, Analyses, and Value. December 1888, January, 1889.

FIGURE 1.3. Composting by the Indore method in South Africa as described by Van Vuren, 1939–1942. (Photograph by Mr. R. Nicholson, Stellenbosch, from Van Vuren, 1949.)

compost to be used in agriculture and horticulture. In 1976 the author visited the VAM, Wijster, facility. This method appeared to be well suited for developing countries since it was a relatively low technology utilizing easily available equipment. During a visit to South Yemen on behalf of United Nations Developing Programs, the author recommended that a modification of the VAM system be used in lieu of a high tech mechanical system originally proposed for that country. This low technology process was implemented.

Breidenbach (1971) indicated that there were more than 30 composting systems in 1969. Table 1.1 lists several major systems in the world at that time.

The only major research conducted on composting in the 1950s and 1960s was conducted at the Richmond Station of the University of California under the leadership of Drs. Gottas and Goulke.

In the 1960s U.S. Public Health Service, forerunner of USEPA, initiated two major research and demonstration projects on composting MSW with biosolids. One location was at Gainesville, Florida, and the other at Johnson City, Tennessee (Breidenbach, 1971). In addition to process and economic

TABLE 1.1. Some Major Composting Systems in 1969.

System	Description	Location
Bangalore (Indore)	Pits or heaps, 2 to 3 ft deep, alternate layers of refuse, night soil, and other organics; detention time 120 to 180 days	India, South Africa
Caspari	Refuse compressed into bricks, stacked for 30 to 40 days; curing; blocks ground; product refined	Schweinfurt, Germany
Dano	Rotating drum; 9–12 ft diameter and up to 150 ft long; 1 to 5 days digestion followed by windrowing	Europe
Earp-Thomas	Silo with 8 stacked decks; ground refuse moved from deck to deck by ploughs; air passed upward through a silo; digestion for 2 to 3 days followed by windrowing	Heidelberg, Germany; Turgi, Switzerland; Verona, Italy; Thessaloniki, Greece
Fairfield-Hardy	Circular tank with rotating arm containing vertical screws; forced aeration through bottom; retention time 4 to 5 days	Altoona, Pennsylvania; U.S.; San Juan, Puerto Rico
Frazer-Eweson	Ground refuse placed in vertical bin with 4 to 5 perforated decks with special arms to force composting material through perforations; air forced through bin; 4 to 5 days' retention	None in operation
Jersey (John Thompson system)	Structure with 6 floors; ground refuse moved from floor to floor; detention time of 6 days	Jersey, Channel Islands, Great Britain; Bangkok, Thailand
Metrowaste	Open bins, 20 ft wide × 10 ft deep × 200–400 ft long; machine turns 1 to 2 times in 7 days; air forced through perforations in bottom of bin	Houston, Texas; Gainesville, Florida
Naturizer or International	Five 9-ft wide steel conveyor belts arranged to pass material from one belt to another; air passes upward through digester; retention time 5 days	St. Petersburg, Florida

(continued)

TABLE 1.1. (continued).

System	Description	Location
T. A. Crane	Two cells with 3 horizontal decks; ribbon screws transfer refuse from deck to deck; air introduced in bottom of bin	Kobe, Japan
Tollemache	Similar to Metrowaste system	Spain; Southern Rhodesia, Africa
Triga	Sets of 4 towers or silos; refuse ground; forced aeration; 4 days' detention	France; Moscow, Russia; Buenos Aires, Argentina
Windrow	Open windrows turned by different types of machines	Mobile, Alabama; Boulder, Colorado; Johnson City, Tennessee; Europe; Israel
van Maanen (VAM)	Unground refuse in piles turned by overhead cranes; 120 to 180 days	Netherlands

Source: Breidenbach, 1971.

studies, there was considerable research on plant growth and effects on soil by the University of Florida and The Tennessee Valley Authority. Connecticut Agricultural Experiment Station conducted considerable research on compost utilization in the 1940s, 1950s, and 1970s (Maynard, 1994).

In 1973 the U.S. Department of Agriculture (USDA) at the Beltsville, Maryland, Agricultural Research Center initiated a major research effort on composting of biosolids. U.S. Environmental Agency (USEPA), Maryland Environmental Service (MES), and Washington, D.C. Council of Governments supported much of the research. During 1975 the USDA research team developed the aerated static pile method (Epstein et al., 1976). The research encompassed process and engineering aspects, pathogen and bioaerosol studies, heavy metal uptake studies, plant growth studies and microbiological studies. The basis of this research provided USEPA with data necessary to formulate the regulations in 40CFR257. University of Maryland's Agronomy and Horticulture departments cooperated with USDA and conducted independent research on plant growth. USDA is still conducting composting research primarily in the biological area. This research led to the rapid growth of biosolid composting in the United States.

In 1976, Rutgers University in New Jersey initiated studies of biosolids at the university and in Camden, New Jersey. Considerable microbiological, process engineering, economics, and utilization research was conducted (Bolan et al., 1980; Kasper and Derr, 1981; Singley et al., 1982). In the late 1970s and early 1980s when both USDA and Rutgers began cutting back their research efforts, the University of Ohio began conducting studies on plant growth with emphasis on composting as related to plant diseases.

A major boost to composting research in the United States occurred following the formation of the Composting Council in 1989 and the generous funding by The Procter and Gamble Company. Today the major institutions carrying multiple phases of composting research include USDA, the Universities of Ohio and Florida in the United States, as well as their counterpart institutions in Germany, Italy, Spain, and Israel. The Composting Council in Washington, D.C. supports considerable research. Hyatt (1995) indicates that over 50 research projects are currently under way.

As a result of the early research emphasizing biosolids composting, major European firms entered the American market in the 1980s. Table 1.2 lists the composting systems in the United States and abroad which E&A Environmental Consultants, Inc. investigated and visited. The growth of biosolid

TABLE 1.2. Composting Systems Evaluated and Visited by E&A Environmental Consultants, Inc.

Aerated Static Pile	Windrow
Dano, U.S., Europe	Daneco, U.S., Italy
Bedminster	Heidelberg silo
OTV/OTVD, France, U.S.	Buhler Miag, U.S., Europe
Fairfield hardy	PLM Selbergs, Sweden
Ebara, Japan	Japan Steel Works
Enadisma, Spain	Recomp, U.S.
Environment Recovery Systems	BAV, Europe
Ashbook Tunnel, U.S., Germany	Agripost, U.S.
Purac, U.S.	Seerdrum, England, U.S.
VAM, Holland	Organic Bioconversion, U.S.
Inge Brikolare, U.S.	Ag-Renu, U.S.
Gicom Tunnel, U.S., Holland	

NUMBER OF OPERATING FACILITIES

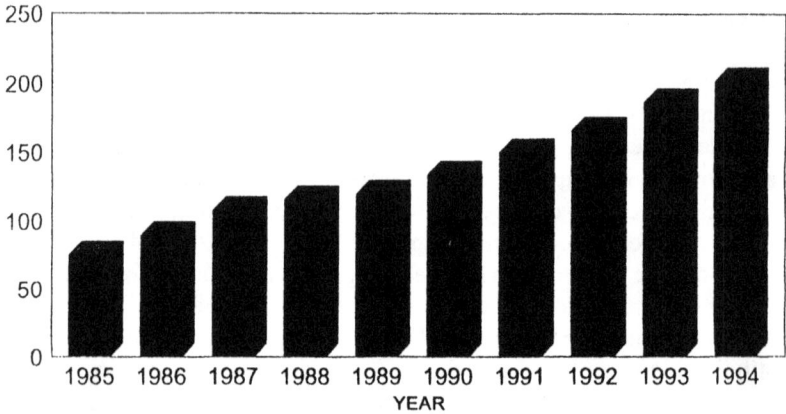

Source: BioCycle

FIGURE 1.4. Growth of biosolid composting facilities in the United States.

composting in the United States is shown in Figure 1.4. This growth is continuing.

The ban by many states on landfilling of yard wastes, brought on by a lack of landfill space near urban areas and the difficulty in siting and permitting landfills forced many communities to compost yard wastes. As a result over 3000 yard waste composting facilities sprang up (Figure 1.5). Many of the operators had no knowledge of composting or the ability to handle different

NUMBER OF FACILITIES

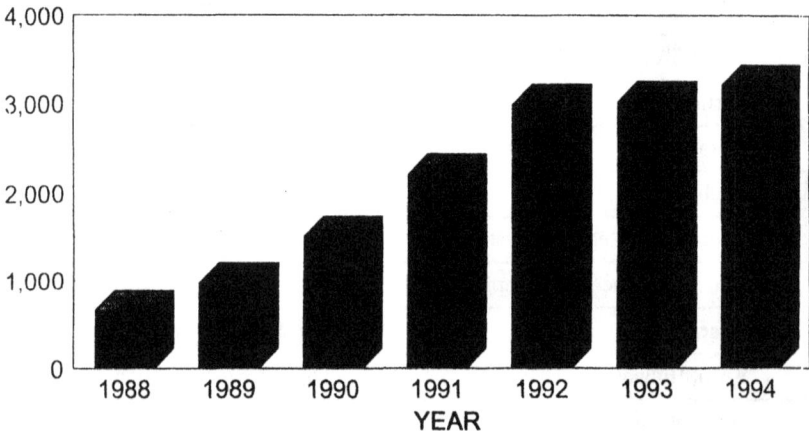

Source: BioCycle

FIGURE 1.5. Growth of yard waste composting facilities in the United States.

TABLE 1.3. *Municipal Solid Waste Composting Facilities in the United States in 1969.*

Location	Company	System	Capacity, tons per day	Waste	Start Date	Status in 1969
Altoona, Pennsylvania	Altoona FAM, Inc.	Fairfield-Hardy	41	MSW, paper	1951	Operating
Boulder, Colorado	Harry Gorby	Windrow	91	MSW	1965	Operating intermittently
Gainesville, Florida	Gainesville Municipal Waste Conversion Authority	Metrowaste Conversion	136	MSW, raw biosolids	1968	Operating
Houston, Texas	Metropolitan Waste Conversion Corp.	Metrowaste Conversion	327	MSW, raw biosolids	1966	Operating
Houston, Texas	United Compost Services, Inc.	Snell	273	MSW	1966	Closed, 1966
Johnson City, Tennessee	Joint USPHS-TVA	Windrow	47	MSW, digested biosolids	1967	Operating
Largo, Florida	Peninsular Organics, Inc.	Metrowaste Conversion	45	MSW, digested biosolids	1963	Closed, 1967
Norman, Oklahoma	International Disposal	Naturizer	32	MSW	1959	Closed, 1967
Mobile, Alabama	City of Mobile	Windrow	273	MSW, digested biosolids	1966	Operating intermittently

(continued)

11

TABLE 1.3. *(continued)*.

Location	Company	System	Capacity, tons per day	Waste	Start Date	Status in 1969
New York, New York	Ecology, Inc.	Varro	136	MSW		Under construction
Phoenix, Arizona	Arizona Biochemical Co.	Dano	273	MSW	1963	Closed, 1965
Sacramento County, California	Dano of America, Inc.	Dano	36	MSW	1956	Closed, 1963
San Fernando, California	International Disposal Corp.	Naturizer	64	MSW	1963	Closed, 1964
San Juan, Puerto Rico	Fairfield Engineering Co.	Fairfield-Hardy	136	MSW	1969	Operating
Springfield, Massachusetts	Springfield Organic Fertilizer Co.	Frazer-Eweson	18	MSW	1954 1961	Closed, 1962
St. Petersburg, Florida	Westinghouse Corp.	Naturizer	95	MSW	1966	Operating intermittently
Williamston, Michigan	City of Williamston	Riker	3.6	MSW, raw bio-solids, corn cobs	1955	Closed, 1962
Wilmington, Ohio	Good Riddance, Inc.	Windrow	18	MSW	1963	Closed, 1965

Source: Breidenbach, 1971.

feedstocks. During certain seasons, large quantities of green wastes were delivered to sites which were often unprepared to handle large volumes or a waste that rapidly decomposed and produced odors. Consequently, odor problems occurred. Although odor is not a health problem, it is a nuisance to the communities surrounding composting facilities and has raised public concern regarding the health impacts of composting. Recently, this concern has been focused on bioaerosols (see Chapter 9).

MSW composting had a dismal start in the United States. In 1969 there were 18 facilities, as shown in Table 1.3 (Breidenbach, 1971). From 1969 until the 1980s, there was no growth and many facilities that operated in 1969 were closed. Today there are 18 facilities. Although many European systems operated well in Europe, the same systems or others in the United States developed problems. MSW composting had to compete with landfill tipping fees. Low tipping fees in many areas made it uneconomical to build good composting facilities. As a result poor facilities were designed, or vendors who had no knowledge of composting saw an opportunity to make money. These vendors often underbid projects, which resulted in low fees and the design of poor facilities or the inability to meet debt payments. Odor problems and poor products resulted and public confidence waned. Multi-million dollar facilities closed in Portland, Oregon, and in Dade County, Florida, because of poor design and odors. Recently several major European companies withdrew from the American market.

PHILOSOPHICAL ASPECTS AND THE FUTURE OF COMPOSTING IN THE UNITED STATES

Growth of biosolids reuse will continue as a result of USEPA's philosophy of encouraging beneficial use of this material. The public has become more comfortable with the concept of beneficial reuse. Pressures from the compost market to produce high quality, consistent products will drive facility design and operational standards. Biosolids have also improved in quality; the Clean Water Act of 1977 reduced contamination by mandating industrial pretreatment and identifying point sources of pollution. USEPA, in the recent 40CFR503 regulations (USEPA, 1993) has tightened up the pathogen reduction requirements which will force composting operations to do a better job of operations and monitoring. Unlike MSW, most of the biosolids are produced by publicly operated wastewater treatment plants. The public agencies involved often have a greater respect for public opinion and also are not motivated by the need to profit from their operations.

Unfortunately, some regulators and environmental organizations have attached a stigma to biosolids, regardless of their chemical constituency. For example, the state of Minnesota has established a classification system for compost whereby a Class A compost cannot have biosolids as a feedstock. As a result, there is not a single biosolids composting facility in the state. Furthermore, Minnesota allows the application of biosolids to agricultural lands. This is irrational, since biosolids compost will generally have lower concentrations of heavy metals as a result of the use of bulking agents such as wood chips, sawdust or other material. Also, many toxic organic compounds decompose during composting. Table 1.4 shows the heavy metals in biosolids from Washington, D.C. and the compost made from it. Zn, Cu, Cd, and Pb were lower in compost than in the feedstock.

The potential for metals to enter the food chain from compost is much less than from the direct application of biosolids to land. Compost products generally have a higher market value in non–food chain uses such as horticulture (nurseries, landscaping, sod or turf, and soil reclamation) than in agriculture. Furthermore, composting disinfects the biosolids, whereas biosolids used for direct land application can contain higher levels of pathogens. The compost treatment for pathogen reduction is classified by USEPA as Class A and is much less restricted for land application than most biosolids treatments that are classified as Class B. This is not to suggest that land application of biosolids is unsafe, but rather to illustrate the illogic of overly stringent compost use standards.

The condemnation of a feedstock, as advocated by many environmental organizations, rather than establishing standards for a product can have greater environmental consequences since it reduces the options for biosolid disposal or use. Domestic biosolids may have lower lead levels than some urban yard wastes. The latter often contain soil material containing lead from unleaded gasoline, since lead does not leach out of soil. This is one reason

TABLE 1.4. Comparison of Heavy Metals in Sewage Biosolids and Their Respective Compost.

Concentration (μg/g)				
Heavy Metal	Digested Sewage Biosolid	Compost from Digested Sewage Biosolid	Raw Sewage Biosolid	Compost from Raw Sewage Biosolid
Zn	1760	1000	980	770
Cu	725	250	420	300
Cd	19	9	10	8
Pb	565	320	425	290

that the feedstock should not be regulated. Standards need to be set for the product as USEPA has done in the 40CFR503 biosolids regulations.

What are the options for biosolid disposal? Essentially, we can dispose of any waste in the air (combustion), water, or soil. Of these three media, only the soil affords the opportunity to control and manage the potential consequences. Wastes incinerated even with the best technology available will still release some level of pollutants. These are dispersed into our environment. Once in the air, it is impossible to control or manage their destiny. Pollutants discharged into waters are controlled by federal legislation. Again once a pollutant enters a water course, it is nearly impossible to control and manage its destiny. Pollutants entering a soil system can, to a great extent, be controlled and managed to reduce impacts to human health and the environment. For example, the uptake of most heavy metals from soil to plants can be reduced by increasing the pH. Organic matter can reduce the availability and movement of certain elements through chelation (see Chapter 6 for greater detail). We can avoid applying wastes that contain unacceptable levels of trace elements to food chain crops or use crops which do not accumulate heavy metals in the food portion of the plant. For example, fruit or grain is much less prone to heavy metal accumulation than leaves or vegetative portions of plants.

Another issue that several of the major environmental organizations are espousing is that waste to be composted should be source separated. This is an ambitious goal and, where practical, should be encouraged. However, several major aspects need to be considered. The argument for source-separated organics waste (SOW) composting is based on the higher heavy metal content which may be found in mixed MSW. Both the data by Epstein et al. (1992) and Richard and Woodbury (1992) indicate that with the exception of lead, most of the regulated heavy metals in MSW are significantly lower than USEPA's recommended levels for land application of biosolids (see Chapter 6). In addition there was no difference between SOW compost and mixed MSW compost for many heavy metals. Lead and other heavy metals will be further reduced in MSW as a result of removal of lead acid batteries, recycling of other batteries, and the exclusion of other heavy metal sources from the waste stream. However by categorically excluding mixed MSW from composting, the development of technology to accomplish separation is discouraged. There are many communities in the United States, as well as other countries, where source separation of organics is impractical. This is especially true in many urban areas where societal differences or economics does not favor source separation. Should we discourage composting and thereby promote landfilling or incineration?

Recycling and source separation are very complementary to composting. A good recycling and source-separation program will reduce the size of the composting facility, thereby reducing capital, operational and maintenance costs. Furthermore a cleaner compost will be produced since many of the contaminants (plastics, metals, glass) are removed. This compost would have a higher market value.

ADVANTAGES AND DISADVANTAGES OF COMPOSTING

Composting has many benefits:

- Many community wastes can be composted. Thus a single composting facility can handle municipal and industrial organic biosolids, MSW, yard wastes, food wastes, etc.
- A composting facility can be designed and operated to minimize environmental impacts. Odors and bioaerosols can be controlled.
- Composting can help meet states' landfill reduction and recycling goals.
- Composting can decompose or degrade many organic materials.
- Composting produces a usable product.

The major disadvantages to composting are:

- Odor and bioaerosol emissions can occur during the process. These odors and bioaerosols can be controlled through better facility design and operations management.
- Composting facilities take up more space than some other waste management technologies. Space requirements are often related to storage and market demand.
- A product must be marketed.

CONCLUSION

Composting as a waste management technology in the United States and elsewhere can expand and play a much greater role. In the United States, biosolid composting is growing and should continue to grow. The growth of MSW composting facilities will depend on the development of good composting facilities and the economics of waste management. Community protection is needed to insure better facilities and reliable systems. States may need to regulate or provide a process of review to insure that communities will have good facilities which can meet environmental requirements with respect to odors and bioaerosols. The financial institutions must be convinced that the technology is reliable and that the systems can be designed and operated in an environmental and economical manner. Without financing, there will not be growth.

Uniform standards are needed for compost products that will not hinder distribution and marketing. Many states are specifying standards for MSW and biosolid compost, but there is a wide variation between states. Standards should be based on sound scientific basis and related to public health and the protection of the environment.

REFERENCES

Beccari, G. 1922. Apparatus for working garbage and refuse of towns. U.S. Patent 1,329,105 January 27, 1920, Reissue 15,417, July 25, 1922.

Bolan, M. P., G. H. Nieswand, M. E. Singley. 1980. *Sludge Composting and Utilization: Statewide Applicability for New Jersey.* N.J. Agric. Expt. Sta., Rutgers U., New Brunswick, NJ.

Breidenbach, A. W. 1971. *Composting of Municipal Solid Wastes in the United States.* Pub. SW-47r. U.S. Environmental Protection Agency.

Epstein, E. 1976. Personal visit.

Epstein, E., G. B. Willson, W. D. Burge, D. C. Mullen, and N. K. Enkiri. 1976. A forced aeration system for composting wastewater sludge. *J. Water Pollut. Control Fed.* 48:688–694.

Epstein, E., R. L. Chaney, C. Henry, and T. J. Logan. 1992. Trace elements in municipal solid waste compost. *Biomass and Bioenergy.* 3(3–4):227–238.

Howard, A. 1935. The manufacture of humus by the Indore process. *J. of the Royal Society of Arts.* 84:26–29.

Howard, A. 1943. *An Agricultural Testament.* Oxford University Press. London.

Hyatt, G. W. 1995. Economic, scientific, and infrastructure basis for using municipal composts in agriculture. In *Agriculture utilization of urban and industrial by-products.* Amer. Soc. of Agronomy Special Pub. no. 58. Madison, WI.

Kasper Jr., V. and D. D. Derr. 1981. *Sludge Composting and Utilization: An Economic Analysis of the Camden Sludge Composting Facility.* NJ Agric. Expt. Sta., Rutgers U., New Brunswick, NJ.

Martin, D. L. and G. Gershuny. 1992. *The Rodale Book of Composting.* Rodale Press, Emmaus, PA.

Maynard, A. A. 1994. Seventy years of research on waste composting and utilization at the Connecticut Agricultural Experiment Station. *Compost Sci. & Util.* 2(2):13–21.

Richard, T. L. and P. B. Woodbury. 1992. The impact of separation on heavy metal contaminants in municipal solid waste composts. *Biomass and Bioenergy.* 3(3–4):195–211.

Singley, M. E., A. J. Higgins, and M. Frumkin-Rosengaus. 1982. *Sludge Composting and Utilization: A Design and Operating Manual.* N.J. Agric. Expt. Sta., Cook College, Rutgers, New Brunswick, NJ.

USEPA. 1989. *Decision-Makers Guide to Solid Waste Management.* EPA/530-SW-072. United States Environmental Protection Agency, Washington, D.C.

USEPA. 1992. *Characterization of Municipal Solid Waste in the United States: 1992 Update.* EPA/530-S-92-019. United States Environmental Protection Agency, Washington, D.C.

USEPA. 1993. The standards for the use or disposal of sewage sludge. Title 40 of the *Code of Federal Regulations* (CFR) 503. *Federal Register* 58FR9248 to 9404. United States Environmental Protection Agency, Washington, D.C.

USEPA. 1994. *A Plain English Guide to the EPA Part 503 Biosolids Rule.* EPA/832/R-93/003. U.S. Environmental Protection Agency, Washington, D.C.

Van Vuren, J. P. J. 1949. *Soil Fertility and Sewage.* Faber and Faber Ltd., London.

CHAPTER 2

Basic Concepts

INTRODUCTION

Composting is the biological decomposition of organic matter under controlled aerobic conditions. In contrast, fermentation is the anaerobic decomposition of organic matter. Many factors affect the composting process; "So many factors are involved, nearly all interrelated, that this complex ecological process is unlikely to succumb to rigorous scientific analysis for many years" (Gray and Sherman, 1969). Some of these factors play a major role in the process while others can influence its direction or extent. Since 1969 new chemical and physical techniques have provided scientists with the tools to examine and manipulate these factors and to delve into the composting process in a rigorous manner.

The basic composting process is depicted in Figure 2.1. The major factors affecting the decomposition of organic matter by microorganisms are oxygen and moisture. Temperature is an important factor in the composting process; however, temperature is the result of the microbial activity. Other important factors that could limit the composting process are nutrients and pH. Nutrients, especially carbon and nitrogen, play an important part in the process as they are essential for microbial growth and activity. Carbon is the principal source of energy, and nitrogen is needed for cell synthesis. Phosphorus and sulfur are also important, but less is known about their role in composting. Microorganisms require the same micronutrients as plants and compete for available micronutrients (Stevenson, 1991). Micronutrients such as Cu, Ni, Mo, Fe, Mg, Zn, and Na are necessary for enzymatic functions, but little is known about their importance to the composting process.

Most of the self-heating of organic matter is the result of microbial

```
                    ┌──────────────────┐
                    │ MICROORGANISMS   │
                    └──────────────────┘
┌──────────────┐         │              ┌──────────────┐
│    WATER     │         ▼              │   OXYGEN     │
└──────────────┘    ORGANIC            └──────────────┘
        FAST │      MATTER
                  ─────────────
RATE OF            CARBOHYDRATES       ┌──────────────────┐
DECOMPOSITION      SUGARS              │ DECOMPOSITION    │
                   PROTEINS            │ PRODUCTS         │
                   FATS                │ CARBON DIOXIDE   │
                   HEMICELLULOSE       │ WATER            │
                   CELLULOSE           └──────────────────┘
        SLOW ▼     LIGNIN
                   MINERAL MATTER
        ┌──────────────┐      ┌──────────────┐
        │   COMPOST    │      │    HEAT      │
        └──────────────┘      └──────────────┘
```

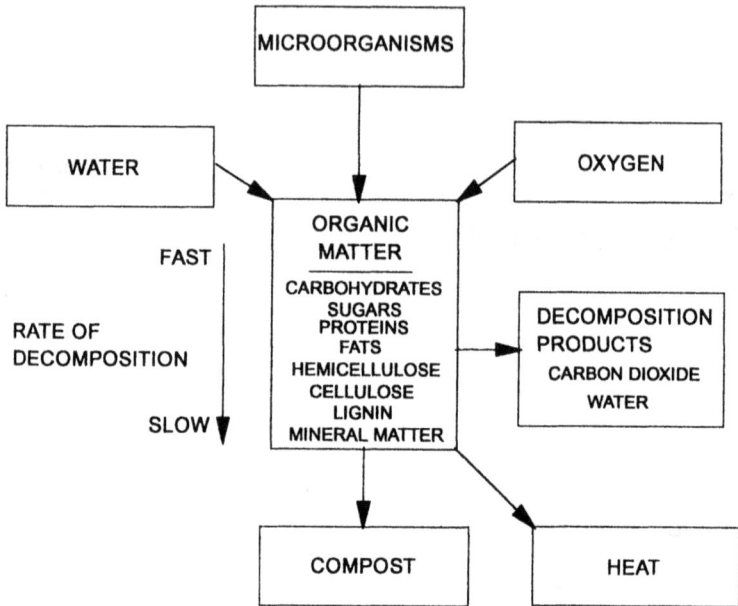

FIGURE 2.1. The composting process.

respiration (Finstein and Morris, 1975). That is, when the mass is insulated, the heat generated increases the temperature of the mass. An increase in temperature affects the microbial population through changes in mesophilic and thermophilic organisms, which in turn affects the rate of decomposition (see Chapter 3). Microbial respiration can therefore be used as an indicator of decomposition and the stability of a compost product. (The microbiological aspects are discussed in greater detail in Chapter 3.) During the process oxygen is consumed, and CO_2 and water are released. In the early days of composting, it was very time consuming or difficult to monitor CO_2 or O_2 continuously during large-scale composting. Consequently, most of the data in the literature are from small-scale or laboratory processes.

In addition to CO_2, ammonia and other volatile compounds are emitted to the atmosphere. In comparison to CO_2 and H_2O these represent very small amounts, however. Wiley and Pierce (1955) represented the aerobic composting process in the following chemical equation:

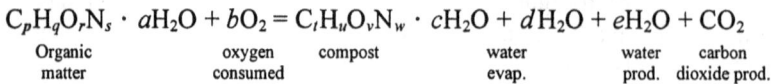

$$C_pH_qO_rN_s \cdot aH_2O + bO_2 = C_tH_uO_vN_w \cdot cH_2O + dH_2O + eH_2O + CO_2$$

| Organic matter | oxygen consumed | compost | water evap. | water prod. | carbon dioxide prod. |

(The small letters represent constants for different conditions.)

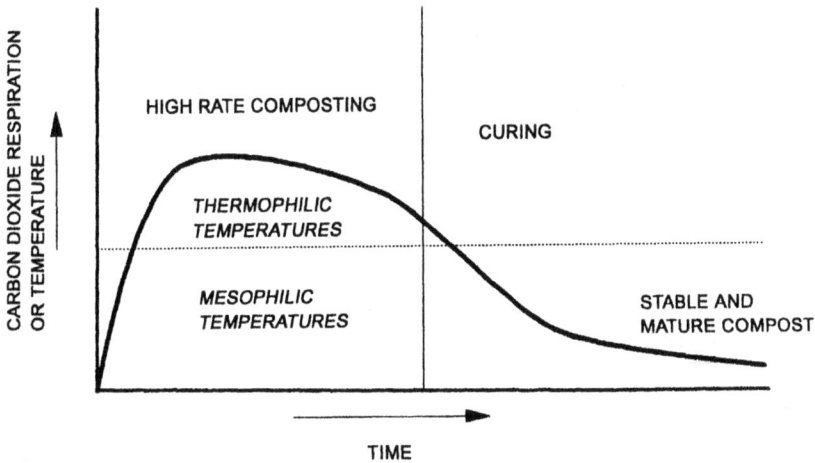

FIGURE 2.2. Phases during composting as related to carbon dioxide respiration and temperature.

The composting process can be depicted in many ways. Figure 2.2 shows a general relationship for respiration and temperature as a function of composting time. The shape of the curve varies with the feedstock being composted and the composting method. In this figure there are two major stages, active composting and curing. The same curve can be subdivided into temperature zones, mesophilic (<45°) and thermophilic (>45°C). During the active composting phase easily decomposable and putrescible compounds are broken down, and pathogens can be eliminated. During the curing phase the compounds less susceptible to carbon mineralization (i.e., trans-

TABLE 2.1. The Susceptibility of Organic Compounds Found in Compost Feedstock to Mineralization.

Organic Compound	Susceptibility to Mineralization
Sugars	Very susceptible
Starches, glycogen, pectin	
Fatty acids, glycerol, lipids, fats, phospholipids	
Amino acids	
Nucleic acids	

(continued)

TABLE 2.1. (continued).

Organic Compound	Susceptibility to Mineralization
Protein	Usually susceptible
Hemicellulose	
Cellulose	
Chitin	
Low molecular weight aromatics and aliphatic compounds	Resistant
Lignocellulose	
Lignin	

formation of C to CO_2) are broken down along with fatty acids. Fatty acids can be phytotoxic, and the use of incompletely processed compost can result in plant injury (see Chapter 5). The breakdown of organic matter during composting is a stepwise reduction of complex substances to simpler compounds. Table 2.1 indicates the principal compounds in organic matter and their susceptibility to decomposition.

In this chapter the basic aspects of the composting process will be described. Specifically, the effect of oxygen, aeration, moisture, temperature, nutrients and pH will be discussed. The carbon and nitrogen cycles during the composting process will also be covered. (The carbon and nitrogen cycles in the soil as a result of compost application are discussed in Chapter 12.)

OXYGEN AND AERATION

Oxygen is essential for the microbial activity in composting since it is an aerobic process. Three principal aeration methods provide O_2 during composting: physical turning of the mass, convective air flow, and mechanical aeration. Windrow methods use the former two ways, whereas static systems provide O_2 by using blowers or through convective air flow. The latter, often called passive aeration, is highly dependent on the porosity of the matrix.

Several studies have shown that in un-aerated windrows oxygen is depleted in the lower parts of the windrow in very short time after turning, as seen in Figure 2.3 (Wiley and Spillan, 1962; E&A Environmental Consultants, Inc., 1994). The lack of oxygen results in anaerobic conditions. Consequently, putrescible compounds are formed. When the windrow is turned, these compounds can cause offensive odors.

Most of the developers of European windrow systems recognized this problem and provided for negative aeration to the windrow, whereby the withdrawn air is scrubbed in a biofilter. This can be seen both in uncovered outdoor facilities such as Fredrikssund, Denmark, or covered facilities such as Falkenburg, Sweden. In Fredrikssund the air from the trenches is directed into a compost biofilter for removal of odorous compounds (Figure 2.4).

Caballero (1984) reported that with a municipal biosolids windrow, oxygen demand was more than doubled immediately after turning than before turning. This could be the result of breaking up the biosolid particles and exposing a greater surface area to microbial activity. Reduction in moisture with an increase in free air space would also increase microbial activity.

A similar consumption of oxygen is seen in static piles; oxygen reaches very low levels within 20 minutes after blower shutdowns (Thompson, 1984). Hence in aerated static pile composting, blowers are regulated by both temperature and time. The interval between on/off periods should be kept to approximately 15 minutes.

Schultz (1960) studied the relationship between O_2 and temperature in a small reactor. He found a linear relationship when O_2 was plotted on a log scale versus time (Figure 2.5). Oxygen consumption rate was expressed by the following equation:

$$Y = a \cdot 10^{KT}$$

where a is a constant = 0.1; $K = 0.28$ for temperatures in the range of 20 to 70°C.

PERCENT OXYGEN

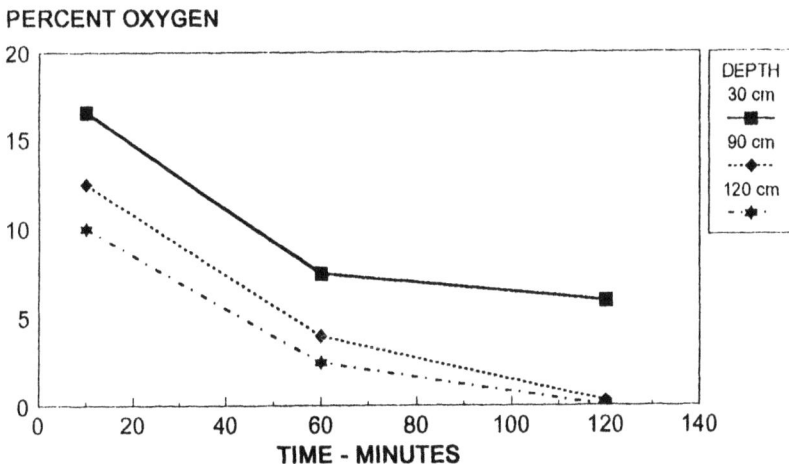

FIGURE 2.3. Oxygen depletion in a yard waste windrow following turning.

FIGURE 2.4. Aerated windrow composting in Fredrikssund, Denmark. Windrow placed over aerated trench which is under negative air.

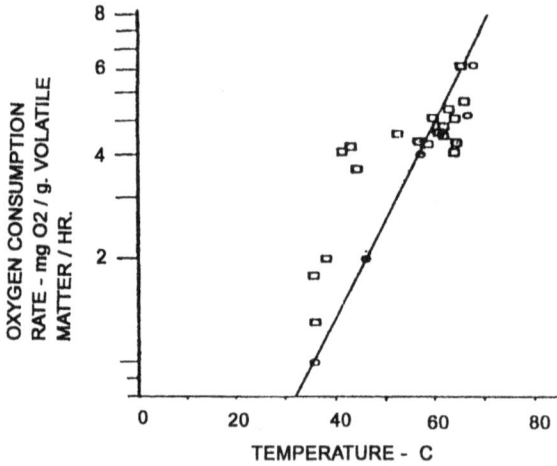

FIGURE 2.5. The relationship between oxygen consumption and temperature during composting in a laboratory reactor. (Data from Schultz, 1960.)

The data were only for the first seven days of composting. The relationship shown is understandable since this is the period of greatest microbial activity. In the same publication Shultz (1960) showed that O_2 consumption decreased with time after the seventh day with a concomitant decrease in temperature as shown in Figure 2.6 (Schultz, 1960). In Schultz's small-scale rotating laboratory drum composter, microbial activity essentially ceased between the ninth and tenth days. This needs to be understood in relation to the nature of the experiment. Such data can be used in determining aeration rate requirements for static systems for specific feedstocks during the most active period of microbial activity.

Jeris and Regan (1973a) reported that in one study of MSW composting, O_2 increased with temperature from 30°C to 66°C when the composted material had a moisture content of 45%. At higher moisture contents, the O_2 uptake rate peaked at 45°C. Wiley and Pierce (1955) evaluated different aeration rates in relation to temperature in a small agitated vessel. Low aeration rates of 2.6–4.3 mg O_2/hr/g volatile solids (VS) resulted in a temperature delay and incomplete decomposition; medium aeration rates of 6.1–19.6 mg O_2/hr/g · VS provided for peak temperatures in four days and temperature declined slowly thereafter; high aeration rates of 21.8–51.8 mg O_2/g · VS resulted in peak temperatures in four days followed by a rapid decline.

Kaibuchi (1961) reported that peak temperatures during composting of MSW in bins with forced aeration were obtained with an aeration rate of 4.3 mg O_2/hr/g · VS. Finally, Viel et al. (1987) showed that O_2 consumption rate for three different composting feedstocks increased during the first day and then decreased for the subsequent nine days. It is evident from these studies that oxygen consumption and aeration rates vary with the feedstock and its method of preparation.

The term "respiratory quotient" (RQ) is the ratio of CO_2/O_2 consumed. If one molecule of CO_2 is produced for every O_2 molecule consumed, RQ equals 1.0. Gray et al. (1973) indicated that different organic compounds have different RQ values. For example, during the oxidation to CO_2 and H_2O, starch has an RQ of 1.0, proteins have an RQ of 0.81, and fats have an RQ of approximately 0.71. According to Wiley and Pierce (1955) and Schultz (1960), RQ for composting was in the range of 0.87 to 0.91. This shows that less CO_2 is produced in relation to oxygen. However, chemical reactions releasing O_2 also result in a lower RQ.

One of the few and best examples of the relationship between oxygen and carbon dioxide for an aerated static pile is shown in Figure 2.7 (Singley et al., 1982). The O_2 and CO_2 were highly correlated, $R^2 = 0.951$ (Figure 2.8). The 22 sets of data fit the regression equation $Y = 19.15 + -0.857X$.

Oxygen consumption is affected by moisture content since the latter

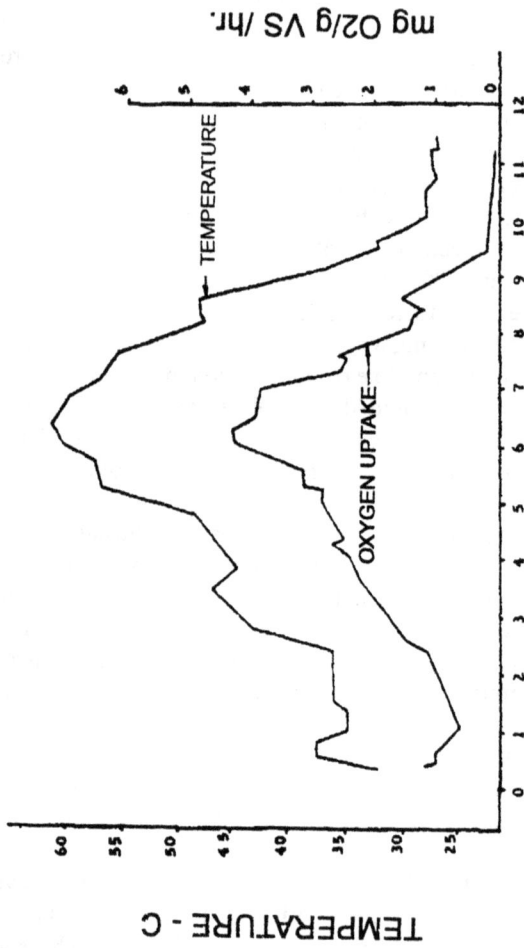

FIGURE 2.6. Changes in oxygen uptake and temperature during composting of artificially prepared MSW in a small laboratory reactor. (Data from Schultz, 1960.)

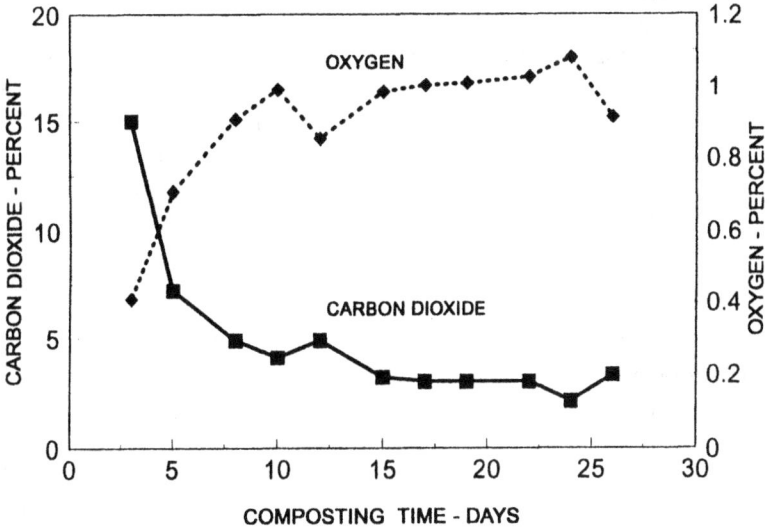

FIGURE 2.7. Changes in carbon dioxide and oxygen during composting of biosolids in an aerated static pile. (Data from Singley et al., 1982.)

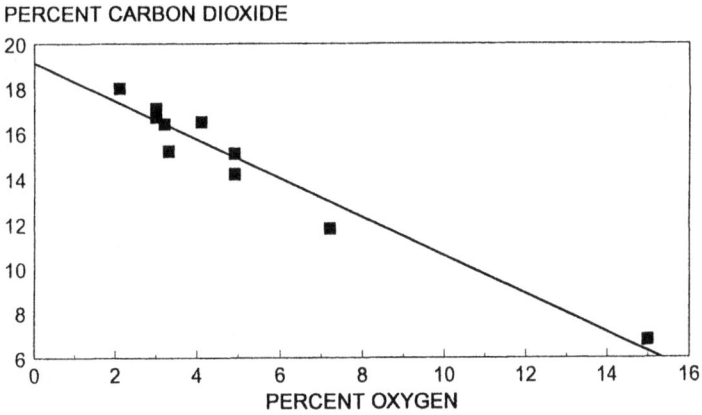

FIGURE 2.8. The relationship between oxygen and carbon dioxide during static aerated pile composting. (Data from Singley et al., 1982.)

impacts microbial activity. Regan and Jeris (1970) found that oxygen consumption was greater at 56% moisture than at 85% (Figure 2.9). The higher O_2 uptake at the lower moisture content is probably the result of greater microbial activity since more free air space (FAS) at the lower moisture content is more favorable for composting. The term "free air space" was suggested by Schultz (1961) based on soil aeration concepts of free pore space (Buckingham, 1904). Free air space is that portion of the total pore space that is not occupied with water (Figure 2.10). It is usually calculated by the following equation, which relates the bulk density (BD) to the specific gravity (SG).

$$\text{Free air space (FAS)} = 100(1 - BD/SG) \times \text{dry mass}$$

The pore space allows air to diffuse through the media and provide oxygen to the microorganisms. Since different materials have different densities and particle sizes, the relationship between moisture and FAS varies slightly (Jeris and Regan, 1973b). The optimum moisture content ranged from 53% to 65% and the corresponding FAS ranged from 32% to 36%. The relationship between free pore space, moisture, and oxygen consumption rate during the composting of MSW is shown in Figure 2.11.

As shown, maximum oxygen consumption (maximum microbial activity) occurred at approximately 65% moisture and a free pore space of 30%. This moisture content is higher than what has been found in field studies to be the optimum moisture content. The data also show that as composting proceeded, oxygen consumption decreased. For municipal biosolids the optimum moisture content is near 55% and should not exceed 60%. As mentioned, the optimum moisture content varies with the composting technology used. The 65% moisture content suggested by Jeris and Regan (1973b) can be applied to dynamic systems (windrow type) as the moisture loss is greater in these systems.

The relationship of free air space, moisture, and oxygen to the potential production of odors is shown in Table 2.2. Increasing the ratio of yard debris to produce waste increased the amount of O_2, decreased bulk density, and reduced the production of total mercaptans. Grinding, which reduced particle size, resulted in an increase in bulk density, reduced O_2 concentration, and increased generation of total mercaptans which can result in odors.

Total porosity and pore size as well as free air space is important. Figure 2.12 shows the relation between porosity and air flow resistance after compaction (Singley et al., 1982). The more compaction and less pore space, the greater the air flow resistance. A good analogy is observed in soils. A clay soil has a greater total pore space than sand, but the pores are very small and water permeability is restricted. In the sand, on the other hand, the larger

Oygen uptake - mg/hr./g VM

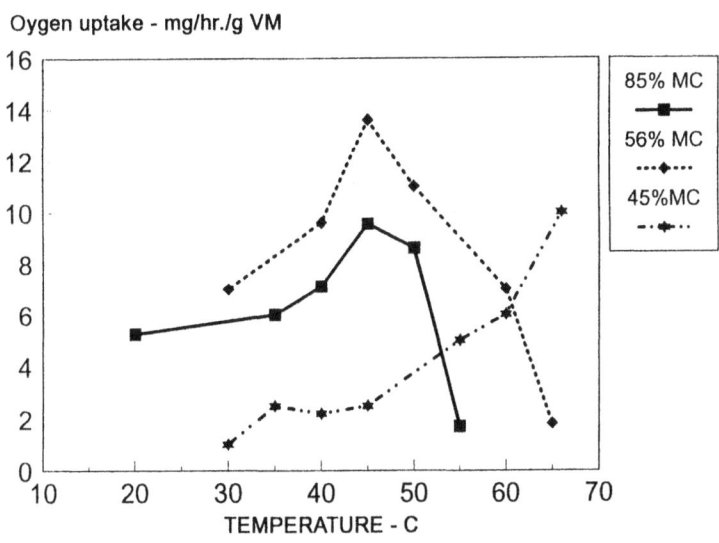

FIGURE 2.9. The effect of temperature on oxygen uptake in composting. (Reprinted with permission from Regan and Jeris, 1970.)

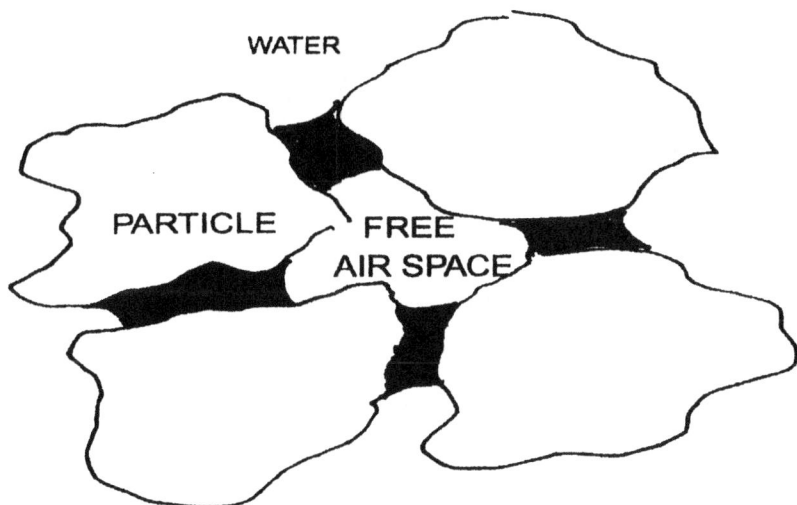

FIGURE 2.10. The relationship of free air space to water and particles in a composting media.

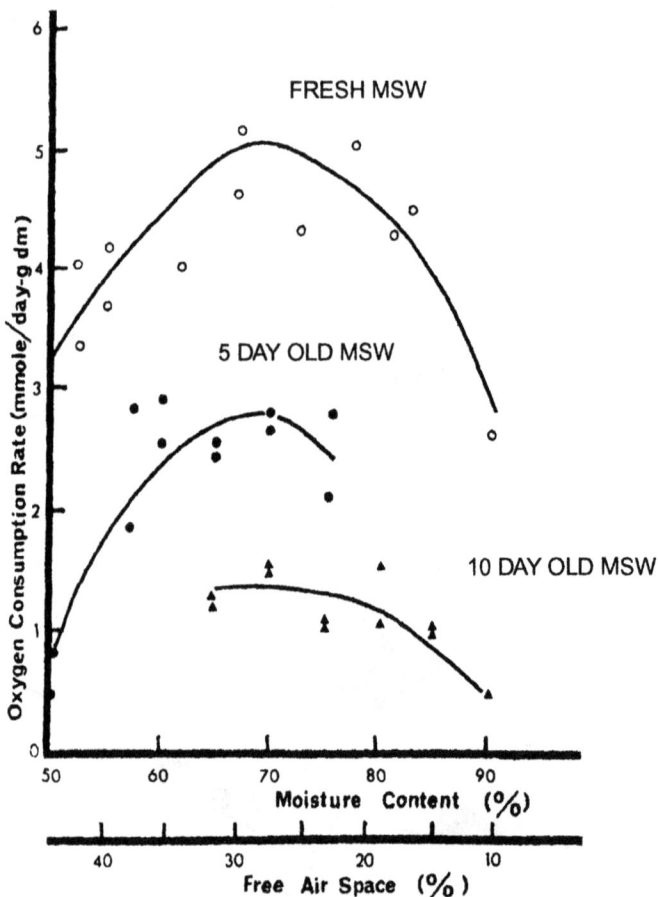

FIGURE 2.11. The relationship of oxygen consumption rate to moisture and free air space (Reprinted with permission from Jeris and Regan, 1973.)

TABLE 2.2. *Effect of Bulk Density on Oxygen Concentration and Total Mercaptans.*

Pile ID	1	2	3	4
Initial mix ratio: Yard debris:produce waste	4:0	4:1	4:1	2:1
Preprocessing	None	None	Ground	Ground
Bulk density, kg/m³ (lb/cy)	178 (300)	297 (500)	891 (1500)	832 (1400)
Oxygen concentration	19.9	18.8	0.3	0
Total mercaptans	0.2	0.5	25	100

Source: E&A Environmental Consultants, Inc., 1993b.

pores transmit water more readily although the number of pores is lower. Adding compost to a clay soil increases pore size and provides for greater infiltration and permeability (see Chapter 11).

Field data showing oxygen uptake during the composting period are very limited. Halley et al. (1980) measured decrease in oxygen concentration from ambient oxygen concentrations and the daily specific aeration rate (Figure 2.13). Early in the process, oxygen demand and consumption are high. As

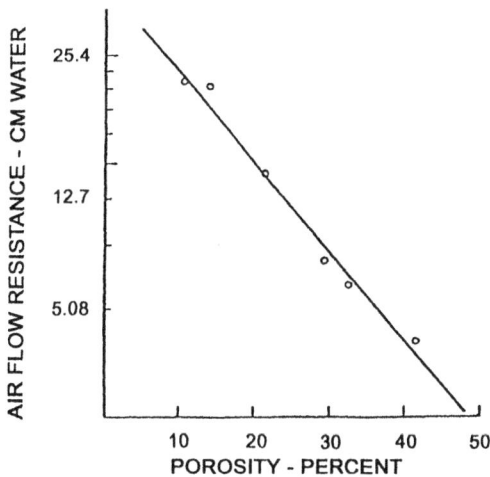

FIGURE 2.12. The relationship between porosity and air flow resistance to porosity. (Data from Singley et al., 1982.)

OXYGEN UPTAKE RATE - gO_2/hr/kg - DS

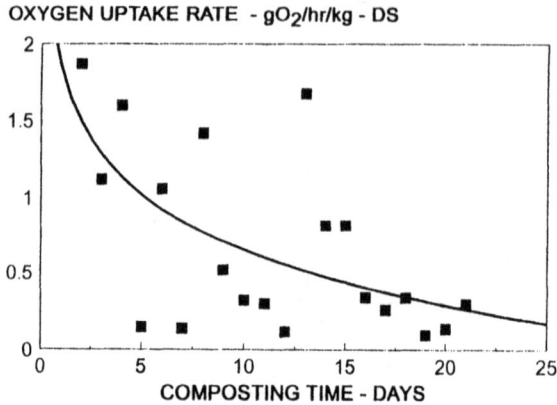

FIGURE 2.13. Changes in the oxygen uptake rate during composting. (Data from Halley and Shea, 1980.)

the easily decomposable organic material is utilized, oxygen consumption decreases. Temperatures during the initial period were highest.

In static systems the rate of aeration can be important for moisture removal primarily for ease of material handling. This is particularly true when dealing with wet materials such as municipal biosolids and food wastes. Dynamic systems often release sufficient moisture during turning. Although overdrying can occur in both systems, it is more likely in dynamic systems. When this happens, the composting process can slow down or cease.

MOISTURE

Moisture in the composting process can affect microbial activity and thus influence temperature and rate of decomposition. In addition, moisture can affect the composition of the microbial population (Coppola et al., 1983). Moisture is produced as a result of microbial activity and the biological oxidation of organic matter. In addition water is lost through evaporation. Based on work using a laboratory composter, Viel et al. (1987) reported that water released through microbial activity was greater than water lost through evaporation.

In a bench scale study maintaining temperature at 55°C in which biosolids and straw were composted, Witter and Lopez-Real (1987) found that water loss, CO_2 production, and aeration demand followed the same patterns. Initially, there was an increase in these parameters but after two days there was a decrease; at the 10th to 12th day the decrease leveled off. However, Manios et al. (1987) found that moisture content in piles increased over the 70 days during the composting of olive leaves.

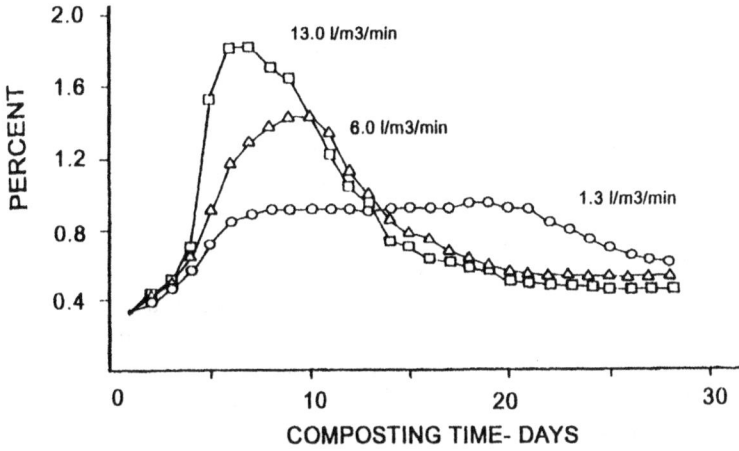

FIGURE 2.14. The effect of aeration on water release during composting of biosolids:wood chips mixtures. (Data from Coppola et al., 1983.)

In a fairly large pilot study using biosolids and wood chips, Coppola et al. (1983) showed that water release was a function of aeration rate (Figure 2.14). The greatest water loss was observed at an aeration rate of 13.0 $l/m^3/min$. Moisture content decreased between the 7th and 15th day, and stabilized at a release rate of approximately 0.4%. At 1.3 $l/m^3/min$ the release rate of moisture was essentially constant. The authors related these data to the microflora, concluding that 10 $l/m^3/min$ of aeration was most favorable to the microflora. If the percent water release is indicative of microbial activity, an aeration rate of 13 $l/m^3/min$ appears to be better than 10 $l/m^3/min$.

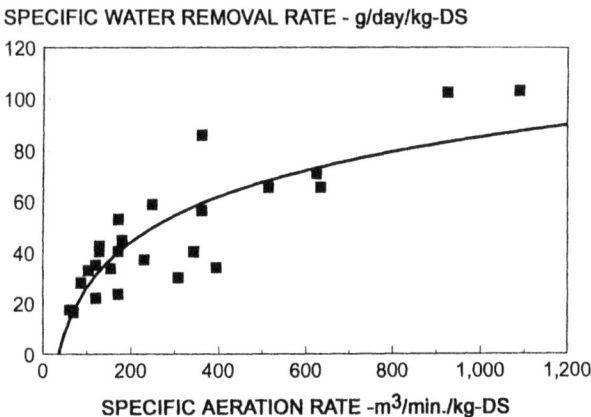

FIGURE 2.15. The relation between specific aeration rate and specific water removal in aerated static pile composting. (Data from Halley et al., 1980.)

Field data during aerated static pile composting of biosolids by Halley et al. (1980) showed that increasing the specific aeration rate increased specific water removal (Figure 2.15). A similar trend was reported by Willson (1977), although the magnitude was higher due to higher temperatures.

The optimum moisture content is between 50 and 60% (Schultz, 1962; Poincelot, 1975). Below 40%, microbial activity decreases and above 60% anaerobic conditions can exist because of blocked pore space. Wiley and Pierce (1955) evaluated three moisture regimes during the composting of MSW. The highest temperature was achieved when the moisture content ranged from 55 to 69%. Higher moisture, 72 to 77%, produced the lowest temperatures and a moisture content of 40 to 53% yielded an intermediate temperature. The experiment, conducted in small laboratory units over a six-day period, showed a linear relationship between temperature and H_2O production and evaporation. Using a full-scale windrow, Reinhart et al. (1993) found that moisture had a significant effect during composting of yard waste and municipal biosolids. Peak temperatures and time to reach stable temperatures were lower when the final moisture content was below 30%. Peak temperatures were highly correlated with moisture content (Figure 2.16).

The changes of moisture during composting vary with the method of

PEAK TEMPERATURE - C

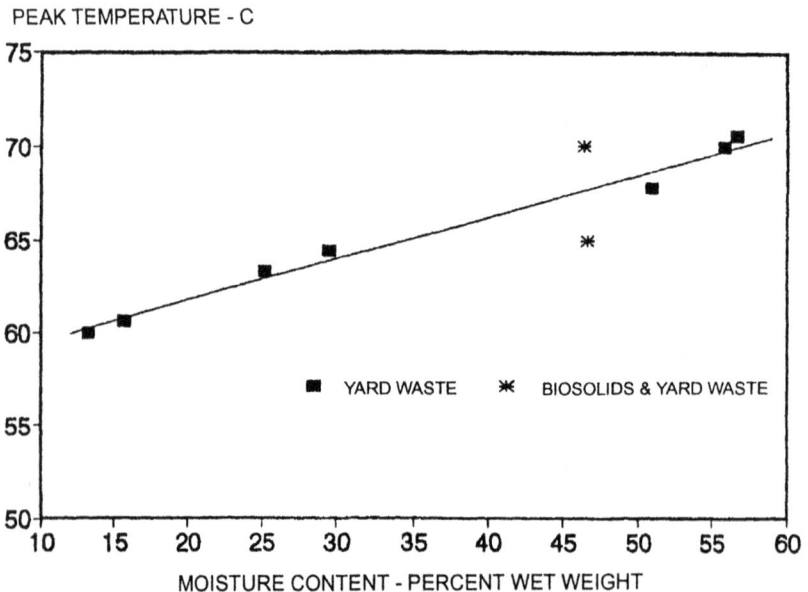

FIGURE 2.16. The relationship between moisture content and peak windrow temperature during composting. (Reprinted with permission from Reinhart et al., 1993.)

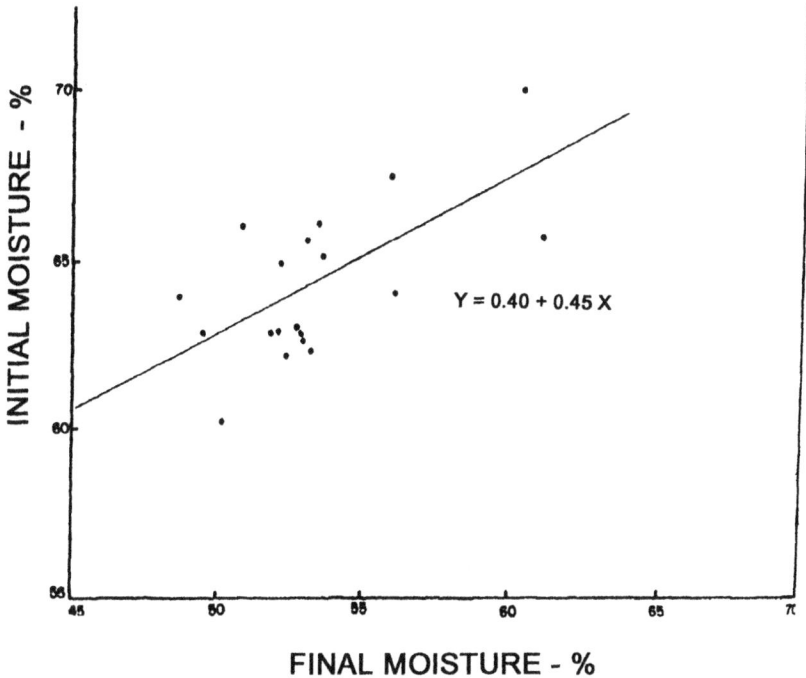

FIGURE 2.17. The effect of initial moisture content of a mixture of biosolids and wood chips on the final moisture content. (Data from Willson and Thompson, 1976.)

composting, the bulking agent, and the feedstock. De Bertoldi et al. (1982) measured the moisture content variation for three composting systems in a large pilot study. The initial moisture content of MSW and municipal biosolids mixture was 67.3%. Turning reduced the moisture content to 55% in 15 days and to 43% in 30 days. Further positive aeration decreased the moisture content of the mixture to 48% in 15 days and to 29% in 30 days. Negative (suction) aeration decreased the moisture content 57% and 45% in 15 and 30 days, respectively. As long as the moisture content remains in the 45% to 55% range during the active composting period, moisture is not a limiting factor.

Moisture removal is important for efficient material handling. For example, in the composting of biosolids, screening is often employed immediately after the active composting period. It is difficult to efficiently screen compost when the moisture content exceeds 45%. At higher moisture contents the rate of bulking agent recovery decreases. Therefore, in the design of aerated static piles, different aeration rates are employed at different times. In the initial phases, sufficient aeration must be supplied to support biological activity; however aeration is kept low to avoid overdrying or cooling the material. The aeration rate for biological activity is approximately 1/6 to 1/10 of that needed for water removal. Later in the process when it is desirable to remove water for efficient screening and materials handling, the aeration rate is increased.

A relationship between the initial moisture content of the biosolids and wood chip mixture and the final moisture content of the compost is shown in Figure 2.17 (Willson and Thompson, 1976). Willson and Thompson (1976) indicated that the initial moisture content of a compost mix should be less than 60% in order to obtain a moisture content less than 50% for efficient screening. The 60% moisture value is also the upper limit for adequate free air space.

TEMPERATURE

Temperature plays a major role in the composting process. At the same time, it is also a function of the process. Probably the most important aspect of temperature is its impact on the microbiological community. Thus, other vital reactions and elements of the composting process are affected and change with temperature. Temperature also affects the moisture relationships, which in turn affect microbiological activity. The interaction between various parameters and temperature often makes it difficult to separate cause and effect.

The composting process is often depicted in terms of time-temperature relationship. In a well-managed system temperature is regulated or manipulated to achieve the objectives desired. For example, if wastes contain human pathogens, disinfection becomes a primary objective. In these cases, USEPA and the states regulate the temperature that must be achieved before a product can be declared safe for use (see Chapter 14). For wastes that may not contain human pathogens such as animal manures, food processing wastes, and certain green materials, temperature control may be needed for plant pathogen destruction or weed seed destruction.

The time-temperature relationship affects the rate of decomposition of the organic matter and therefore is important for the production of a stable and mature product for consumer use. The effect of temperature on the microbial system results in major changes in the kinds and amounts of organisms (see Chapter 3). This effect is so important that often the time-temperature relationship is described in the broad terms of the organisms: mesophilic and thermophilic (see Chapter 3).

The type of process and the degree of control have a major effect on temperature and its consequences. The highest degree of temperature control is usually obtained in static or dynamic enclosed systems (classified by USEPA as in-vessel). Static systems offer the next highest degree of temperature control, and windrow systems offer the least. Temperature, in any system, is rarely uniform throughout the composting mass. The center of the mass tends to be hotter and the outer edges cooler. Greater pile surface area results in greater heat loss. The larger the mass or volume, the more heat is generated and the higher the potential temperature in the center.

Temperature generally rises initially and then levels off. At some point when microbial activity slows, temperatures begin to descend. Figures 2.18 and 2.19 show time-temperature curves for a static pile and a windrow system. In both cases temperatures exceeded the 55°C as required by USEPA 40CFR503 regulations for biosolids (see Chapter 13).

Table 2.3 shows a comparison between an aerated static pile and a windrow for several combinations of yard waste, wood waste, food waste, and mixed waste paper (E&A Environmental Consultants, Inc., 1993a, 1993b). Temperatures in the aerated static pile were consistently higher than in the windrow. Temperatures were lowest with wood wastes since this material had less available C for microbial activity.

There has been some debate regarding the optimum temperature for decomposition of organic matter. One reason for this controversy is that different feedstocks or materials decompose more rapidly at different temperatures. Most data in the literature indicate that the optimum temperature lies between 50–60°C. Wiley and Pierce (1955) indicated that maximum CO_2 production occurred at temperatures between 60–65°C for MSW (mixed garbage and refuse). Schultz (1961) reported that maximum decomposition of MSW occurred at temperatures between 65–70°C. Schultz (1960) and Regan and Jeris (1970) reported data by others showing that maximum oxygen uptake rates occurred between 45°C and 66°C.

Since oxygen uptake is a function of microbiological activity, the highest oxygen uptake rate should indicate the most optimum decomposition temperature. Jeris and Regan (1973c) reported that optimum temperatures for the decomposition of MSW was near 60°C. Bach et al. (1984) also reported that for biosolids composting 60°C was the optimum temperature.

It should be kept in mind that temperatures exceeding 55°C must be maintained for several days if waste contains pathogens. It is easier to achieve the high temperatures required for disinfection earlier in the process when easily digestible carbon for maximum microbial activity is available. In static piles and agitated bed systems, the center of the mass must be at a temperature higher than 55°C to ensure that the entire mass is meeting temperature requirements since the outer extremities have lower tempera-

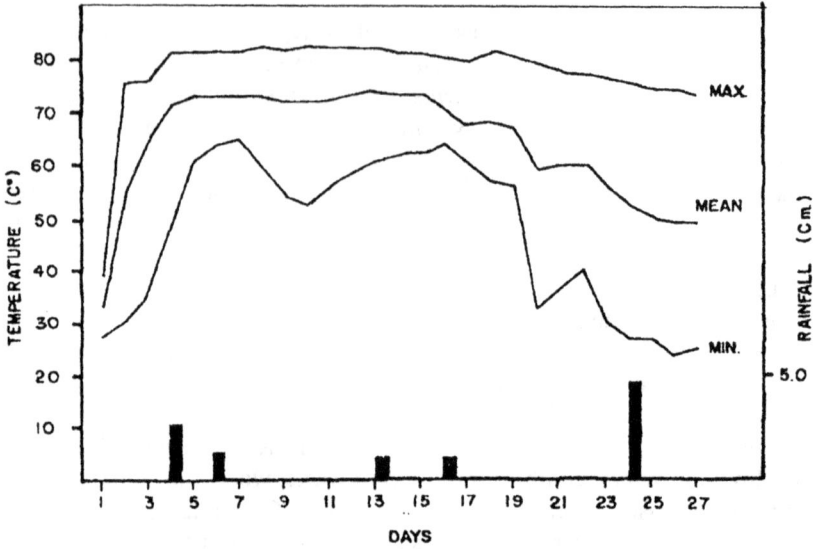

FIGURE 2.18. Time-temperature curves for biosolids and woodchips.

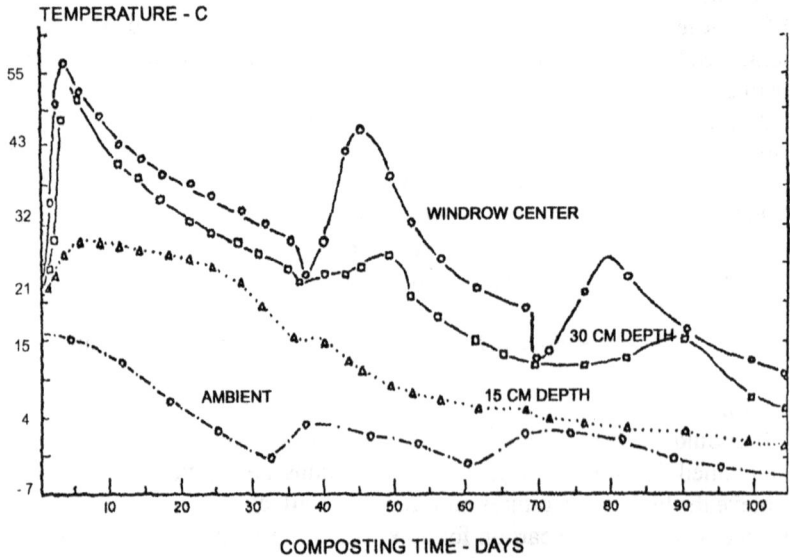

FIGURE 2.19. Windrow temperatures during composting of biosolids and bark. (Data from Walke, 1975.)

38

TABLE 2.3. Number of Consecutive Days at Indicated Temperatures for Several Food Waste Mixtures with Yard Waste (YD), Wood Waste (WW), Food Waste (FW), and Mixed Waste Paper (MWP).

Pile Type	Composition	Temperature, °C			
		>50	>55	>60	>70
ASP	YD/FW	49	30	17	2
	YD/MWP/FW	35	35	32	4
	WW/FW	48	16	7	2
ATW	YD/FW	34	17	8	5
	YD/MWP/FW	26	11	8	4
	WW/FW	13	8	6	4

Source: E&A Environmental Consultants, Inc. 1993b.

tures than the center. If there is no need to disinfect the waste, lower temperatures may achieve faster stabilization. Yard wastes or wastes containing weed seeds or plant pathogens may need to achieve high temperatures to produce a marketable product, however.

NUTRIENTS

Carbon

The two most important nutrients are carbon and nitrogen. Few other inorganic chemical reactions have been studied. The C to N ratios during composting affect the process and the product. As indicated earlier, the important parameter is the carbon available to microorganisms, not the total carbon in the material. During microbial growth, approximately 25 to 30 parts of C are needed for every unit of N (University of California, 1953; Gotass, 1956; Waksman, 1938).

Carbon is provided to the microbial community from decomposing plants and wastes from animals and humans. The C is utilized for cellular growth. Some of the microbial biomass returns C to the cycle. During microbial activity, respiratory CO_2 is evolved and emitted to the atmosphere. The readily available C is utilized initially. As the composting process continues, however, the rate of CO_2 evolution decreases as a result of decreased metabolic activity and the decrease of available carbon.

The change in CO_2 evolution as the composting process progresses has been studied by several researchers (Wiley and Pierce, 1955; Bach et al., 1984; Sikora and Sowers, 1985; Michel et al., 1993; Singley et al., 1982). Wiley and Pierce (1955) indicated that CO_2 and released moisture increase to a peak nearly simultaneously with peak temperature and then decrease. Carbon dioxide output exceeds H_2O output initially, but as the more available C is utilized the rate of CO_2 decreases and moisture output exceeds CO_2 output. The relationship between CO_2 evolution and volatile solids (VS) during the composting of biosolids and rice hulls is shown in Figure 2.20 (Bach et al., 1984). Bach et al. found a direct relationship between VS and CO_2. The higher the VS content of organic material, the greater the production of CO_2. Since VS represents the total C, rather than the available C in materials, this relationship indicates that all of the C in rice hulls is readily available to microorganisms.

An example of the changes in CO_2 during composting is found in Figure 2.21. The data show the rate of CO_2 evolution during composting of different ratios of leaves and grass in a small laboratory reactor under controlled temperature conditions (Michel et al., 1993). Three ratios, 100% leaves, 2/3:1/3 leaves to grass, and 1/3:2/3 leaves to grass, were studied. Higher grass contents in relation to leaves resulted in greater evolution of CO_2. The C/N ratios for the three mixtures studied were 48, 30, and 22. Carbon dioxide evolution rate increased for the first eight days and then decreased. The rate of decrease was greater for the two grass mixtures than with the leaves. This indicated that the grass provided more readily available C and N than the leaves. Available nutrients were consumed more quickly, resulting in a more

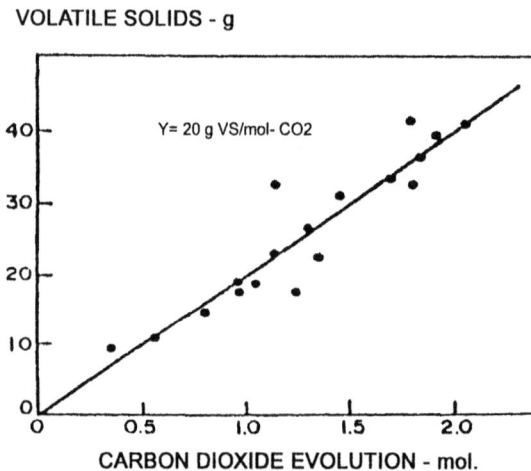

FIGURE 2.20. The relationship between carbon dioxide evolution and volatile solids during composting of biosolids and rice hulls. (Data from Bach et al., 1984.)

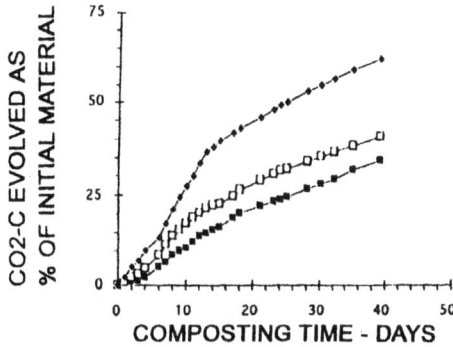

FIGURE 2.21. Cumulative carbon dioxide–carbon for three yard waste mixtures. (Reprinted with permission from Michel et al., 1993.)

drastic drop in activity. After 16 to 22 days, the rate of CO_2 was relatively constant.

Different compost feedstocks are more susceptible to C mineralization than others (Table 2.1). However, some feedstocks such as paper and paper products can be easily composted when prepared properly by size reduction and pulping. Microbial decomposition of organic matter takes place on surfaces of particles. The greater the surface area, the more rapid the decomposition. Reducing particle size of paper greatly increases its rate of decomposition (University of California, 1953).

E&A Environmental Consultants, Inc. (1993a, 1993b) composted milk cartons, bags used to ship diapers, and other paper products with food waste and yard waste in less than 20 days. Figure 2.22 shows the decrease in weight of paper bags used to hold diapers (courtesy of International Paper Company). The paper bags were shredded and composted with stable yard waste and nitrogen was added to bring the mixture up to a C/N ratio of 30:1. In 18 days, 87% of the paper was degraded and in 31 days 93% was degraded. Figure 2.23 shows the changes in paper and compost (E&A Environmental Consultants, Inc., 1994). In a large pilot study using yard waste, produce waste and ground wax-coated cardboard (WCC), 49% to 68% of the WCC was lost in 38 days and 87% to 89% was lost in 85 days (E&A Environmental Consultants, Inc., 1993a).

In nature, the decomposition rate of organic matter continues at a very slow rate since there is a slow release of C. This C comes from more recalcitrant decomposable compounds such as lignin. A similar situation can occur when composting MSW that is high in cellulose and low in N (Figure 2.24). Temperatures may remain high for extensive periods of time as a result of continual low-rate decomposition of the organic matter and a supply of C.

PERCENT DRY WEIGHT

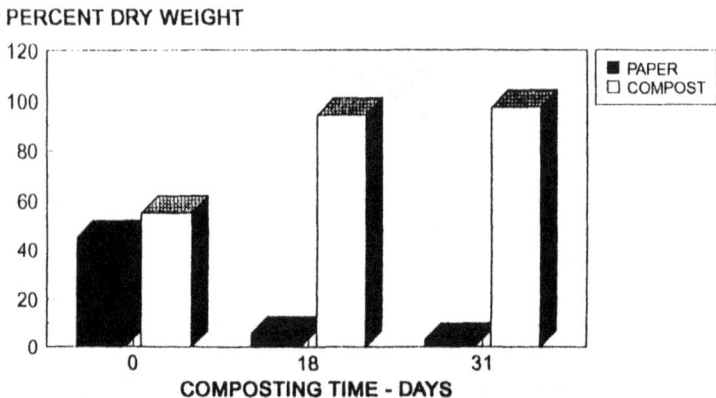

FIGURE 2.22. Decomposition of kraft paper used to hold diapers. (From E&A Environmental Consultants, Inc., 1995. Courtesy of International Paper Company.)

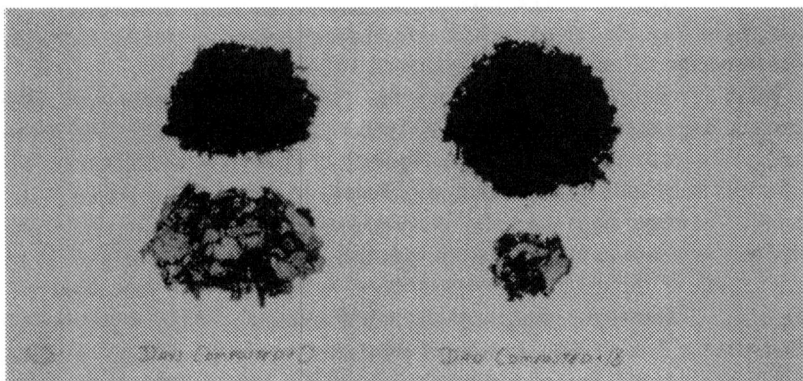

FIGURE 2.23. Decomposition of kraft paper used to package diapers during composting with yard waste and nitrogen. (From E&A Environmental Consultants Inc., 1975. Courtesy of International Paper Company.)

TEMPERATURE - C

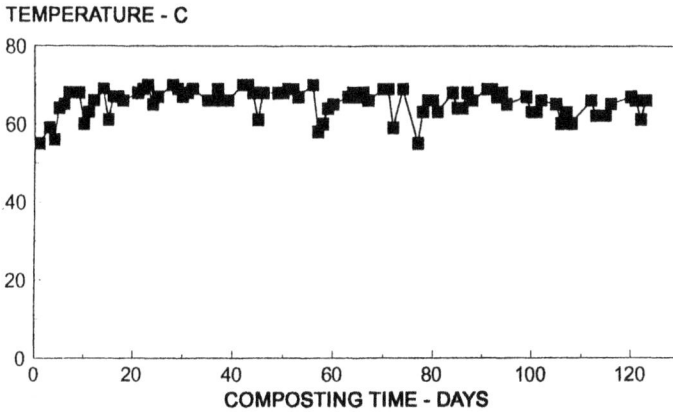

FIGURE 2.24. Average temperature during windrow composting of MSW. (Data from Wu and Epstein, 1994.)

Nitrogen

Microorganisms need N for protein synthesis. Bacteria may contain 7% to 11% N on a dry weight basis and fungi from 4% to 6% (Anderson, 1956). The amount of N in wastes varies with the type of waste. For example, food wastes and biosolids have higher N content than yard waste. The N content and C/N of several feedstocks used in composting are shown in Table 2.4. Microorganisms utilize C and N at a ratio of 30:1. Low C/N ratios in feedstocks result in nitrogen volatilization in the form of ammonia. This is particularly true under alkaline conditions. The imbalance of C/N is illustrated by the problem encountered by many facilities that receive large volumes of grass in the summer and do not have a sufficient carbon source to offset the low C/N ratio. Anaerobic or partially aerobic conditions can result in ammonia release to the atmosphere (Knuth, 1970). The loss of N reduces the value of compost as a fertilizer.

At C/N ratios exceeding 50:1, the composting process slows because of rapid cell growth and depletion of available N, resulting in reduced cellular growth. As cells die, their stored N becomes available to living cells (Bishop and Godfrey, 1983).

Cappaert et al. (1975) found that the addition of mineral N fertilizer to bark increased the rate of decomposition. The optimum addition of N was between 0.5% and 0.8% depending on the moisture content. Cappaert et al. (1976) found that the amount of N needed for maximum O_2 consumption varied with the type of bark. Respiration for hardwood was highest with the addition of 1.5% N. With softwood bark, maximum respiration occurred at

TABLE 2.4. Nitrogen Content and C/N Ratios of Several
Feedstocks Used in Composting.

Compost Feedstock	Nitrogen, Percent Dry Weight	C/N Ratio	Data Source
MSW, USA	0.2–3	15	Numerous data sources and experience
MSW, Japan	1.2–2.7	13–31	Inoko et al., 1979
Biosolids	<0.1–17.6		Sommers, 1977
Biosolids digested	0.5–3.4	15.7	Parker and Sommers, 1983 Poincelot, 1975
Fruit wastes	1.52	34.8	Poincelot, 1975
Yard wastes	0.19–1.17 1.95	— 22.8	Lisk et al., 1992 Kayhanian and Tchobanoglous, 1992
Paper	0.25	173	Poincelot, 1975
Sawdust	0.11	511	Poincelot, 1975
Grass clippings	2.46–5.0	10–20	Michel, 1993 E&A Environmental Consultants, Inc., 1993a
Leaves	0.93	48	Michel, 1993
Produce waste	1.8–2.5	15–25	E&A Environmental Consultants, Inc., 1993a
Food wastes	3.2	15.6	Kayhanian and Tchobanoglous, 1992
Pharmaceutical wastes	2.55	19	Poincelot, 1975
Wood (pine)	0.07	723	Poincelot, 1975
Seaweed	1.9	19	Gotaas, 1956
Oat straw	1.05	48	Gotaas, 1956
Wheat straw	0.3	128	Gotaas, 1956

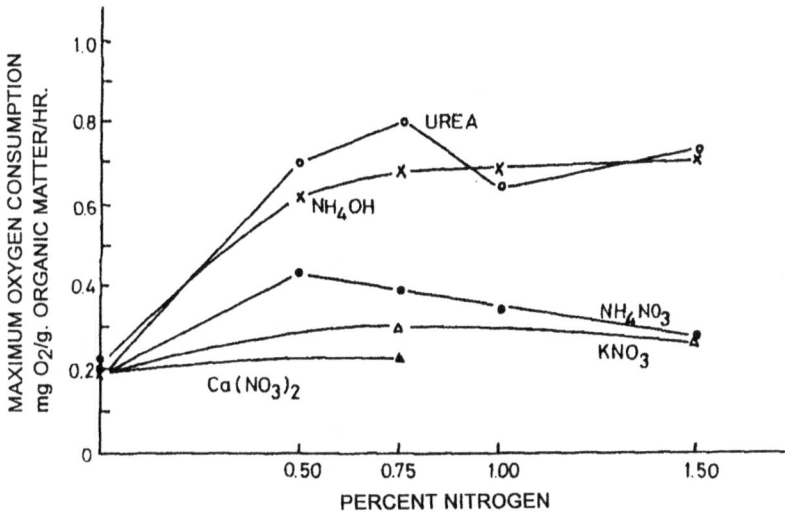

FIGURE 2.25. Effect of nitrogen source on oxygen consumption during composting of softwood bark. (Reprinted with permission from Cappaert et al., 1976.)

N levels between 0.5% and 0.75%. At the higher N values respiration was depressed. Oxygen consumption varied with the N source (Figure 2.25). Urea provided the highest respiration rate.

Changes in concentration of nitrogen species can vary with aeration rate and bulking agent ratio during the composting of biosolids, as seen in Figures 2.26 and 2.27 (Bishop and Godfrey, 1983). In aerated static piles, total N decreased from 1.6% to 1.2%. A much greater decrease occurred without aeration. Most of the organic N was lost or mineralized during the first 14 days. The greatest reduction in total and organic N occurred with an aeration rate of 8 m^3/min/m^3, indicating that aeration enhanced microbial mineralization and volatilization. Both NO_3-N and NH_3-N initially increased and then decreased. This could be the result of an increase in N-fixing bacteria.

Increasing the bulking agent ratio from 1:1 to 3:1 greatly reduced total and organic N (Figure 2.27). A higher bulking agent ratio results in better aerobic conditions due to greater porosity, which enhances mineralization and volatilization. The decrease in total and organic N occurred during the first seven days. Nitrate N and NH_3-N again increased initially and then decreased.

Aoyama (1985) determined the distribution of various N compounds in four fractions of four composts. Three of the composts were from animal wastes and one was MSW compost. In the MSW compost, the highest amount of organic N was found in the water-soluble fraction; the least

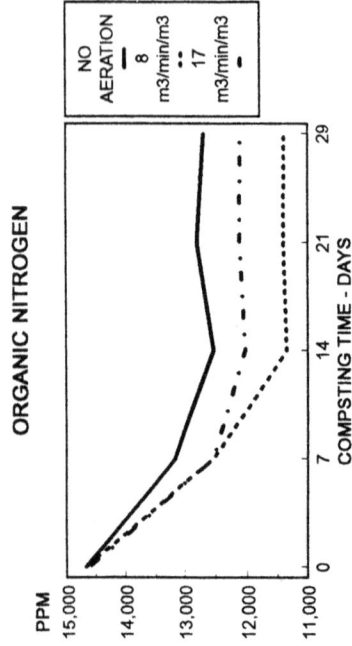

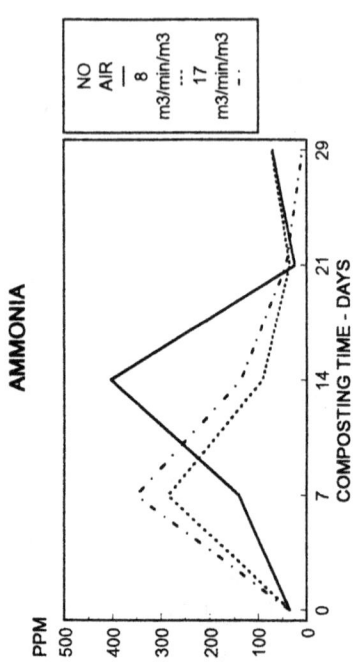

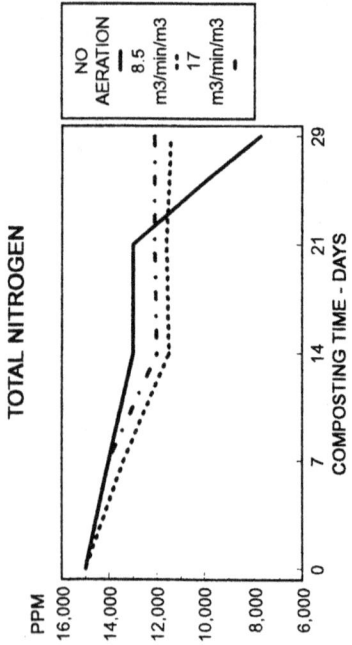

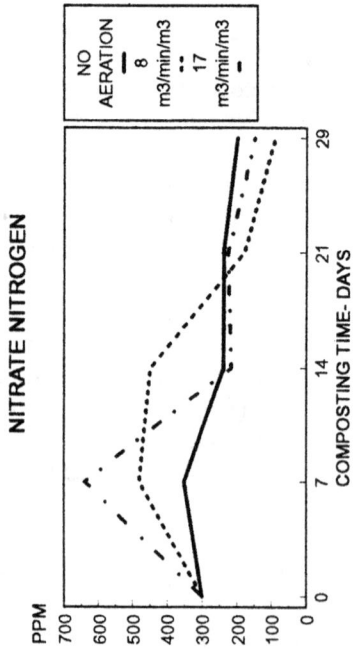

FIGURE 2.26. Changes in nitrogen species as affected by aeration during biosolids composting. (Reprinted with permission from Bishop and Godfrey, 1983.)

46

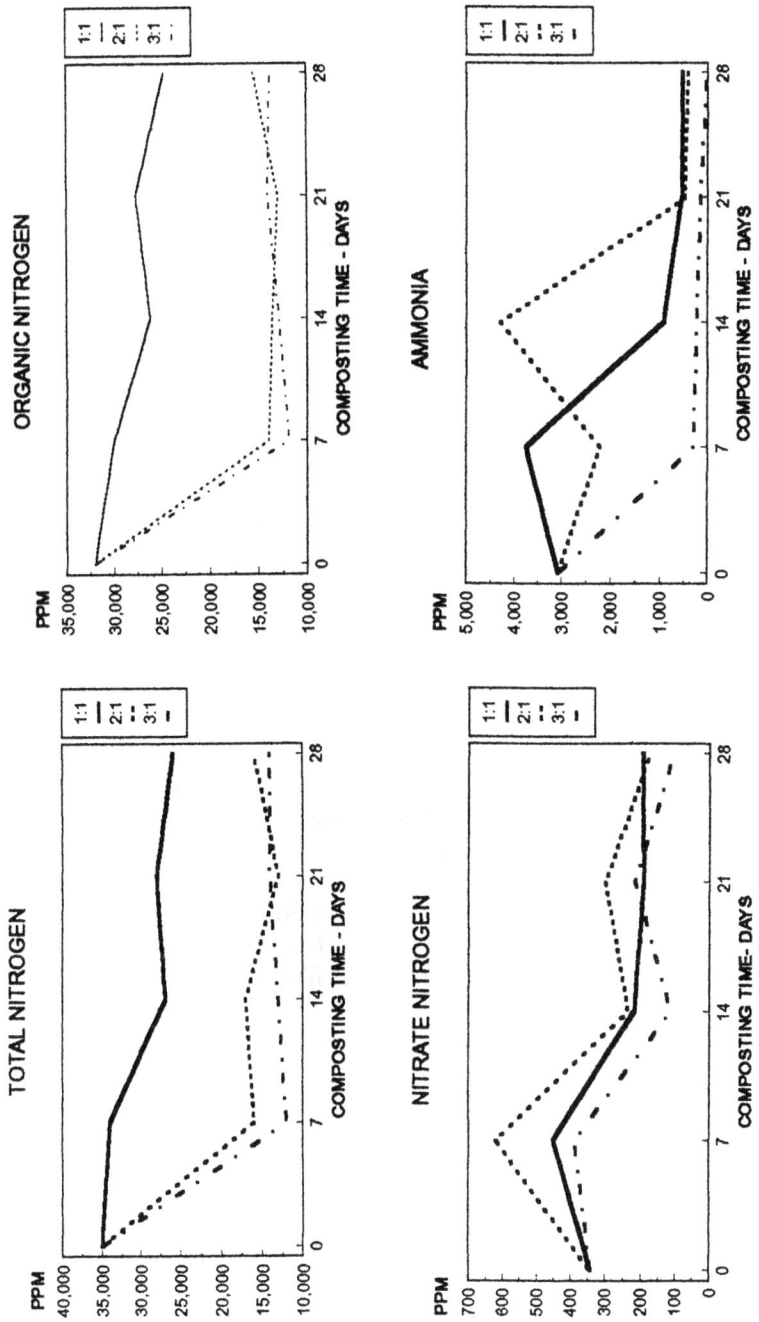

FIGURE 2.27. Changes in nitrogen species as affected by bulking agent ratio during composting of biosolids. (Reprinted with permission from Bishop and Godfrey, 1983.)

amount of organic N was in the coarse solid fraction. This fraction represented the sediment after shaking a 100 g sample with water and allowing it to settle for 16 hours. Based on Stokes' law, even this fraction would contain a significant amount of fine particles. Mineralized N ranged from 2.0% to 15.7% of the organic N for the various fractions, with an average of 7% to 8%. Total organic N was 1712 mg N/g dry matter. Apparently, different organic wastes and the different amounts of carbonaceous amendment affected the N mineralization rate.

De Bertoldi et al. (1982) found that N loss was 18% with turning, 11% with negative (suction) aeration, and 5% with positive aeration. The lower loss of N by forced aeration was attributed to lower pH and temperature.

PH

pH is a measure of the acidity or alkalinity of a medium. Although measuring pH is very simple, there can be a considerable discrepancy as a result of different compost-to-water ratios used (Carnes and Lossin, 1970). In the measurement of pH in soil, a slurry of 1:1 ratio of soil to water is often used. Carnes and Lossin (1970) indicated that due to the complex nature of buffers, a high dilution is better. They recommended a dilution ratio of 1:50 compost:water.

pH affects the growth response of organisms. Bacteria that require pH of 5 or less for maximal growth are called "acidophiles." Their optimum pH is usually 2 or 3. Bacteria that grow best at a pH range of 7 to 12 are termed "alkalophiles." Their optimum value is usually 9.5. Finally, organisms preferring a pH near neutrality are called "neutrophiles." Compost feedstocks can vary from a low pH of 3 to 4 as in the case of grape pomace or have a high pH as a result of lime addition. The pH curve for a biosolid and bark mixture is shown in Figure 2.28 (Walke, 1975). Initially, the pH decreased from approximately 5.5 to near 5.1. This decrease is short and the result of formation of organic acids. Thereafter, as the temperature increases, the pH increases.

Jeris and Regan (1973c) found that maximum thermophilic composting occurred at a pH range of 7.5 to 8.5. Stabilization efficiency decreased at lower or higher pH values. The effect of pH on temperature during the composting of raw biosolids is found in Table 2.5 (Epstein et al., 1977). As shown, the highest temperatures for the longest period of time were obtained with a pH range of 6.5 to 9.6. With a pH below 6.0 and above 9.6, temperatures were below the 55°C level necessary to meet U.S. regulations.

However, Site II in Maryland composts biosolids that have a pH above 12 and achieves temperatures above 55°C for several days. Also, during the

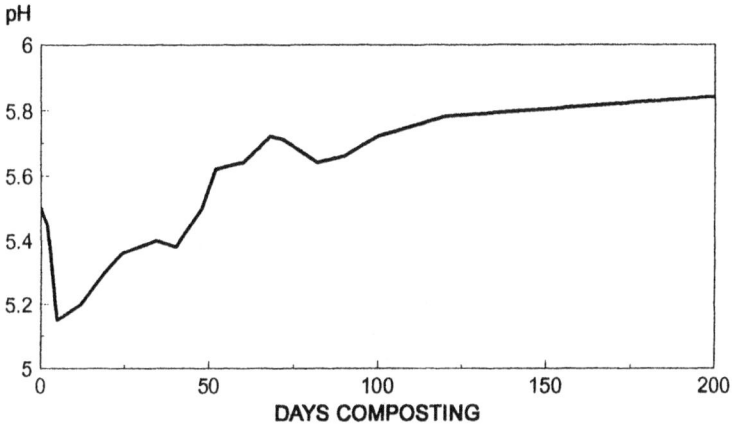

FIGURE 2.28. Changes in pH during windrow composting of sewage biosolids and bark. (Data from Walke, 1975.)

composting of low pH grape pomace in Napa Valley, California, high temperatures are achieved. Several explanations may be given for the difference by the data cited by Epstein et al. (1977) and the results in Site II and Napa Valley. For example, factors such as moisture can affect the temperature at various pH levels and the insulation can also affect temperature.

TABLE 2.5. Effect of pH on Temperatures Attained during 21 Days of Composting Raw Biosolids with Wood Chips (Epstein et al., 1977).

Biosolid pH	Maximum Temperature (°C)	Number of Days Minimum Temperature Exceeded		
		40°C	50°C	60°C
5.3	79	3	1	0
6.0	69	6	0	0
6.5	89	18	8	6
7.1	84	11	11	5
9.6	82	19	15	9
10.7	73	2	0	0
11.6	66	1	0	0

SUMMARY

Composting is the biological decomposition of organic matter under controlled aerobic conditions. During the composting process, microorganisms decompose organic compounds, which consist of carbohydrates, sugars, proteins, fats, hemicellulose, cellulose, and lignin. Carbohydrates are more easily decomposed whereas lignin is more resistant to decomposition. Many factors affect the composting process. Aerobic microorganisms need oxygen, water and nutrients for their metabolism and cell synthesis. As a result of microbial activity, heat is liberated and, if contained within the composting mass, the temperature rises. Temperature increases through the mesophilic phase into a thermophilic phase and then decreases back into the mesophilic phase. During the course of these transitions, the microbial population changes, thereby affecting the rate of organic matter decomposition. Knowledge of the basic aspects of composting is important to achieve effective and maximum decomposition in the shortest time with minimal negative impacts to the environment.

REFERENCES

Anderson, M. S. 1956. Compost as a means of garbage disposal. *The Soil and Crop Sci. Soc. of Florida Proc.* 16:134–144.

Aoyama, M. 1985. Properties of fine and water-soluble fractions of several composts. I. Micromorphology, elemental composition and nitrogen mineralization of fractions. *Soil Sci. Plant Nutr.* 31(2):189–198.

Bach, P. D., M. Shoda, and H. Kubota. 1984. Rate of composting of dewatered sewage sludge in continuously mixed isothermal reactor. *J. Ferment. Technol.* 62(3):285–292.

Bishop, P. L. and C. Godfrey. 1983. Nitrogen transformations during sludge composting. *BioCycle.* 24(4):34–39.

Buckingham, E. Contribution to our knowledge of the aeration of soils. *U.S. Dept. of Agr. Bull. Soils Bull.* 25.

Caballero, R. 1984. *Experience at a Windrow Composting Facility: Los Angeles County Site.* Technology Transfer, U.S. EPA, Municipal Environ. Res.: Cincinnati, OH.

Cappaert, I., O. Verdonk, and M. De Boodt. 1975. Composting of hardwood bark. *Compost Sci.* 16(4):12–15.

Cappaert, I., O. Verdonck, and M. De Boodt. 1976. Composting of bark from pulp mills and the use of bark compost as a substrate for plant breeding. *Compost Sci.* 17(4):6–9.

Carnes, R. A. and R. D. Lossin. 1970. An investigation of the pH characteristics of compost. *Compost Sci.* 11(5):18–21.

Coppola, S., S. Dumontet, and P. Marino. 1983. Composting raw sewage sludge in mixture with organic or inert bulking agents, pp.125–147. In E. I. Stentiford (ed.). *Proc. of the Int'l. Conf. on Composting of Solid Wastes and Slurries.* The Univ. of Leeds, England.

De Bertoldi, M., G. Vallini, and A. Pera. 1982. Comparison of three windrow compost systems. *BioCycle.* 23(2):45–50.

E&A Environmental Consultants, Inc. 1993a. *Composting Produce Waste and Wax Coated Cardboard Using a Low Technology Approach: Pilot Study Results.* Final report. The Clean Washington Center, Seattle, WA.

E&A Environmental Consultants, Inc. 1993b. *Food Waste Collection and Composting Demonstration Project for City of Seattle Solid Waste Utility.* Final report. Seattle, WA.

E&A Environmental Consultants, Inc. 1994. In-house data.

Epstein, E., G. B. Willson, and J. F. Parr. 1977. The Beltsville aerated pile method for composting sewage sludge. pp. 201–213. In *New Processes of Waste Water Treatment and Recovery.* Soc. of Chem. Ind. London.

Finstein, M. S. and M. L. Morris. 1975. Microbiology of solid waste composting. *Advances in Applied Microbiology.* 19:113–149.

Gotaas, H.B. 1956. *Composting—Sanitary Disposal and Reclamation of Organic Wastes.* World Health Organization, Geneva, Switzerland.

Gray, K. R. and K. Sherman. 1969. Accelerated composting of organic wastes. *Chem. Eng.* 20(3):64–74.

Gray, K. R., A. J. Biddlestone, and R. Clark. 1973. Review of composting—Part 3: Processes and products. *Process Biochem.* 8(10):11–30.

Halley, Jr., E. M., T. G. Shea, and L. W. Braswell. 1980. Process dynamics in aerated static-pile composting. Paper presented at the *53rd Annual Water Pollution Control Fed.* Las Vegas, Nevada.

Inoko, A., K. Miyamatsu, K. Sugahara, and Y. Harda. 1979. On some organic constituents of city refuse composts produced in Japan. *Soil Sci. Plant Nutr.* 25(2):225–234.

Jeris, J. S. and R. W. Regan. 1973a. Controlling environmental parameters for optimal composting. Part I. *Compost Sci.* 14(1):10–15.

Jeris, J. S. and R. W. Regan. 1973b. Controlling environmental parameters for optimum composting. Part II. Moisture, free air space and recycle. *Compost Sci.* 14(2):8–15.

Jeris, J. S. and R. W. Regan. 1973c. Controlling environmental parameters for optimal composting. Part III. *Compost Sci.* 14(3):16–22.

Kaibuchi, Y. 1961. Research on composting of city refuse and night soil. *J. Sanitary Eng. Div. ASCE.* SA6. 87:101–139.

Kayhanian, M. and G. Tchobanoglous. 1992. Computation of C/N ratios for various organic fractions. *BioCycle.* 33(5):58–60.

Knuth, D. T. 1970. Nitrogen-cycle ecology of solid waste composting. *Develop. Ind. Microbiol.* 11:387–395.

Lisk, D. J., W. H. Gutenmann, M. Rutzke, H. T. Kuntz, and G. Chu. 1992. Survey of toxicants and nutrients in composted waste materials. *Arch. Environ. Contam. Toxicol.* 22:190–194.

Manios, V. I., P. E. Tsikalas, H. I. Siminis, and O. Verdonck. 1987. Phytotoxicity of olive tree leaf compost, pp. 296–308. In M. De Bertoldi et al. (ed.). *Compost: Production Quality and Use.* Elsevier Applied Sci. London.

Michel Jr., F. C., C. Adinarayana Reddy, and L. J. Forney. 1993. Yard waste composting: Studies using different mixes of leaves and grass in a laboratory scale system. *Compost Sci. & Util.* 1(3):85–96.

Parker, C. F. and L. E. Sommers. 1983. Mineralization of nitrogen in sewage sludges. *J. Environ. Qual.* 12(1):150–156.

Poincelot, R. P. 1975. The biochemistry and methodology of composting. *Bull. 754.* The Connecticut Agr. Expt. Station, New Haven, CT.

Regan, R. W. and J. S. Jeris. 1970. A review of the decomposition of cellulose and refuse. *Compost Sci.* 11(1):17–20.

Reinhart, D. R., A. R. deForest, S. J. Keely, and D. R. Vogt. Composting of yard waste and wastewater treatment plant sludge mixtures. *Compost Sci. & Util.* 1(2):58–64.

Schultz. K. L. 1960. Rate of oxygen consumption and respiratory quotients during aerobic decomposition of a synthetic garbage. *Compost Sci.* 1:36–40.

Schultz, K. L. 1961. *Aerobic Decomposition of Organic Waste Materials.* Final report. Project RG-4180 (C5R1). National Institutes of Health, Washington, DC.

Schultz, K. L. 1962. Continuous thermophilic composting. *Appl. Microbiol.* 10:108–122.

Sikora, L. J. and M. A. Sowers. 1985. Effect of temperature control on the composting process. *J. Environ. Qual.* 14(3):434–439.

Singley, M. E., A. J. Higgins, and M. Frumkin-Rosengaus. 1982. *Sludge Composting and Utilization—A Design and Operating Manual.* New Jersey Agr. Expt. Sta., Rutgers U., New Brunswick, NJ.

Sommers, L. E. 1977. Chemical composition of sewage sludges and analysis of their potential use as fertilizers. *J. Environ. Qual.* 6:225–232.

Stevenson, F. J. Organic matter-micronutrient reactions in soil, pp. 145–186. In J. J. Mortvedt et al. (ed.). *Micronutrients in Agriculture.* 2nd ed. Soil Sci. Soc. Am. Madison, WI.

Thompson, J. 1984. *Experiences at Static Pile Composting Operations.* Technology Transfer, U.S. EPA, Municipal Environ. Res., Cincinnati, OH.

University of California. 1953. *Reclamation of Municipal Refuse by Composting.* Tech. Bull. No. 9, Univ. of California, Berkeley.

Viel, M., D. Sayag, and L. Andre. 1987. Optimization of agricultural industrial wastes management through in-vessel composting, pp. 231–237. In M. De Bertoldi et al. (ed.). *Compost: Production Quality and Use.* Elsevier Applied Sci., London.

Waksman, S.A. 1938. *Humus.* Williams and Wilkins Co., Baltimore, MD.

Walke, R. H. 1995. The preparation, characterization and agricultural use of bark-sewage compost. Ph.D. Dissertation, Univ. of New Hampshire, Durham, NH.

Wiley, J. S. and G. W. Pierce. 1955. A preliminary study of high rate composting. *Proc. Am. Soc. Civil Eng.* Paper No. 846. 81:1–28.

Wiley, J. S. and J. T. Spillan. Refuse-sludge composting in windrows and bins. *Compost Sci.* 2(4):18–25.

Willson, G. B. 1977. Equipment for composting sewage sludge in windrows and in piles. In *Composting of Municipal Residues and Sludges. Proc. Nat'l Conf. on Composting of Municipal Residues and Sludges.* Hazardous Materials Control Res. Inst., Silver Spring, MD.

Willson, G. B. and J. L. Thompson. 1976. Dewatering of sludge compost piles, pp. 46–54. In *Proc. Nat'l. Conf. on Municipal & Indust. Sludge Composting.* Hazardous Materials Control Res. Inst., Silver Spring, MD.

Witter, E. and J. M. Lopez-Real. 1987. Monitoring the composting process using parameters for compost stability, pp. 351–358. In M. De Bertoldi et al. (ed.). *Compost: Production Quality and Use.* Elsevier Applied Sci., London.

Wu, N. and E. Epstein. 1994. Making the desert bloom. Paper presented at *The Composting Council Annual Meeting,* Alexandria, VA.

CHAPTER 3

Microbiology

INTRODUCTION

As a biological process, composting involves a myriad of microorganisms. These organisms decompose organic matter and organic compounds. Several important factors affect the microbiological population. These include oxygen, moisture, temperature, nutrients and pH as discussed in Chapter 2. Because of the complex nature of organic matter and many organic compounds, both natural and xenobiotic, many microbes and other organisms are involved in the decomposition process.

Some more resistant organic compounds have been shown to be biodegraded. *Phanerochaete chrysosporium* is known to degrade lignin (Jeffries, 1987). Similarly, Crawford (1978) reported that *Streptomyces* decomposed lignocellulose. Janshekar and Fiechter (1982) evaluated several bacteria that used the carbon source of lignin-related aromatic compounds such as benzoic, *p*-OH-benzoic, vanillic, veratric, syringic, and *p*-coumaric acids. The white rot fungus, *Phanerochaete chrysosporium,* is known to biodegrade chlorinated biphenyls and chlorinated dibenzodioxins, recalcitrant xenobiotic compounds.

Knowledge, understanding, and identification of microorganisms that are capable of biodegrading specific compounds is very useful in the composting process and the biodegradation of pollutants and soil contaminants through composting. Attempts have been made to accelerate the composting process through seeding or optimizing the growth of certain organisms.

The microbial population during composting affects its own destiny by producing heat that in turn affects microbial populations. Krueger et al.

(1973) indicate that microorganisms are classified according to the temperatures they can tolerate and grow.

Microorganisms	Temperature Range
Cryophiles or psychrophiles	0–25°C (32–77°F)
Mesophiles	25–45°C (77–104°F)
Thermophiles	>45°C

Organisms associated with composting fall into the two classes: mesophiles and thermophiles. Although we are primarily concerned with the growth of organisms at mesophilic and thermophilic temperatures, it is of interest to know that many organisms survive and grow at low or high temperatures (Marsh and Simpson, 1976).

For example, Singer (1954) stated that the fungi of the arctic are no different from fungi of other regions. Similarly, Kelly (1954) showed that there are no distinct arctic bacteria. Thus, compost stored outdoors in cold climates can support microbial growth. The lowest temperature at which the growth of any microorganism has been found is −7.5°C. Table 3.1 lists several microorganisms that can tolerate and grow at low temperatures. Many of these species are found in composting media.

TABLE 3.1. Some Microorganisms That Grow at Temperatures below 0°C.

Organism	Growth Temperature (°C)	Reference
Alternaria radicina	−0.5	Lauritzen (1926)
Aplanobacter insidiosum	−1.7	Jones and McCulloch (1926)
Bacterium spp.	−7.5	Tarr (1954)
Cladosporium herbarum	−6	Brooks and Hansford (1923)
Cladosporium sp.	−6.7	Berry and Magoon (1934)
Lactobacillus sp.	−6	Berry and Magoon (1934)
Penicillium sp.	−6	Berry and Magoon (1934)
Pseudomonas fluorescens	−4	Berry and Magoon (1934)
Rizoctonia carotae	−3	Rader (1948)
Sporotrichum sp.	−6.7	Berry and Magoon (1934)

TABLE 3.2. Some Microorganisms That Grow at
Temperatures Exceeding 75°C.

Organism	Growth Temperature (°C)	Reference
Actinomyces thermofuscus	65	Waksman et al. (1939)
Bacillus kaustrophilus	73–75	Prickett (1928)
Bacillus terminalis var. *thermophilus*	73–75	Prickett (1928)
Bacillus thermophilus	78	Georgevitch (1910)
Bacillus Losanitichi	78	Georgevitch (1910)
Chemolithotrophic bacteria	>90	Tansey and Brock (1978)
Clostridium sp.	75 (optimum 60)	Tansey and Brock (1978)
Sulfolobus acidocaldarius	85–90 (optimum 70–75)	Tansey and Brock (1978)
Desulfovibrio thermophilus	85	Tansey and Brock (1978)
Thermomicrobium roseum	85 (optimum 70–75)	Tansey and Brock (1978)

Microorganisms have also been found to grow at very high temperatures. Table 3.2 lists some microorganisms that grow at temperatures exceeding 75°C. Although historically it has often been stated that composting ceases or is extremely reduced at temperatures exceeding 60°C (McKinley et al., 1985), this is not the case. Tansey and Brock (1978) indicated that there can be potential benefit in favoring thermotolerant organisms in composting; for example, temperatures of 70 to 80°C are required for productive compost in mushroom production. Tansey and Brock (1978) cited the beneficial activity of thermophilic fungi in the retting of guayule, which is a source of natural rubber. Retting is analogous to composting, as the latex is improved by microbial decomposition of the plant material resulting in rubber with greater tensile strength.

In composting, the primary interest is in the organisms that decompose organic matter at the most efficient temperature and other conditions. However, when composting wastes that are contaminated with human pathogens, destruction of these pathogens is an important secondary objective. This occurs at thermophilic temperatures exceeding 55°C.

Waksman and Cordon (1939) indicated that plant residue decomposition is influenced by material preparation. According to these authors, the most important factors are (1) the nature of the composting feedstock, especially the proportion of nitrogenous compounds to carbohydrates; (2) decomposition temperature; and (3) the microbial population of the compost. The latter two factors are very interactive. As pointed out in Chapter 2, the most important factors affecting composting include oxygen, C/N ratio, temperature, and moisture.

Although microbiology is the single most important facet of composting, relatively little research has been conducted in this area. Unlike physical and chemical factors that are easy to measure, biological aspects are more difficult to evaluate. This chapter provides information on the factors that affect the microbial population, rates of microbial degradation, and the manifestations that occur because of changes in these factors.

MICROBIAL POPULATIONS

Both the variety of microbes and the microbial populations fluctuate throughout the composting process. Table 3.3 shows the wide range of organisms that have been isolated from compost (Walke, 1975; Poincelot, 1975). Figure 3.1, based on data by Walke (1975), shows the fluctuations of actinomycete, fungi, bacteria, and total microbial populations as a function of time during windrows composting of biosolids and bark.

TEMPERATURE

Temperature is the single most important parameter affecting the number and types of microorganisms in a composting pile. As temperatures increase, the growth of organisms accelerates. While many composting organisms thrive at temperatures near 50°C, numerous organisms grow at temperatures exceeding 50°C.

Waksman and Cordon (1939) found that introduction of thermophilic populations to a compost pile brought about greater decomposition at higher temperatures than did the introduction of mesophilic populations. Waksman et al. (1939) noted that at 50°C, thermophilic fungi, bacteria, and actinomycetes were all active in compost; however, at 65°C the fungi were rare whereas bacteria and actinomycetes were predominant. At 75°C, spore-forming bacteria were the predominant, or perhaps the only organisms. Waksman and Cordon (1939) reported that little growth of thermophilic actinomycetes occurred at 28°C. A temperature of 50°C proved to be optimum, with no growth at 65°C.

TABLE 3.3. Microorganisms Identified in Composting.

Bacteria	**Actinomycetes**
Aerobacter (aerogenes)	*Nocardia brasiliensis*
Bacillus megatherium	*Thermomonospora viridis*
B. stearothermophilus	*T. curvata*
B. cereus	*Micromonospora parva*
B. mycoides	*M. vulgaris*
Pseudomonad sp. (seven isolates)	*Thermoactinomyces vulgaris*
Flavobacterium sp.	*Actinoplanes* sp.
Micrococcus sp.	*Thermopolyspor polyspora*
Sarcina sp.	*Pseudonocardia*
Cellumonas folia	*Streptomyces violaceoruber*
Chondrococcus exiguus	*S. thermoviolaceus*
Mycococcus virescens	*S. rectus*
M. fulvus	*S. thermofuscus*
Thibacillus thiooxidans	*S. thermovulgaris*
T. denitrificans	*Thermomonospora fusca*
Proteus sp.	*T. glaucus*

Fungi	
Rhizopus nigricans	*Absidis orchidis*
Rhizoctonia sp.	*Rhizopus arrhizus*
Geotrichum candidum	*Candida (parapsilosis)*
Mucor pusillus	*Cladosporium herbarum*
Penicillium digitatum	*Rhodotorula rubra*
Mucor racemosus	*Aspergillus tamarii*
Torulopsis sp.	*Zygorhynchus vuilleminii*
Aspergillus flavus	*Trichosporon cutaneum*
Absidia (ramosa)	*Verticillium* sp.
Saccharomyces sp.	*Synecephalastrum* sp.
Pulluloria sp.	*Pichia* sp.
Pythium sp.	*Cylindrocaron* sp.
Hanisenula sp.	*Chaetomium (thermophile)*
Trichoderma koningi	*Lipomyces* sp.
Talaromyces (Penicillium) duponti	*Sporotrichium thermophile*
Stysanus stemonitis	*Fusarium moniliforme*
Glibotrys (alaboviridis)	
Humicola insolens	
Humicola griseus var. *thermoideus*	

Protozoans	**Algae**
Chilomonas (paramecium)	*Hormidium (nitens)*
Cyathomonas (truncata)	*Vaucheria (terrestris)*
Lycogala epidendrum	*Euglena mutabilis*
Cercomonas (crassicanda)	*Protococcus vulgaris*
	Dactylococcus (bicandatus)
	Chlorococcum humicola
	Microcoleus vaginatus
	Porphyridium (cruentum)
	Kentrosphaera sp.
	Diatoms (unidentified)

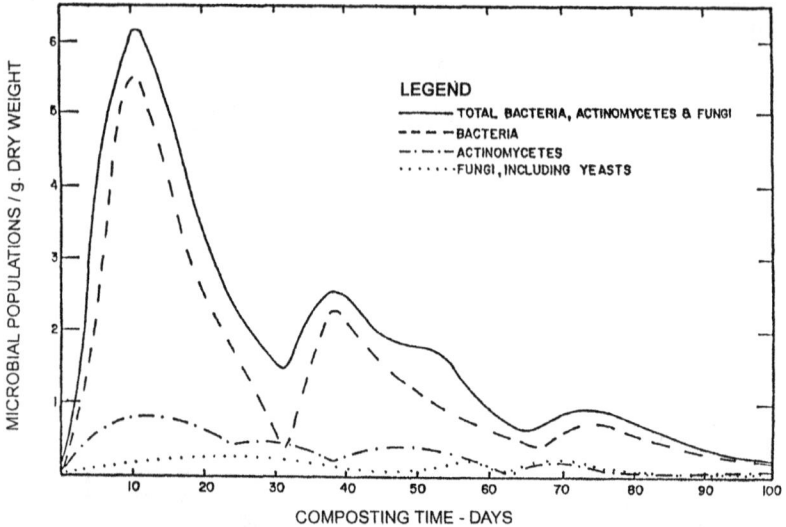

FIGURE 3.1. Changes in microbial population during windrow composting of biosolids and bark. (Data from Walke, 1975.)

One of the earliest studies of the effect of temperature on microorganisms during composting was carried out by Webley (1947). During the composting of grass and straw, the numbers of aerobic mesophilic bacteria were reduced at high temperatures (55–65°C). This was followed by regrowth when the temperature fell below 40°C. As the temperature decreased below 40°C, the rate of recovery of bacteria resulted in much higher numbers than the initial count. For grass compost the maximum number of bacteria was 2.6×10^{10}/g of fresh weight whereas for the straw compost the maximum count was 7.8×10^9/g of fresh weight.

Webley (1947) could not account for this "striking phenomenon," whereby the final numbers were higher than the initial. Most likely, bacteria at the cooler regions of the container survived and repopulated the compost. Further, the high temperatures eliminated the competitive organisms that normally would have inhibited their growth. Earlier research at USDA in Beltsville, Maryland showed that when sterile compost was inoculated by pathogenic microorganisms, the levels of salmonellae in the compost were much higher than normally found in unsterile material. Forsyth and Webley (1948) noted that the development of thermophilic organisms at high temperatures is rapid and takes place in aerobic zones of compost piles.

The upper limit for fungi has been found to be 60°C (Cooney and Emerson, 1964). Thus, it is expected that fungi would not be found in the center of composting piles where temperatures often exceed 65°C. However, as the temperature in the center of a compost pile drops, fungi from the cooler regions of the pile recolonize into the center.

Chang and Hudson (1967) studied the succession of fungi in straw and grass compost (Figure 3.2). They found that both mesophilic and thermophilic fungi decreased in population as temperature of compost reached 70°C. However, as soon as the temperature decreased below 65°C, the thermophilic fungi resumed growth and their population peaked at approximately 45°C.

In other studies with wheat straw compost, mesophilic bacteria increased in population during the first two days of composting and then decreased when the temperature rose to 70°C. As the temperature descended to 55°C, populations increased. Populations of both thermophilic bacteria and thermophilic fungi grow as the temperature increases. Peak populations of bacteria occurred at 77°C. Above 70°C, thermophilic fungal growth was limited and contributed to an increase in temperature. Above 77°C, the activities of thermophilic bacteria and actinomycetes continued to generate heat.

Thermophilic fungi were found to have recolonized once the temperatures were between 50–60°C, whereas mesophilic fungi did not recolonize until temperatures were at 20–45°C. *Humicola insolens* appeared to be the most thermotolerant fungus, present even at the maximum heat phase, which exceeded 70°C.

Other fungi also present at high temperatures were *Chaetomium thermophile, Humicola lanuginosa, Malbranchae pulechella* var. *sulfurea* and *Talaromyces duponti. Aspergillus fumigatus, Mycelia sterila* C.t.6, and

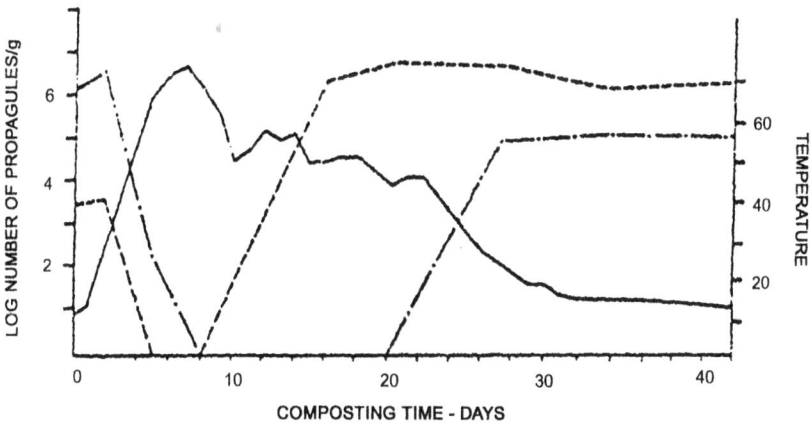

FIGURE 3.2. Changes in fungal populations during the composting of wheat straw. (Data from Chang and Hudson, 1967.)

Mucor pusillus were not found during peak temperature (4 to 14 days). *Aspergillus fumigatus* recolonized after 14 days but *Mucor pusillus* did not.

Chang (1967) hypothesized that the reason why *Mucor pusillus* did not recolonize was nutritional factors rather than temperature. Thus, he found that this fungus could only utilize simple carbon compounds such as glucose and cellobiose and could not utilize cellulose and hemicellulose as a carbon source. Therefore, as the compost process proceeded and the easily available carbon compounds were utilized by other microorganisms, the lack of carbon source precluded the growth and development of *Mucor pusillus*.

Walke (1975) studied the fluctuations in microbial populations during windrow composting of biosolids and bark (Figure 3.3). He concluded that during the initial stages of composting, soluble compounds such as sugars, alcohols, acids and proteins were available, and bacteria dominated as the principal microbes in the compost. Turning the windrows on the 36th and 68th day resulted in an increase in bacteria. This could be the result of improved aerobic conditions. However, fungi and actinomycetes decreased. Although bacterial activity was reduced towards the end of the composting period, bacteria remained the dominant organism. Windrow temperatures in the center increased to 55°C during the first five days and then decreased to 27°C until the 36th day when the windrow was turned. The temperature then rose to approximately 45°C and then decreased until the windrow was turned on the 68th day, when it rose again to approximately 27°C.

Microbial activity was measured using the triphenylformazan (TPF) test, an enzyme assay (Figure 3.4). An excellent relationship was obtained between microbial counts and transmittance and a standard curve of TPA and transmittance. Walke indicated that the method was faster, easier, and less expensive than the plate count method. This method could be used to determine optimum windrow management.

Microbial changes during the decomposition of leaves were reported by Hankin et al. (1976). Compost was produced primarily from deciduous leaves and composted in a pile 6 m long, 3 m wide, and 2 m high, which was turned once over a 100-day period. Bacteria increased during the first 10 days from 10^8 cells/g to 10^{10} cells/g. During that period, temperatures ranged from 17°C to 36°C. After 10 days numbers ranged from 10^7 to 10^9 cells/g. Many mesophilic bacteria were killed at temperatures of 40°C to 58°C. Fungi numbers were lower (approximately 10^6 cells/g) and declined during the thermophilic phase. Actinomycetes were recovered in large numbers (10^8/g of sample) during the thermophilic stage. However, their numbers declined as the pile cooled.

The authors also reported on the microorganisms that produced extracellular degradative enzymes (Table 3.4). Lipase producers fluctuated during the entire composting period. Numbers ranged from 10^4 to 10^{10} cells/g and

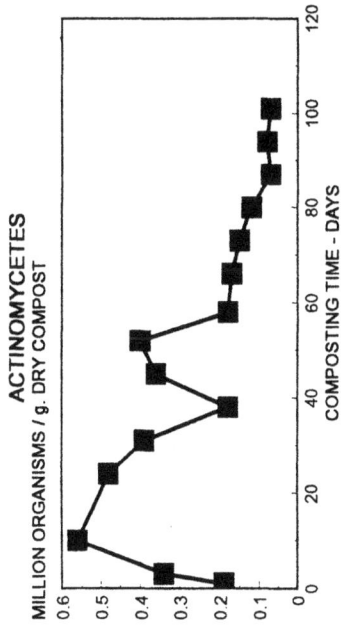

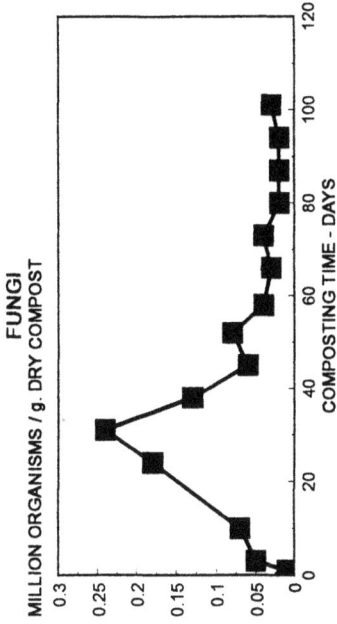

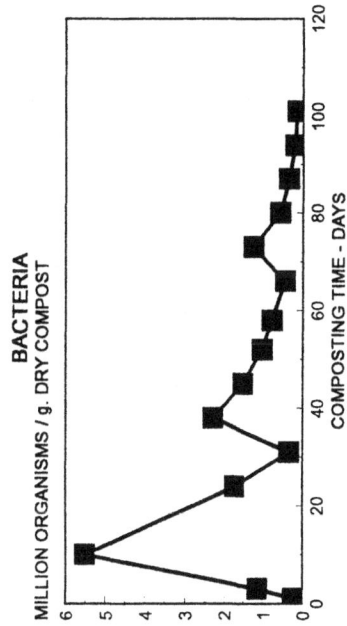

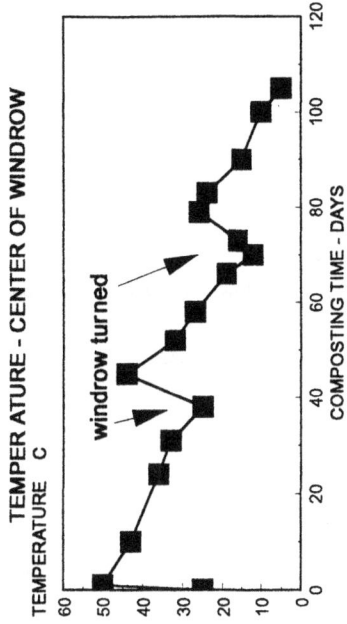

FIGURE 3.3. Changes in microbial population and temperature in center of windrow during composting of biosolids and bark. (Data from Walke, 1975.)

61

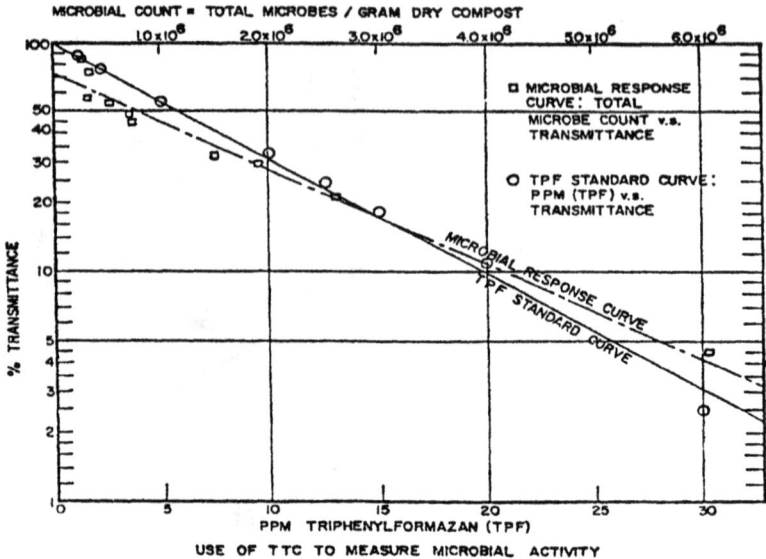

MICROBIAL COUNT = TOTAL MICROBES / GRAM DRY COMPOST

□ MICROBIAL RESPONSE
CURVE : TOTAL
MICROBE COUNT v.s.
TRANSMITTANCE

O TPF STANDARD CURVE :
PPM (TPF) v.s.
TRANSMITTANCE

% TRANSMITTANCE

PPM TRIPHENYLFORMAZAN (TPF)

USE OF TTC TO MEASURE MICROBIAL ACTIVITY

FIGURE 3.4. Measurement of microbial activity in biosolid and bark compost using triphenyl-formazan. (Data from Walke, 1975.)

followed patterns similar to the total bacteria. Fungi-producing lipase enzymes ranged from 10^3 to 10^6 cells/g.

Finally, the authors examined alkane utilizers since alkanes constitute from 10% to 50% of the total surface lipids (Kolattukudy, 1968). The numbers of alkane producers essentially followed the same pattern as total bacteria, but represented only about 10% to 20% of the bacterial population. Other enzyme producers studied include protease, urease, cellulose (Cx), amylase and pectate transeliminase.

De Bertoldi et al. (1980) studied the populations of bacteria, fungi and actinomycetes during windrow and forced-air composting of biosolids with the organic fraction of MSW. The total number of bacteria increased during the first 14 days to a level of 10^9 and then decreased to 10^7 (Figure 3.5). A similar pattern at lower magnitude was observed for the ammonia, proteolytic, pectinolytic, and cellulolytic bacteria. The nitrogen-fixing bacteria increased within the first seven days and then decreased. The nitrogen-fixing bacteria were identified as *Azomonas* sp., *Klebsiella,* and *Enterobacter* (aerobes).

Bacillus sp. (facultative aerobes) and *Clostridium* sp. (anaerobes) generally followed the thermophilic phase. These organisms followed the temperature curve that reached a peak of 70°C at seven days and then decreased.

Thermophilic temperatures existed for the first 23 days followed by mesophilic temperatures.

Both mesophilic and thermophilic fungal populations initially increased and then decreased (Figure 3.6). Two separate peaks were observed. For the thermophilic fungi, the peak occurred after three to four days of composting. For the mesophilic fungi, the peak occurred after 35 days. Cellulose-degrading fungi were more numerous than the other fungi. The changes in actinomycetes populations are shown in Figure 3.7. As expected, the mesophilic actinomycetes decreased during the first seven days and then reappeared in the 10th day, reaching a maximum at 30 days although the temperatures from the 10th to the 23rd days exceeded mesophilic temperatures.

Safwat (1980) investigated the changes in the microbial populations during the composting of cottonseed waste (Figure 3.8). Mesophilic bacteria were higher than the populations of thermophilic bacteria throughout the composting period, although thermophilic temperatures exceeding 68°C were achieved. From the 7th to the 120th day, temperatures exceeded 52°C. Cellulose decomposing bacteria increased with increasing temperature for the first 60 days and then decreased to undetectable amounts during the next

TABLE 3.4. *Extracellular Degradative Enzymes Identified during Composting of Leaves.*

	Minimum	Maximum
Total Bacteria:	10^7	10^{10}
Enzyme		
Lipase producers (bacteria)	10^4	10^{10}
Lipase producers (fungi)	10^3	10^6
Alkane utilizers	10^4	10^9
Protease producers (bacteria)	10^6	10^{10}
Protease producers (fungi)	10^3	10^6
Urease producers	low, variable	
Cellulose (Cx) producers	<17% of total bacteria	
Amylase producers	<20% of total bacteria	
Pectin degraders (bacteria)	10^4	10^8
Pectin degraders	10^3	10^8

Adapted from Hankin et al., 1976.

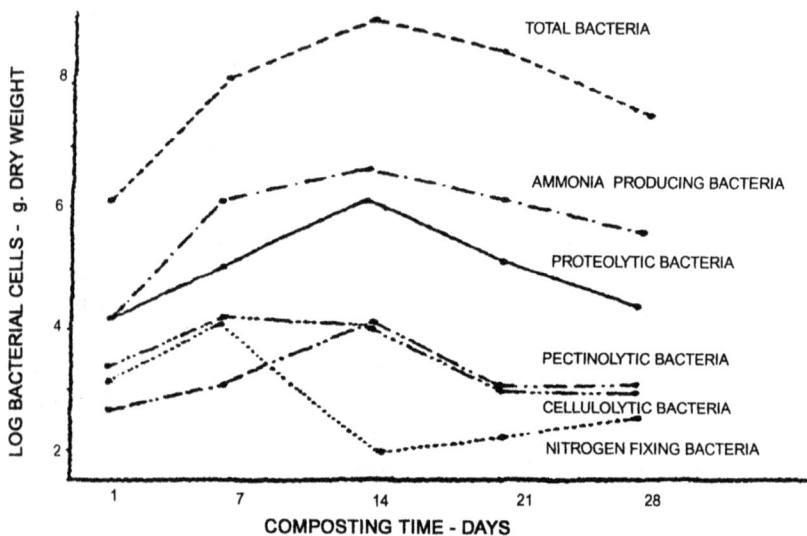

FIGURE 3.5. Changes in major physiological bacterial populations during the composting of biosolids. (Reprinted with permission from De Bertoldi et al., 1980.)

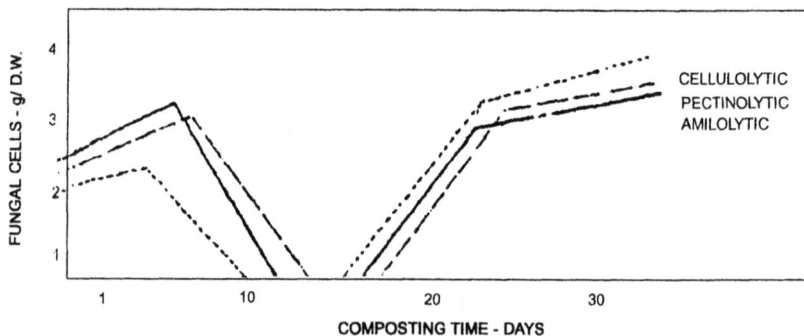

FIGURE 3.6. Changes of major physiological fungal populations during composting of biosolids. (Reprinted with permission from De Bertoldi et al., 1980.)

64

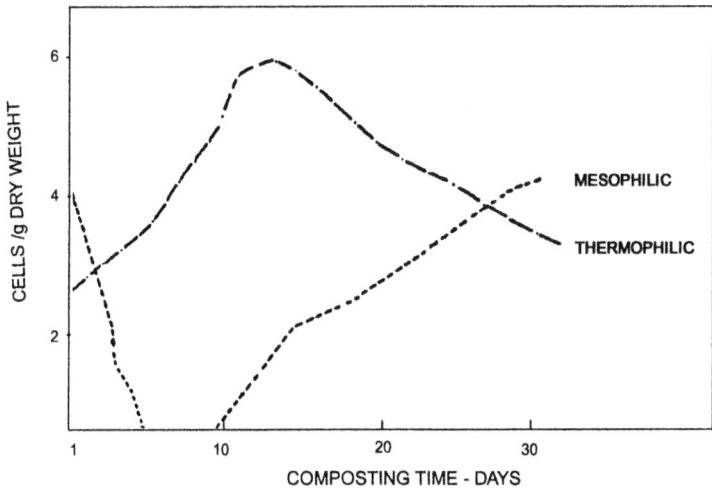

FIGURE 3.7. Changes in actinomycetes population during the composting of biosolids. (Reprinted with permission from De Bertoldi et al., 1980.)

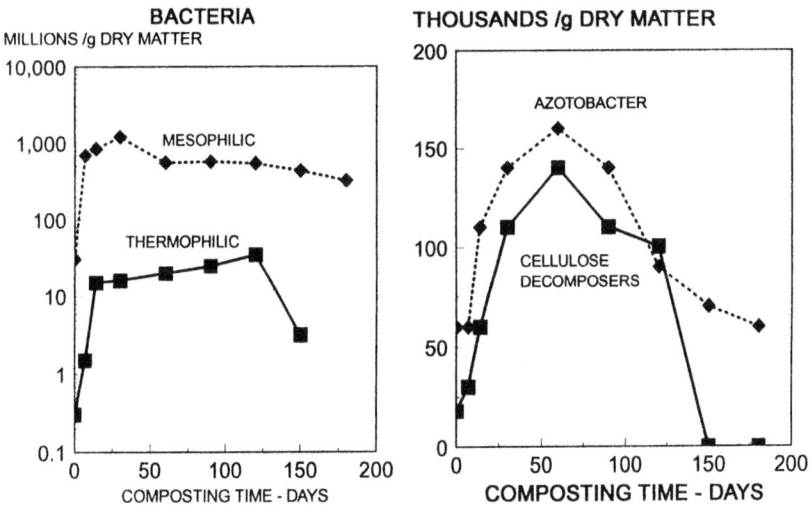

FIGURE 3.8. Changes in bacteria, cellulose decomposers, and azotobacter populations during composting of cottonseed waste. (Data from Safwat, 1980.)

90 days. Azotobacter, a nitrogen fixer and utilizer, also increased rapidly during the first 60 days. Even at 180 days it was found at a level of 60,000 organisms/g dry matter.

Using a small, air-induced and insulated vessel, Chino et al. (1983) determined the microbial populations during composting of biosolids and rice hulls (Figure 3.9). The highest counts of aerobic bacteria, fungi, ammonifiers, and denitrifiers occurred in the middle stages of composting (35 days). Actinomycetes increased for the first 35 days and then decreased. This is similar to the pattern found by Walke (1975) and De Bertoldi et al. (1980) for thermophilic actinomycetes.

The reverse was true for the mesophilic actinomycetes. Data on nitrogen bacteria are shown in Table 3.5 (Chino et al., 1983). Ammonia and denitrifier bacteria increased during the first 35 days and then decreased. Ammonia and nitrite oxidizers decreased during the entire period of composting. The high numbers of denitrifiers (9900×10^5 organisms/g) seem erroneous in contrast to the total aerobic bacteria (1920×10^5 organisms/g). Temperatures in the system exceeded 60°C and large quantities of ammonia were liberated. No data are presented on oxygen. It is possible that conditions were anaerobic, which would explain the relatively low aerobic bacteria counts. Alternatively, the number of denitrifiers may have been reported erroneously. A denitrifier count of 9900 would correlate better with data collected in other studies.

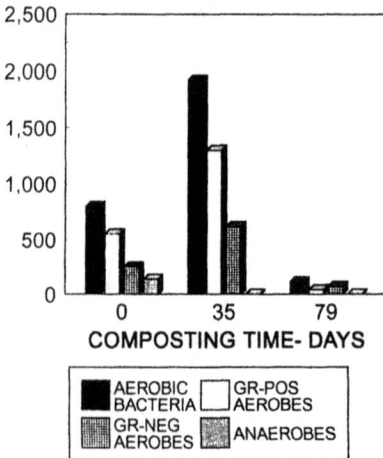

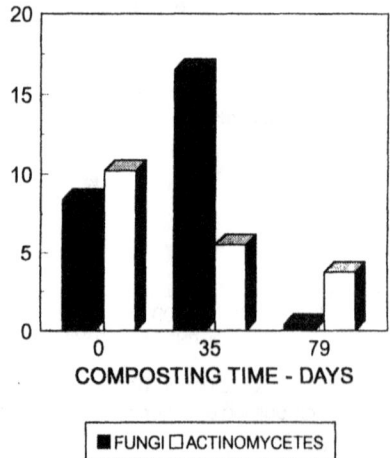

FIGURE 3.9. Changes in microbial populations during composting of biosolids and rice hulls in a small vessel. (Data from Chino et al., 1983.)

TABLE 3.5. *Changes in Nitrogen Bacteria during Composting of Biosolids and Rice Hulls.*

Organism	Number of Organisms/g Solids at Different Composting Times		
	0 Days	35 Days	79 Days
Ammonifiers	3.4×10^5	2.4×10^7	4.4×10^6
Denitrifiers	1.3×10^8	9.9×10^8	2.0×10^7
Ammonia oxidizers	4.3×10^6	1.4×10^6	3.7×10^4
Nitrite oxidizers	8×10^3	3×10^2	3×10^2

Source: Chino et al., 1983.

Diaz-Ravina et al. (1989) found significantly lower numbers (approximately 3700) of denitrifiers in MSW compost (Figure 3.10). Ammonifiers and proteolytic were the most numerous of the nitrogen bacteria. The amylolytics and aerobic cellulolytic represented the largest number of the carbon assimilators, and among the sulfur metabolizers, the sulfur reducers and elementary sulfur oxidizers were the most numerous.

Strom (1985a, 1985b) studied thermophilic bacteria in MSW composting in a small laboratory reactor and also collected samples from an MSW composting facility at Altoona, Pennsylvania and an MSW-biosolids facility in Leicester, England. Of the 652 randomly selected colonies, 87% were identified as *Bacillus* spp. Other isolates included two genera of unidentified non-spore-forming bacteria, the actinomycetes *Streptomyces* spp. and *thermoactinomyces* sp., as well as the fungus *Aspergillus fumigatus*. All of the samples recovered from Altoona and Leicester except for one were also found in the laboratory study.

Strom (1985b) concluded that maximum microbial diversity was obtained at 60°C and that in the laboratory the diversity dropped sharply at incubation temperatures between 60°C and 65°C. However, his data show that the percentage of isolates among taxa were higher at 60°C to 65°C and 65°C to 69°C. This was due to the presence of *Bacillus stearothermophilus*. There was no indication that microbial diversity was related to decomposition.

In summary, during aerobic composting three temperature phases affect microbial populations. The first, the mesophilic (ambient to 45°C) phase, is followed by a thermophilic (45°C to 70°C) phase. Finally, the third stage is a return to mesophilic (45°C to cooler) temperatures. Maximum growth of mesophilic bacteria occurs during the final mesophilic phase. Thermophilic bacteria, fungi, and actinomycetes peak during the thermophilic phase.

TOTAL MICROORGANISMS

LOG /g DRY COMPOST

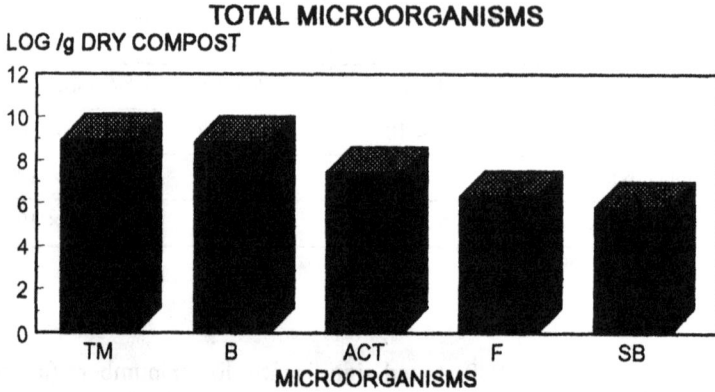

TM = Total microbial population; B= bacteria;
ACT = actinomycetes; F= fungi; SB = aerobic spore-forming bacteria.

CARBON ASSIMILATORS

LOG /g DRY WEIGHT

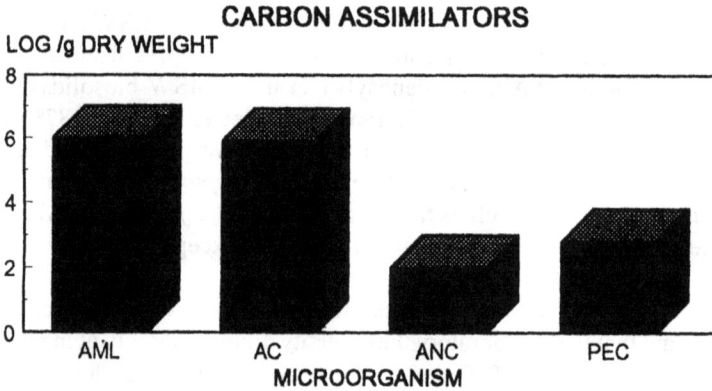

AML =Amylolytics; AC = aerobic cellulolytics;
ANC = anaerobic cellulolytics; PEC = pectolytics.

FIGURE 3.10. Microbial populations in MSW compost. (Data from Diaz-Ravina, 1989.)

NITROGEN MICROORGANISM

LOG /g DRY WEIGHT

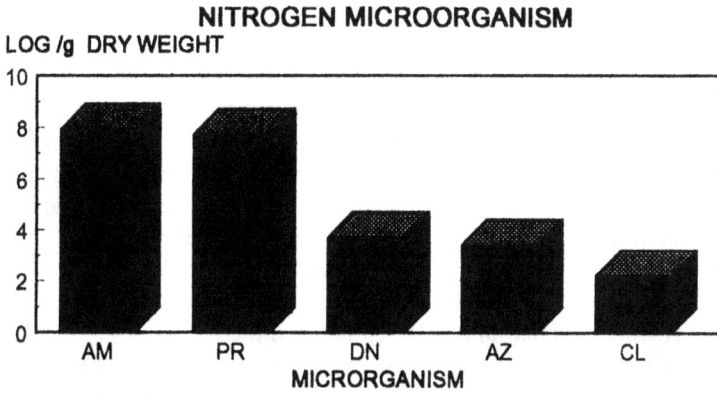

AM = ammonifiers; PR = proteolytics; DN = denitrifiers
AZ = azobacter; CL = cladosporium

SULFUR METABOLIZERS

LOG /g DRY WEIGHT

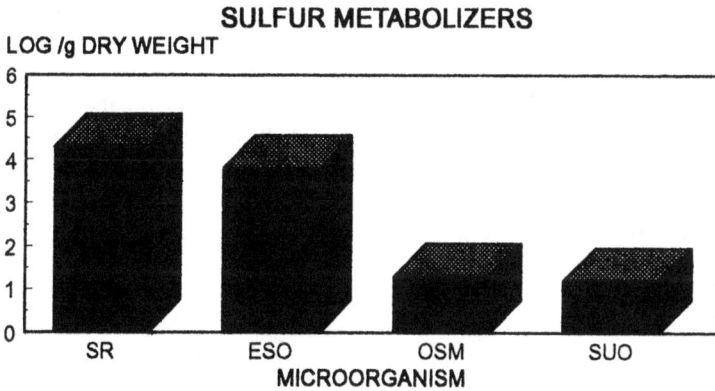

SR = sulfur reducers; elementary sulfur oxidizers
OSM = anaerobic organic sulfur mineralizers; SUO = sulfide oxidizers

FIGURE 3.10 (continued). Microbial populations in MSW compost. (Data from Diaz-Ravina, 1989.)

Mesophilic fungi populations were higher during the initial mesophilic phase.

MOISTURE

Moisture is often a limiting factor in the composting process resulting in microbial inactivation and slowing or cessation of the decomposition process. Numerous studies have tried to determine the effect of moisture on microbial growth (Marsh and Simpson, 1976), but no data are available on the effect of moisture during composting on microbial diversity.

Fungi can grow at relative humidities (RH) below 100%, whereas bacteria grow best at RH of 100% (Finstein and Morris, 1975). Based on indirect measurements of microbial activity using either O_2 or CO_2 respiration, microbial activity begins to decrease at moisture contents of about 40%. As indicated earlier, moisture contents exceeding 60% restrict oxygen in the composting mass. This in turn affects aerobic microbial species and their activity.

NUTRIENTS

Nitrogen can limit microbial activity and the rate of decomposition of organic matter. The effect of different sources of N on the mesophilic bacteria is shown in Figure 3.11 (Safwat, 1980). In this study, nitrogen in the form of ammonium nitrate and urea was added to batches of cottonseed waste and composted in 1 tonne piles. The populations of each essentially followed the temperature curve; however, some differences were found between additives.

Nitrogen seemed to reduce the population of mesophilic bacteria, with ammonium nitrate having the greatest negative effect. In contrast, the addition of N resulted in slight increases in the thermophilic populations (Figure 3.12). Mesophilic fungi were essentially absent during thermophilic temperatures; however, after 90 days when temperatures dropped below 50°C, they reappeared (Figure 3.13). Ammonium nitrate increased the mesophilic fungal populations over urea and the NO-N treatments. Mesophilic actinomycetes grew rapidly for the first 14 days and then decreased to very low levels (Figure 3.14). Little difference was found between the two N sources on the thermophilic actinomycetes (Figure 3.15). In piles where no N was applied, actinomycetes disappeared after seven days; with urea, the numbers continued to grow for 30 days and then disappeared. Finally, with the nitrate N treatment, actinomycetes persisted for 60 days.

MILLIONS/g DRY MATTER   TEMPERATURE - ˚C

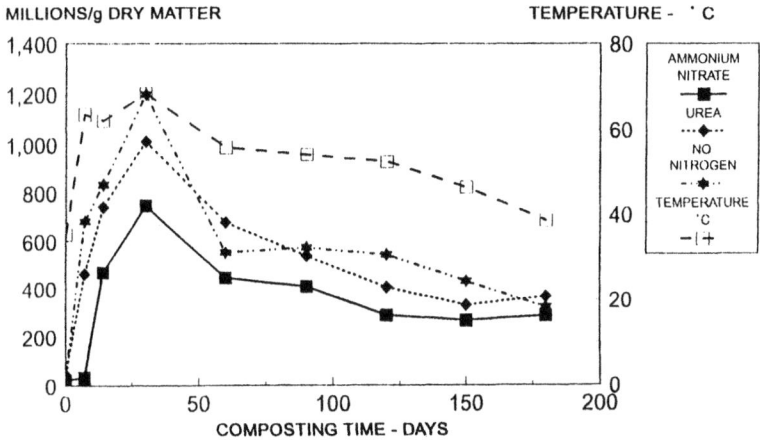

FIGURE 3.11. Effect of nitrogen source on mesophilic bacterial populations during composting of cottonseed waste. (Data from Safwat, 1980.)

THOUSANDS /g DRY MATTER

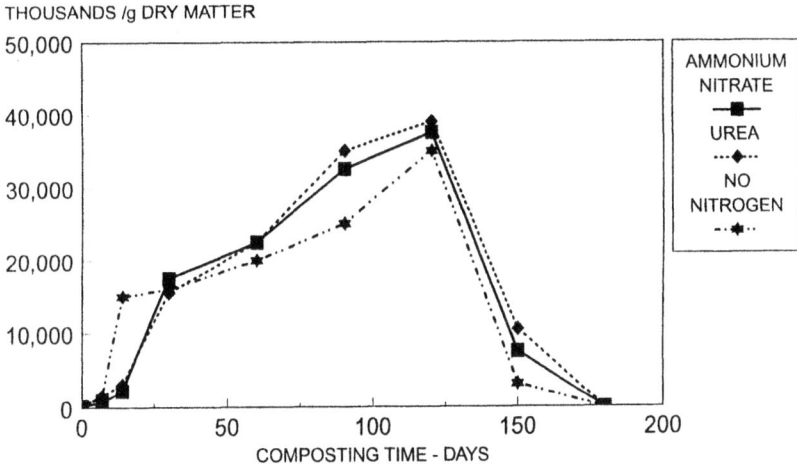

FIGURE 3.12. Effect of nitrogen source on thermophilic bacteria during the composting of cottonseed wastes. (Data from Safwat, 1980.)

71

MILLIONS /g DRY WEIGHT   TEMPERATURE - ° C

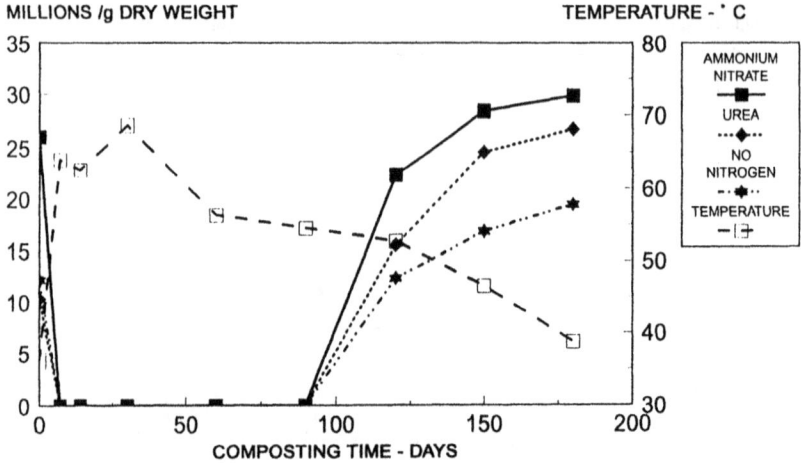

FIGURE 3.13. Effect of nitrogen source on mesophilic fungal populations during composting of cottonseed waste. (Data from Safwat, 1980.)

MILLIONS / g DRY MATTER   TEMPERATURE - ° C

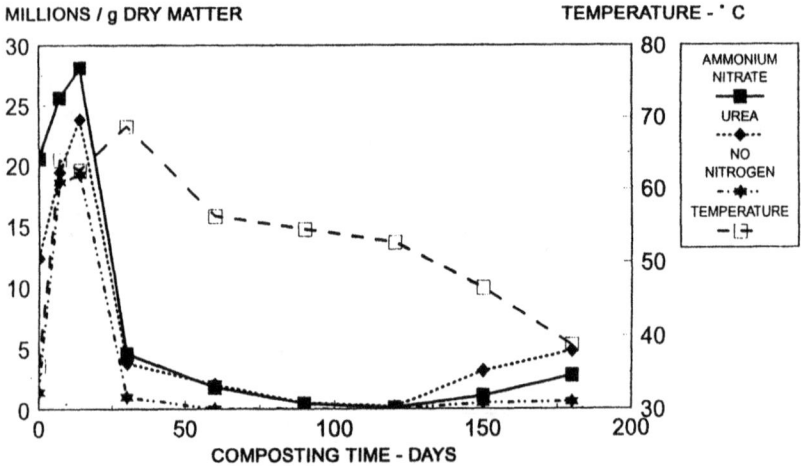

FIGURE 3.14. The effect of nitrogen source on mesophilic actinomycetes during composting of cottonseed wastes. (Data from Safwat, 1980.)

72

THOUSANDS /g DRY MATTER   TEMPERATURE - °C

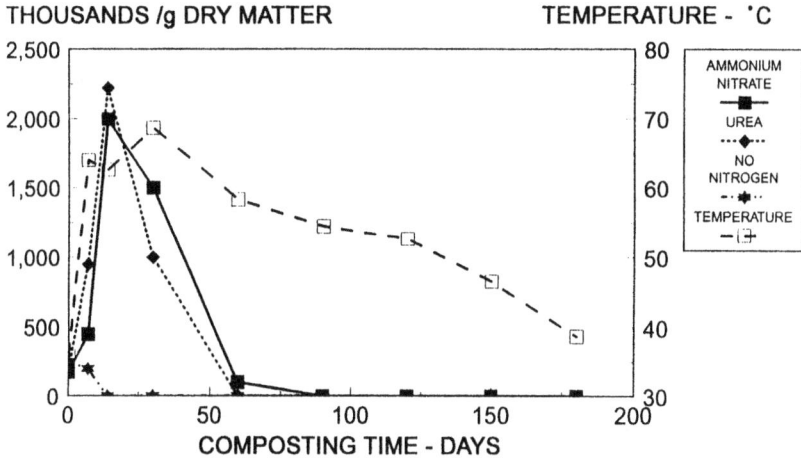

FIGURE 3.15. The effect of nitrogen on thermophilic actinomycetes during the composting of cottonseed waste. (Data from Safwat, 1980.)

INOCULANTS

Golueke et al. (1954) conducted an intensive study of inoculants. Their study did not find any effect of adding inocula on the process or the product even though the inocula tested were rich in bacteria. The inocula tested were garden soil, horse manure, partially decomposed organic matter, and special commercial bacterial cultures. The existing indigenous microflora was adequate, and the addition of other bacteria was superfluous.

Gray et al. (1971) discussed the controversy over the benefits of inocula. Inoculation may be of value if the indigenous microflora is unable to increase in numbers due to certain environmental limitations. At the time of the report, the authors concluded that the value of additives and inocula remains unproven.

Poincelot (1975) indicated that pure culture studies must "be viewed with reservation because of competition and antagonism in mixed populations."

Nakasaki et al. (1992) evaluated the seeding of tofu refuse on the composting reaction in an effort to accelerate the process. They prepared both sterilized and nonsterilized media. Although a change in the pattern of CO_2 evolution rate was observed, they did not find any difference in the final conversion of carbon.

At the present time, no data in the literature indicate that the addition of inoculants, microbes, or enzymes accelerates the composting process. Preparing the feedstock in terms of particle size, optimum moisture, and optimum C/N ratio along with managing the composting system for proper aeration appears to provide the best conditions for optimum composting.

SUMMARY

Microbiology is the heart of the composting process. The microbial populations involved are numerous, often reaching levels of 10^9/g to 10^{10}/g of compost. Composting involves mesophilic and thermophilic bacteria, fungi, and actinomycetes. Temperature is the single most important factor affecting microbial populations, their growth, and activity. Microbiological development progresses in certain temperature phases. The initial rise in temperature from ambient to 45°C is called the mesophilic phase. Temperatures can then ascend to 70°C or higher in the thermophilic phase. Microbial activity decreases as the food source (available C) is utilized, and temperatures decrease and eventually approach ambient temperature. This is a return to the mesophilic stage.

Moisture affects microbial activity in two ways. Like all organisms, microbes require water to function; moisture content below 40% can inhibit biological activity. Further, at moisture contents above 60%, aerobic microorganisms may lack oxygen since the free pore space may be blocked by water, thereby restricting oxygen diffusion.

An understanding of the microbial populations can help identify factors that affect the composting process. Since different organisms are able to utilize different carbon sources, a knowledge of microbial populations and activities may suggest modifications to the composting process that would result in greater decomposition. This aspect is very important in utilizing compost for degradation of organic contaminants in feedstocks.

REFERENCES

Berry, J. A. and C. A. Magoon. 1934. Growth of micro-organisms at and below 0°C. *Phytopathology.* 24:780–796.

Brooks, F. T. and C. C. Hansford. 1923. Mould growth upon cold-store meat. *Brit. Mycol. Soc. Trans.* 8:113–142.

Chang Y. 1967. The fungi of wheat straw II. Biochemical and physiological studies. *Trans. Br. Mycol. Soc.* 50(4):667–677.

Chang, Y. and H. J. Hudson. 1967. The fungi of wheat straw compost. I. Ecological studies. *Trans. Br. Mycol. Soc.* 50(4):649–666.

Chino, M., S. Kanazawa, T. Mori, M. Araragi, and B. Kanke. 1983. Biochemical studies on composting of municipal sewage sludge mixed with rice hull. *Soil Sci. Plant Nutr.* 29(2):159–173.

Cooney, D. G. and R. Emerson. 1964. *Thermophilic Fungi, Eumycota.* W. H. Freeman Publ. Co., San Francisco.

Crawford, D. L. 1978. Lignocellulose decomposition by selected *Streptomyces* strains. *Applied Environ. Microbiol.* 35(6):1041–1045.

De Bertoldi, M., U. Citernesi, and M. Griselli. 1980. Bulking agents in sludge composting. *BioCycle.* 21(1):32–41.

Diaz-Ravina, M. M. J. Acea, and T. Carballas. 1989. Microbiological characterization of four composted urban refuses. *Biol. Wastes.* 30:89–100.

Finstein, M. S. and M. L. Morris. 1975. Microbiology of solid waste composting, pp. 113–149. In D. Perlman (ed.). *Advances in Applied Microbiology.* Academic Press, New York.

Forsyth, W. G. C. and D. M. Webley. 1948. The microbiology of composting II. A study of the aerobic thermophilic bacterial flora developing in grass composts. *Proc. Soc. Appl. Bact.* 3:34–39.

Georgevitch, P. 1910. *Bacillus thermophilus Jivoini* nov. spec. und *Bacillus thermophilus Losanitchi* nov. spec. *Centr. Bakteriol. Parasitenk.* 27:150–167.

Golueke, C. G., B. J. Card, and P. H. McGauhey. 1954. A critical evaluation of inoculums in composting. *Appl. Microbiol.* 2:45–53.

Gray, K. R., K. Sherman, and A. J. Biddleston. 1971. A review of composting—Part 1. *Process Biochemistry.* 6(6):32–36.

Hankin, L., R. P. Poincelot, and S. L. Anagnostakis. 1976. Microorganisms from composting leaves: ability to produce extracellular degradative enzymes. *Microbial Ecol.* 2:296–308.

Janshekar, H. and A. Fiechter. 1982. On bacterial degradation of lignin. *European J. Appl. Microbiol. Biotechnol.* 14:47–50.

Jefferies, T.W. 1987. Physical, chemical and biochemical considerations in the biological degradation of wood, pp. 213–230. In J. F. Kennedy et al. (ed.). *Wood and Cellulosic: Industrial Utilization, Biotechnology, Structure and Properties.* Ellis Horwood Ltd., West Sussex, England.

Jones, F. R. and L. McCulloch. 1926. A bacterial wilt and root rot of alfalfa caused by *Aolanabacter insidiosum. J. Agr. Res.* 33:493–521.

Kelly, C. D. 1954. The cryptogamic flora of the arctic. IV. *Bacteria. Botan. Rev.* 20:417–424.

Kolattuludy, P. E. 1968. Biosynthesis of surface lipids. *Science.* 159:498–505.

Krueger, R. G., N. W. Gillham, and J. H. Coggin, Jr. 1973. *Introduction to Microbiology.* The Macmillan Co., New York.

Lauritzen, J. I. 1926. The relation of black rot to storage of carrots. *J. Agr. Res.* 33:1025.

Marsh, P. B. and M. E. Simpson (1976). *The Limiting Effects of Temperature and Moisture on the Growth of Fungi and Bacteria.* Unpublished report. USDA. Agr. Res. Serv., Beltsville, MD.

McKinley, V. L., J. R. Vestal, and A. E. Eralp. 1985. Microbial activity in composting. *BioCycle.* 26(6):39–43.

Nakasaki, K., A. Watanabe, M. Kitano, and H. Kubota. 1992. Effect of seeding on thermophilic composting of tofu refuse. *J. Environ. Qual.* 21:715–719.

Poincelot, R. P. 1975. *The Biochemistry and Methodology of Composting.* The Connecticut Agr. Expt. Sta., New Haven, CT.

Prickett, P. S. 1928. Thermophilic and thermoduric micro-organisms with special reference to species isolated from milk. *N.Y. Agr. Expt. Sta.* (Geneva) 147:1–58.

Rader, W. E. 1948. *Rhizoctonia carotae* N. sp. and *Gliocladium aureum* N. sp., two new root pathogens of carrots in cold storage. *Phytopathology.* 38:440.

Safwat, M. S. A. 1980. Composting cottonseed wastes. *Compost Sci./Land Util.* 21(3):27–29.

Singer, R. 1954. The cryptogamic flora of the arctic. VI. Fungi. *Botan. Rev.* 20:451.

Strom, P. F. 1985a. Effect of temperature on bacterial species diversity in thermophilic solid-waste composting. *Applied and Environ. Microbiol.* 50:899–905.

Strom, P. F. 1985b. Identification of thermophilic bacteria in solid-waste composting. *Applied Environ. Microbiol.* 50:906–913.

Tansey, M. R. and T. D. Brock. 1978. Microbial life at high temperatures: Ecological aspects, pp. 160–216. In F. J. Kushner (ed.). *Microbial Life in Extreme Environments.* Academic Press, New York.

Tarr, H. L. A. 1954. Microbiological deterioration of fish post mortem, its detection and control. *Bacteriol. Rev.* 18:1–15.

Waksman, S. A. and T. C. Cordon. 1939. Thermophilic decomposition of plant residues in composts by pure and mixed cultures of microorganisms. *Soil Sci.* 47(3):217–225.

Waksman, S. A., W. W. Umbreit, and T. C. Cordon. 1939. Thermophilic actinomycetes and fungi in soils and composts. *Soil Sci.* 47:37–61.

Walke, R. 1975. The preparation, characterization and agricultural use of bark-sludge compost. Ph.D. Dissertation. U. of New Hampshire, Durham, NH.

Webley, D. M. 1947. The microbiology of composting. 1. The behavior of the aerobic mesophilic bacterial flora of composts and its relation to other changes taking place during composting. *Proc. Soc. Appl. Bact.* 2:83–89.

CHAPTER 4

Biochemistry

INTRODUCTION

The basic goal of composting is to produce a stabilized product, compost. The biochemical manifestations as a result of microbial action transform the raw organic fraction of the waste or residue into a humus-like material. Composting does not produce humus. Humus is the result of long-term decomposition of organic matter in soil.

Although different feedstocks can be composted as indicated in Chapter 2, the basic biochemical changes are similar as a result of the heterogeneous microbial populations found in the composting media.

When compost is applied to soils, it undergoes biochemical changes by the indigenous microflora in the soil. Although the soil microflora contains many of the organisms present during composting, the overall population is different and more heterogeneous. The estimated soil microbial populations shown in Table 4.1 are considered conservative. Clark (1967) estimated that the live weight of the soil microflora may vary from 0.5 to over 4 metric tons in the surface 15 cm of 1 hectare. This microflora is not only important in the decomposition of organic matter, it also provides nutrients and biomass to the soil.

There are large numbers of slime molds, viruses or phages of bacteria, insects, arthropods, earthworms, and other organisms as shown in Figure 4.1 (Dindal, 1978). According to Dindal (1978), the generalized food web applies to all types of organic matter deposits. Although Dindal (1978) called this myriad of organisms "the food web of the compost pile," in reality, it is the food web of the soil.

Organisms involved in the decomposition process were classified as 1°,

77

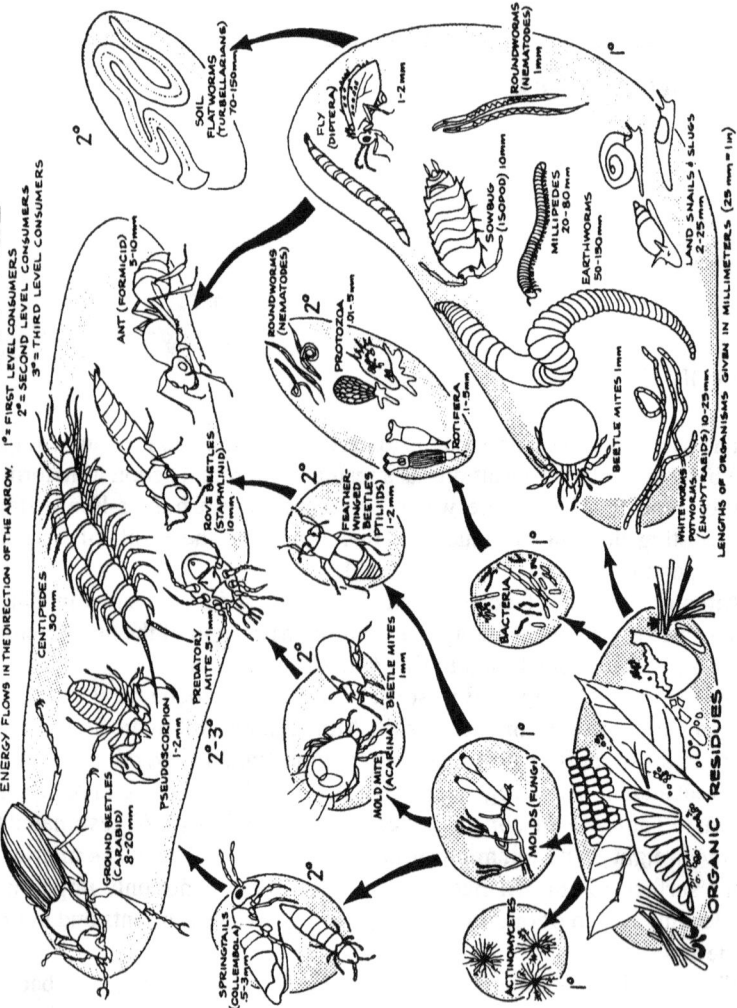

FIGURE 4.1. The food web of soil organisms. (Reprinted with permission from Dindal, 1978.)

TABLE 4.1. *Approximate Number of Organisms in Soils.*

Organism	Estimated Numbers
Bacteria	3,000,000–500,000,000
Actinomycetes	1,000,000–20,000,000
Fungi	5000–900,000
Yeasts	1000–100,000
Algae	1000–500,000
Protozoa	1000–500,000
Nematodes	50–200

Source: Martin and Focht, 1977.

2°, and 3° level consumers. At the first level are the true decomposers of wastes and other organic matter; the second level of organisms feeds on the level 1 decomposers, whereas the third group preys on the second level and upon each other. All of these organisms along with soil organic matter from plants and other residues (applied organic wastes) form the soil biomass. This soil biomass eventually decomposes into humus.

What happens to organic matter during composting and the biochemical changes that occur in the process of humification is presented in this chapter. The pathways of decomposition resulting in humus will also be presented.

ORGANIC MATTER

What is organic matter? Organic matter has been defined as (1) identifiable high-molecular weight compounds such as polysaccharides and proteins; (2) simpler substances such as sugars, amino acids, and other small molecules; and (3) humic substances (MacCarthy et al., 1990). Compost is organic matter and, when added to soils, becomes part of the soil organic matter pool. However, since compost represents a more advanced stage of decomposition than much of the organic matter that is normally applied to soils (i.e., crop residues, manures, leaves etc.), most of the sugars, proteins, simple sugars and amino acids have been metabolized as a source of C and N for the microorganisms. What is principally left is humic substances. As Chen and Inbar (1993) stated: "The organic matter has undergone an initial, rapid stage of decomposition, and is in the process of humification."

Organic matter is heterogeneous, consisting of basic chemical and mineral building blocks. Essentially, seven major elements constitute organic matter:

- carbohydrates and sugars
- protein
- fats
- hemicellulose
- cellulose
- lignin
- mineral matter

The first three constituents are decomposed relatively rapidly whereas hemicellulose, cellulose, and lignin are much more difficult to degrade. Finally, mineral matter does not biodegrade.

The composition of organic matter varies with its source. One of the earliest attempts to characterize organic matter was carried out by Waksman (1936), using what he termed "proximate analysis." Table 4.2 shows the composition of organic matter (Waksman, 1936; Stevenson, 1994). The composition of soil organic matter depends on its source (plants or animal manures) and the length of decomposition in the soil. According to Gray et al. (1971), the manure composition depends on the type of animal and its feed. Similarly, variation in plant composition depends on the species (straw versus grass), age (fresh versus aged), and the environment. The ranges for the various constituents for both plant and manure are shown in Table 4.3.

Organic constituents in compost also vary considerably with their source (Table 4.4). For example, biosolids that have undergone the activated process and food wastes contain considerably lower amounts of cellulose

TABLE 4.2. Waksman's Proximate Analysis of Soil Organic Matter.

Fraction	Chemical Treatment	Percent of Organic Matter
Fats, waxes, oils	Ether extraction	0.5–4.7
Resins	Alcohol extraction	0.3–3.0
Hemicellulose	Hydrolysis (2% HCl)	5–12
Cellulose	Hydrolysis (80% H_2SO_4)	3–5
Protein plus "lignin-humus"	Analysis of final residues for C and N	
Protein	Nitrogen × 6.25	30–35
Lignin-humus		30–50

Source: Stevenson, 1994.

TABLE 4.3. Composition of Organic Matter.

Fraction	Percent, in Dry Weight	
	Plants	Manure
Hot and cold water soluble—sugars, starches, amino acids, aliphatic acids, urea, and ammonium salts	5–30	2–20
Ether/alcohol solubles—fats, oils, waxes, resins	5–15	1–3
Proteins	5–40	5–30
Hemicellulose	10–30	15–25
Cellulose	15–60	15–30
Lignin	5–30	10–25
Inorganic (ash)	1–13	5–20

Source: Gray et al., 1971.

TABLE 4.4. Composition of Several Materials That Are Composted.

Material	Protein (%)	Lipids (%)	Hemicellulose (%)	Cellulose (%)	Lignin (%)
Pine wood[1]	NA	NA	26.0	44.0	27.8
Birch wood[1]	NA	NA	39.0	40.0	19.5
Bagasse[1]	NA	NA	30.0	33.4	18.9
Wheat straw[1]	NA	NA	28.4	30.5	18.0
Refuse[2]	2–8	5–10	NA	35–55	3–5
Biosolids	37.0	4.7	NA	2.6	6.9
Food wastes	12–18	9–15	NA	10	NA

[1]From Poincelot, 1975.
[2]From Lynch, 1986.
NA = not available.

81

and lignin than woody, wheat straw or bagasse, which contain more protein and less fats or cellulose than food wastes. These constituents determine the rate of composting.

Knowledge of the composition of the material to be composted is important in the design of composting systems. For example, combining materials high in protein (nitrogen) with cellulolytic materials reduces the potential for odors as the C/N ratio is more favorable to the microbial population. Materials high in cellulose and lignin take longer to decompose whereas materials high in sugars, other carbohydrates and lipids take less time to form a stabilized compost. Further, combining food waste, biosolids or grass with materials having a high hemicellulose or cellulose reduces the composting time since the readily available carbohydrates and sugars in biosolids, food waste, or grass start breaking down quickly, thereby raising the level of microbial activity.

The relative rate of microbial decomposition for various organic matter constituents is shown in Table 4.5 (Stentiford, 1993). Three major groups are indicated. Group 1 consists of compounds that are readily biodegraded: sugars, starch, fats and proteins. Members of the second group are slower to biodegrade and consist primarily of hemicellulose and cellulose. Finally, the third group is much more resistant to biodegradation, consisting of lignin and lignocellulose.

Lynch (1993) included hemicellulose and cellulose in the lignocellulose classification, whereas Stentiford (1993) separated the hemicellulose and

TABLE 4.5. *The Relative Rate of Microbial Decomposition of Organic Materials (after Stentiford, 1993).*

Readily biodegradable—Group 1
Sugars
Starches, glycogen, pectin
Fatty acid, glycerol
Lipids, fats
Amino acids
Nucleic acids
Protein
Slower to biodegrade—Group 2
Hemicellulose
Cellulose
Chitin
Low molecular weight
Resistant to biodegradation—Group 3
Lignocellulose

cellulose from lignocellulose. Hemicellulose includes a group of branched heteropolysaccharides soluble in diluted alkali. It includes hexose, glucose, mannose and galactose, and the pentoses xylose and arabinose (Lynch, 1993).

Cellulose is a linear, unbranched homopolysaccharide of 10,000 or more D-glucose units connected by b1–4 glycosidic bonds. It is the most abundant organic compound in the world. Approximately 50% to 70% of the dry weight of plants is made of cellulose, which is mostly in the cell wall (Greenland and Oades, 1975). Glucose and cellobiose are the two components of cellulose.

Lignin is the organic constituent most resistant to biodegrade found in compost feedstock. However, several bacteria and fungi degrade lignin. Leszkiewicz and Kinner (1988) indicated that temperature and oxygen concentration greatly influence the biodegradation of lignin. Further, studies have shown that the white rot fungi (Basidiomycetes) are dominant biodegraders. Finally, *Fusarium*, *Aspergillus*, Nocardia, and Streptomycetes can also partially degrade lignin (Higuchi, 1982).

Lignin consists of phenylpropane units represented by coniferyl alcohol (I), *p*-hydroxycinnanyl alcohol (II) and sinapyl alcohol (III) (Stevenson, 1994). According to Lynch (1993), lignocelluloses, which consist of hemicellulose, cellulose, and lignin, are the principal providers of the carbon and energy for microorganisms in the production of compost. Although this may be true for the long-term decomposition of organic materials, during the initial phases of composting the readily available carbon is the important source of carbon.

This is evident during the composting of digested biosolids. Well anaerobically digested biosolids have very little available energy for the microbial population and it is often difficult to achieve temperatures without the addition of an external source of available carbon. At one site, the addition of rice hulls rather than kiln-dried wood chips to well anaerobically digested biosolids resulted in elevated temperatures and efficient composting.

BIOCHEMICAL MANIFESTATIONS OCCURRING DURING COMPOSTING

Ashworth (1942) conducted one of the earlier studies on the changes in the composition of different organic matter materials during composting using Waksman's proximate analysis. Oat straw, grass cuttings mixed with a small proportion of beech leaves, sphagnum peat and a mixture of cotton grass and sphagnum peat were studied in compost piles. Nutrients were applied to all materials to achieve the same nutrient level.

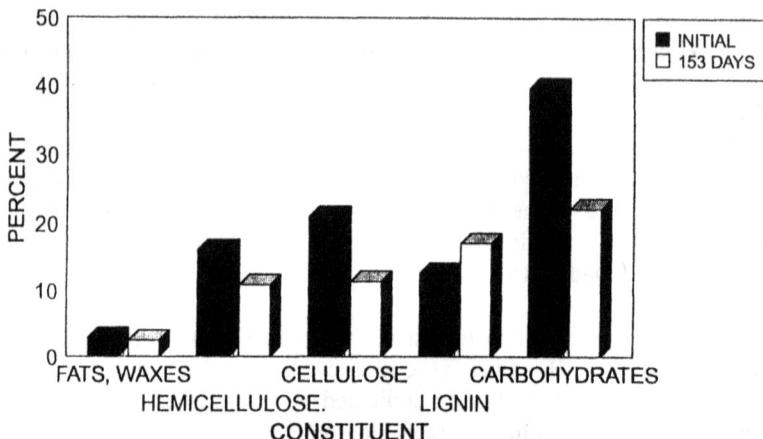

FIGURE 4.2. Changes in composition during composting of grass. (Data from Ashworth, 1942.)

The data for the grass and straw are found in Figures 4.2 and 4.3. As shown, the grass and straw lost approximately 33% of the hemicellulose and 50% of the cellulose. Most of the hemicellulose in the grass was lost during the first 12 days, whereas in the straw it took 60 days. Cellulose in the grass decreased rapidly during the first 30 days with virtually no changes thereafter. With the straw there was no change for the first 24 days, followed by a rapid decrease from the 24th to the 90th day. Changes in fats, waxes, and

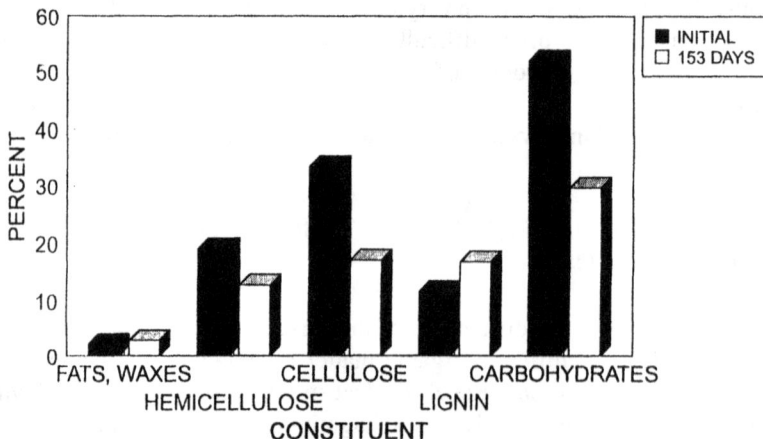

FIGURE 4.3. Changes in composition during composting of straw. (Data from Ashworth, 1942.)

resins were very small. A slight increase was noted following the 12th day. Ashworth (1942) indicated that this could be due to actual synthesis. Previous work did not show such an increase. There were also increases in lignin. These increases were relative due to the breakdown and losses in carbohydrates.

Karim and Chowdhury (1958) evaluated the decomposition of MSW in a laboratory study. Fats and waxes decreased very rapidly during the first 180 days with little change for the next 60 days. Protein decreased in a steady manner during the entire 240 days, and cellulose decreased rapidly from 15 to 90 days with no change thereafter (Figure 4.4). The total losses of each constituent are shown in Figure 4.5. Over a period of 240 days, 73% of the fats and waxes was lost; 49% of the cellulose was lost; 39% of the proteins was lost; and 30% of the lignin was lost. As would be expected, the greatest loss was with the fats and waxes and furfural.

No data were reported on simple carbohydrates. However, Karim and Chowdhury (1958) provided data on furfural, the 2-carboxaldehyde of furan, which is usually found in sugars. The loss of 56% of furfural probably reflects the loss of the simple carbohydrates.

The relatively low loss in proteins is the result of the initial low C/N ratio of 15 and subsequent small change in the C/N ratio. After 90, 180 and 240 days, the C/N ratio was approximately 11. Protein was determined by multiplying the total N by 6.25. No data were provided on temperature. Since this study was conducted in a laboratory setting, it may not have included the entire spectrum of mesophilic and thermophilic organisms because of lower temperatures in the laboratory in comparison to large-scale piles.

Keller (1961) found that over a 115-day period of composting refuse in windrows, lipids decreased by 71%, from a level of 7.7% to 2.2%. The decrease in cellulose as a percent of total organic matter was from 34% to 9%. Both constituents decreased rapidly during the first 40 days (Figure 4.6). After that time, little change took place in lipids and cellulose. Temperatures in the windrows reached 60°C during the first 18 days and began to decrease by the 23rd day. Organic matter decreased from 34% to 9%.

Wiley and Spillans (1961) measured changes in lipids, crude fiber and protein during windrow and bin composting of refuse and biosolids (Figure 4.7). They found 74%, 39%, and 21% reduction in lipids, crude fiber and protein, respectively, over a 30-day period. More nitrogen and, consequently, protein was lost than anticipated. This is surprising since the C/N ratio was 65 in the beginning of composting and 47 at the end of 30 days, indicating that the product was unstable and that there was an excess of C.

The high C/N ratio (65) is not typical of biosolids and refuse mixtures. The high N in biosolids, and consequently low C/N ratio, reduces the high C/N ratio of refuse, resulting in a much more favorable C/N ratio for the

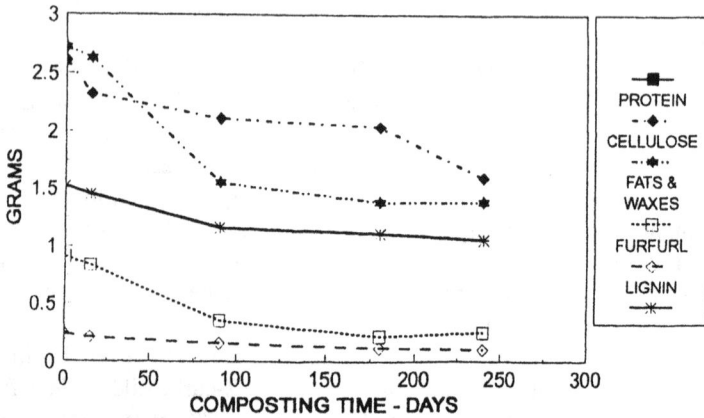

FIGURE 4.4. Changes in organic constituents during MSW composting. (Data from Karim and Chowdhury, 1958.)

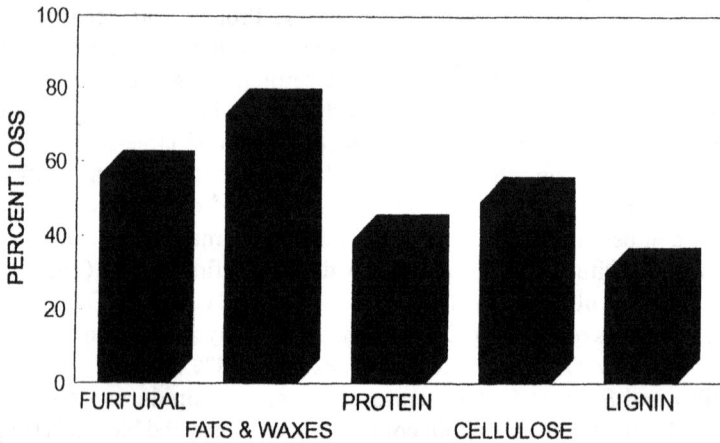

FIGURE 4.5. Losses of several constituents during the decomposition of MSW. (Data from Karim and Chowdhury, 1958.)

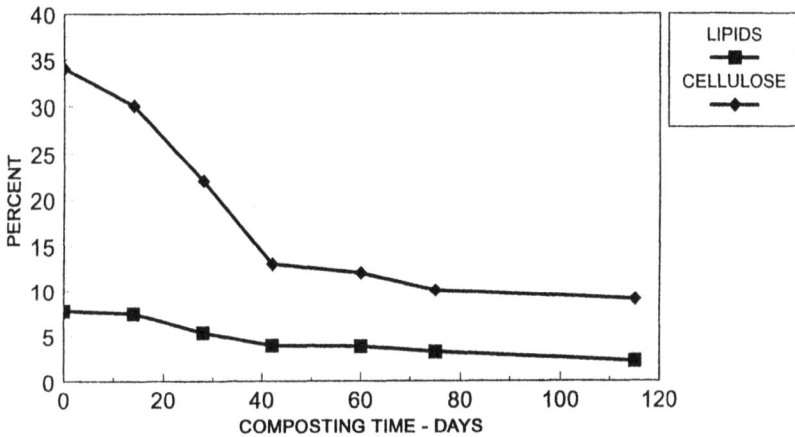

FIGURE 4.6. Changes in percent cellulose and lipids during windrow composting of refuse. (Reprinted with permission from Keller, 1961.)

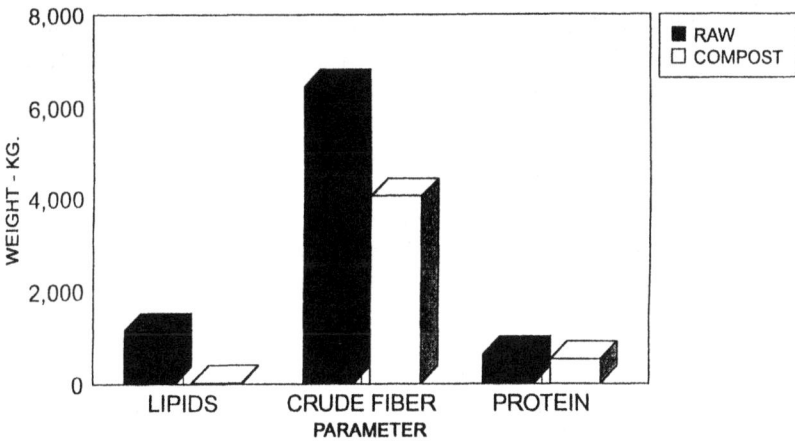

Protein calculated as 6.25N

FIGURE 4.7. Changes in lipids, crude fiber, and protein during windrow and bin outdoor composting of MSW and biosolids. (Data from Wiley and Spillans, 1961.)

TOTAL DRY MATTER
HEMICELLULOSE AND CELLULOSE

LIGNIN, SUGARS AND FATS
AND STARCH

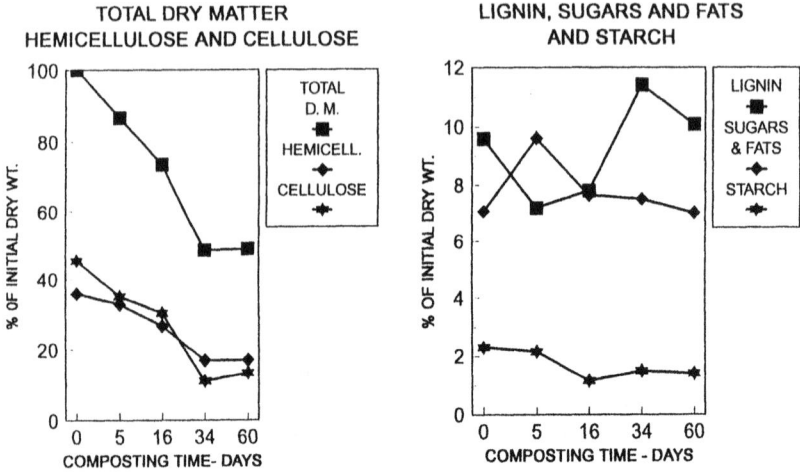

FIGURE 4.8. Biochemical changes during composting. (Data from Chang, 1967.)

mixture. Temperatures reached 68°C to 71°C in approximately seven days and remained near that level until the 24th day when it began to descend.

The pH during composting was measured daily. Initially, it dropped from 6.3 to slightly above 5 during the first day and then increased steadily to over 8 by the 18th day. This high pH could have resulted in ammonia stripping, which accounts for the loss of N. Oxygen levels were between 1.2% and 7.4% at a 30 cm depth and between 0.8% and 1.9% at a 60 cm depth, except right after windrows were turned. Therefore, for most of the composting period anaerobic conditions probably existed.

Chang (1967) studied the biochemical changes that occur during the composting of wheat straw. The changes as a percent of total dry matter are shown in Figure 4.8. As illustrated, cellulose and hemicellulose, which constituted 45.32% and 35.69% of the initial dry weight, decreased to 13.27% and 16.98% of the initial dry weight, respectively, in 60 days. The straw had lost over 50% of its initial dry weight after 60 days of composting, which essentially represents the loss of hemicellulose and cellulose. Hemicellulose and cellulose represented approximately 80% of the total weight of straw. The diatase-soluble fraction represented starch and glycogen and the ethanol-soluble fraction included sugars, glucosides, essential oils, coloring matter and resinous substances. This fraction changed little as a percent of the initial dry weight since these ingredients represented a very small fraction. No changes in protein were noted. Temperatures during the composting period peaked at 68°C in five days and by the 34th day were at 16°C.

Even though the material composted by Chang (1967) and Keller (1961), were very different in composition, their results were similar. As in the study by Keller (1961), cellulose decreased very rapidly during the first 34 days with little change thereafter.

Poincelot (1972) reported that with MSW, sugars and starch were completely metabolized in 5–7 weeks. A 75% loss was noted in lipids or fats and a 60% loss of cellulose and hemicellulose. The data on lipid change during composting were similar to Keller's (1961) findings but much higher for cellulose changes.

During composting other chemical changes can occur that would affect plant growth. For example, DeVleeschauwer et al. (1981) found significant amounts of acetic acid in fresh municipal refuse compared to that in 69-day-old compost (Figure 4.9). There was little difference between the aerated and non-aerated compost. Germination of cress seed was severely impeded when acetic acid concentrations exceeded 300 ppm. Other organic acids including propionic, isobutyric, butyric, and isovaleric, which could also be phytotoxic, were identified in the fresh refuse but were not found after five months of composting.

Roletto et al. (1985) suggested a humification index (HI) as a means of assessing maturity. This index was defined as the percentage of humic acid carbon (CHA) and organic carbon (HI = CHA × 100/C-org.). Sequi et al. (1986) defined HI as the ratio between the organic-C and of the non-humified fraction and that of the humified [humic acid (HA) + fulvic acid (FA)]

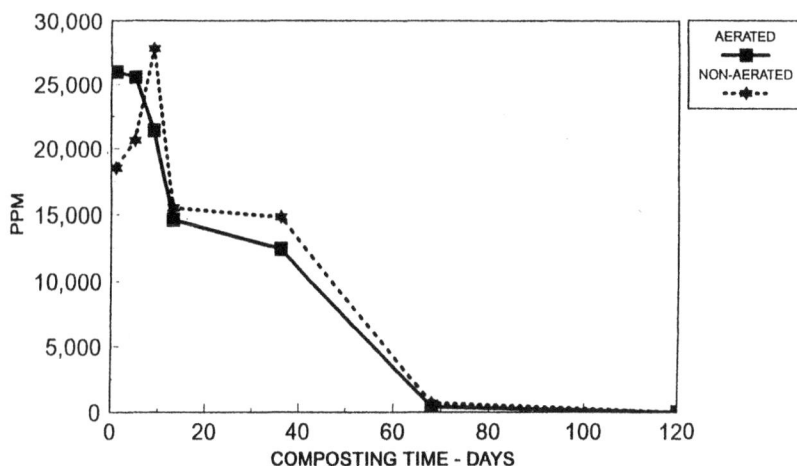

FIGURE 4.9. Acetic acid production during composting. (Reprinted with permission from DeVleeschzuwer et al., 1981.)

fractions. Ciavatta et al. (1993) used the term "degree of humification" (DH), which was determined as follows:

$$\%DH = [(HA + FA)/TEC] \times 100$$

TEC is the total extractable organic carbon.

Saviozzi et al. (1988) measured the changes in several humification parameters during the composting of sludge from a paper-processing factory, straw and refuse. Several parameters were assessed and used as an index of maturity or characterization of state of composting. These included (a) C/N ratio, (b) humification index (HI), (c) humic acid carbon as percentage of organic-C (CHA), (d) fulvic acid carbon as percentage of organic-C (CFA), (e) organic-C as a percent of dry matter, and (f) CHA/CFA ratio, CHA+FA as percent organic-C.

Most of the changes occurred during the first 60 days (Figure 4.10). The humification index decreased within the first 60 days. This parameter was highly correlated with the C/N ratio, which decreased during the same period of time. All of the parameters studied were reported to be linear and to correlate significantly to the C/N ratio. The CHA/CFA ratio increased rapidly during the first 40 days and then increased at a lower rate for the duration of 140 days. CHA (% organic-C) increased for the first 60 days with little change thereafter. The C/N ratio decreased from 26 to 13. The HI

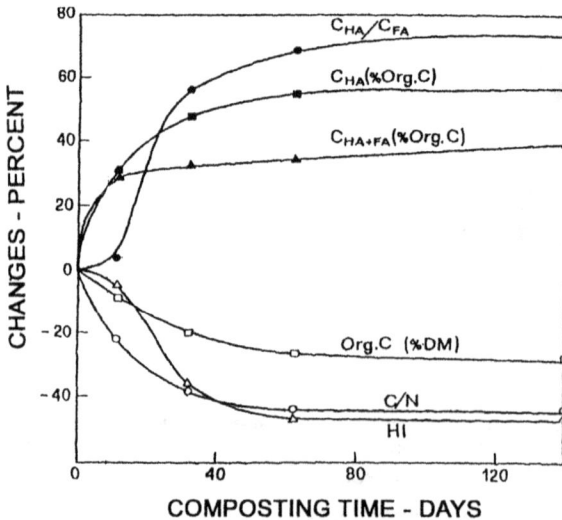

FIGURE 4.10. Relative trend of several humification parameters during 140 days of biosolids-straw composting. (Reprinted with permission from Saviozzi et al., 1988.)

decreased from 3.2 to 0.5 and was highly correlated ($r = 0.86$) with the C/N ratio. Finally, the CHA/CFA ratio increased from 0.3 to 1.8 and also was highly correlated ($r = 0.84$) with the C/N ratio. The authors recommend that before these parameters are used to assess maturity of organic wastes, considerably more tests over a wide range of wastes are needed.

The changes in some of the organic components during composting of manure were studied by Inbar (1989) and Inbar et al. (1989). Changes in the various constituents as related to the initial dry matter are shown in Figure 4.11. As illustrated, the most rapid changes occurred during the first 40 to 60 days. In that time, approximately 50% of the organic matter was oxidized to CO_2 and water. As a percent of the initial dry matter, lignin underwent very little change. Cellulose was reduced by 50% of the original amount. Hemicellulose was reduced to 33% of the original value in the first 40 to 60 days. This was similar to the data given by Ashworth (1942) for grass and straw.

Figure 4.12 shows the changes in the constituents in relation to total organic matter. Lignin content increased from 19.9% to 30.9% on a dry weight basis. As a percent of the total organic matter, it increased over a 140-day period from approximately 29% to 54%.

Using solid-state carbon-13 nuclear magnetic resonance and infrared spectroscopy, Inbar et al. (1989) showed that during the composting of cattle manure, the level of carbohydrates decreased while the level of alkyl C, aromatic C, and carboxyl groups increased (Figure 4.13). At the same time, compounds in the polysaccharide region decreased by 33%.

Throughout the study, appreciable amounts of carbohydrates were the primary source of energy for the microorganisms. The higher rate of decomposition of carbohydrates resulted in accumulation of modified lignin. Through microbial action, the lignin was then degraded to smaller polyphenolic compounds and intermediate phenols. Further, through enzymatic action these in turn were converted to quinones and later condensed to humic acid.

In the same study, Inbar et al. (1989) showed that during the composting of cattle manure, humic substances increased over time (Figure 4.14). Specifically, humic acid increased from 377 to 710 g/kg organic matter. Less increase was found in the fulvic acid and the non-humic fraction.

Inbar (1989) also evaluated the changes in humic substances during the composting of grape pomace. Here humic substances decreased by 20% during the first 70 days of composting and then continued to decrease slightly for the remainder of the composting period. The HA fraction essentially remained unchanged, whereas the FA fraction decreased from 148 to 51 g/kg OM. The non-humic fraction decreased from 9.5 to 3.3% of the total OM during the first 70 days and then remained unchanged. The C/N ratio declined

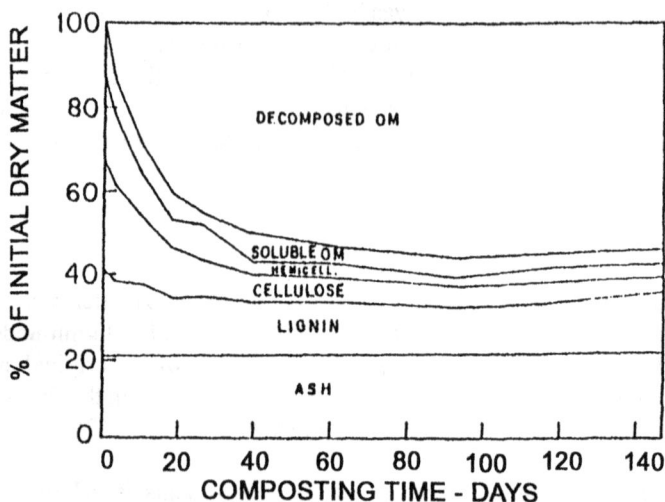

FIGURE 4.11. Changes in relative content of the organic constituents of separated manure during composting. (Data from Inbar, 1989.)

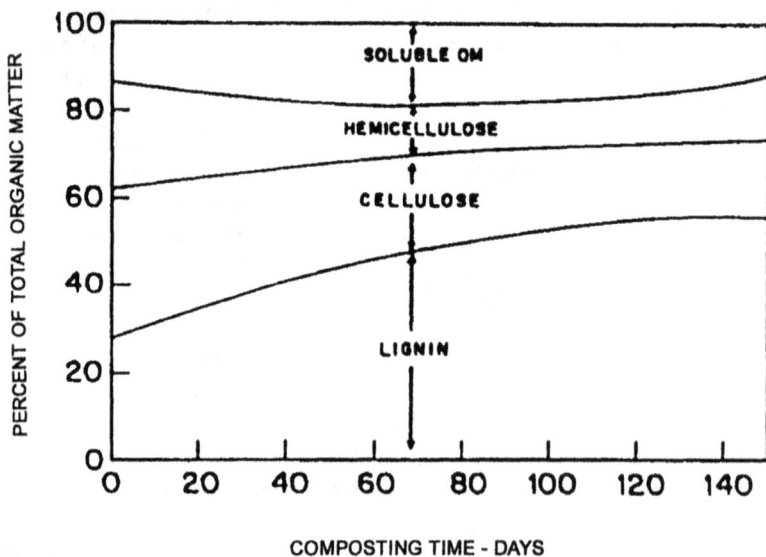

FIGURE 4.12. Changes in organic constituents during composting of separated manure. (Data from Inbar, 1989.)

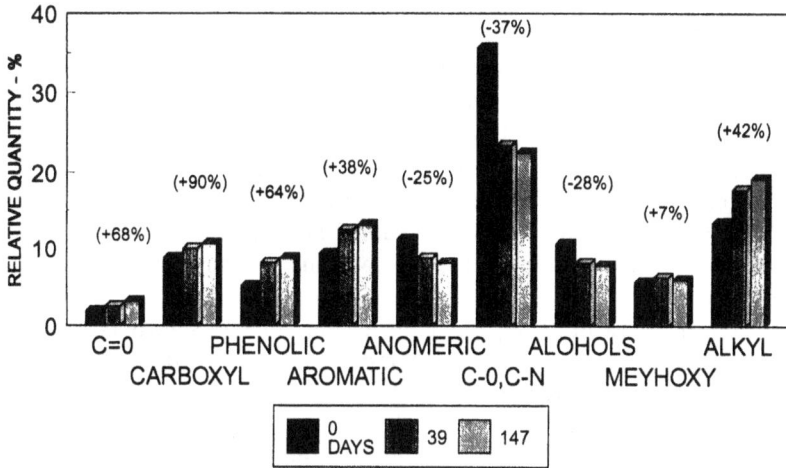

FIGURE 4.13. Distribution of carbon in fresh, partially composted, and mature composted cattle manure and percentage change from fresh (0 days) to mature compost (147 days). (Data from Inbar, 1989.)

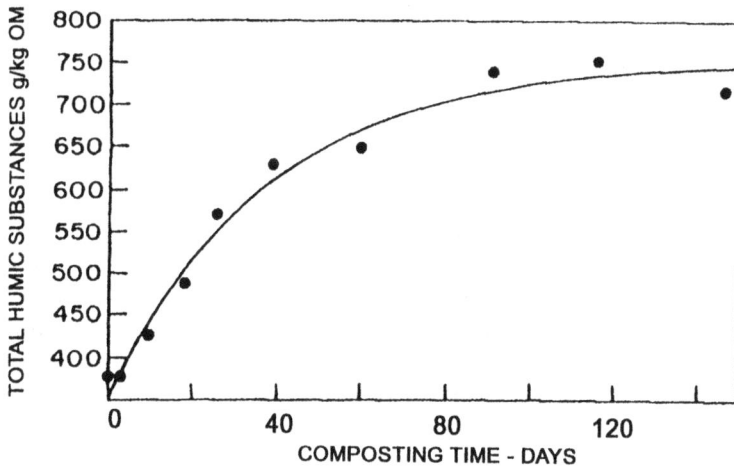

FIGURE 4.14. Changes in the organic matter content of total humic substances during composting of separated cattle manure. (Reprinted with permission from Inbar et al., 1990.)

from an initial value of 35–40 or higher to 18–20 at the end of the composting period. The HI was highly correlated ($R^2 = 0.837$) with the C/N ratio.

Further studies by Inbar et al. (1990) evaluated the changes in humic substances during composting. The total humic substances consisted of two fractions, humic acid and fulvic fractions. The changes in these are shown in Figure 4.15a. Humic acids increased from 184 to 457 g/kg OM and fulvic fraction from 194 to 253 g/kg OM during the 147 days of composting. Changes in fulvic acid and the non-humic fraction are shown in Figure 4.15b. Excellent correlation was obtained between the extracted components of humic substances and the C/N ratio and the composting time (Table 4.6).

At this time, use of these parameters to determine the state of composting and the stability or maturity of the compost is not practical. Methodologies are not uniform, nor are they easy to perform. Furthermore, there is lack of data with respect to a wide range of feedstocks. One of the difficulties in using these parameters is the variability within feedstocks. For example, biosolid composition varies widely depending on whether is raw, aerobically digested, anaerobically digested, limed or unlimed.

Kirchmann and Widen (1994) reported that total fatty acids were higher when composting 100% organic household wastes than when the waste was mixed with park wastes (Figure 4.16). Figures 4.17a and 4.17b show the individual fatty acids determined for both types of wastes. These data give rise to several questions. Why was there a difference between the unmixed organic household wastes and the mixture containing the household waste mixed with park wastes?

Kirchmann and Widen (1994) did not provide an analysis of the two wastes. Possibly the separated organic household wastes had a high nitrogen content as well as more available C, which resulted in oxygen depletion and anaerobic conditions at least for part of the time. According to Kirchmann and Widen (1994), the piles were turned twice each week with a front-end loader. Under these conditions the piles were probably anaerobic within 20 to 60 minutes after turning, making the conditions conducive to the formation of fatty acids. Valerate acid was found in the organic household waste, but not in the mixture of household and park waste. If the toxicity of both composts was essentially the same, as the authors imply, then the presence of valerate acid is not of concern.

Little data are available on the formation of fatty acids under different composting conditions and with different feedstocks. In addition, the level of phytotoxicity of different fatty acids is unknown. Therefore, further research is needed to identify both the conditions that lead to the formation of particular fatty acids and the effect these compounds may have on plant development.

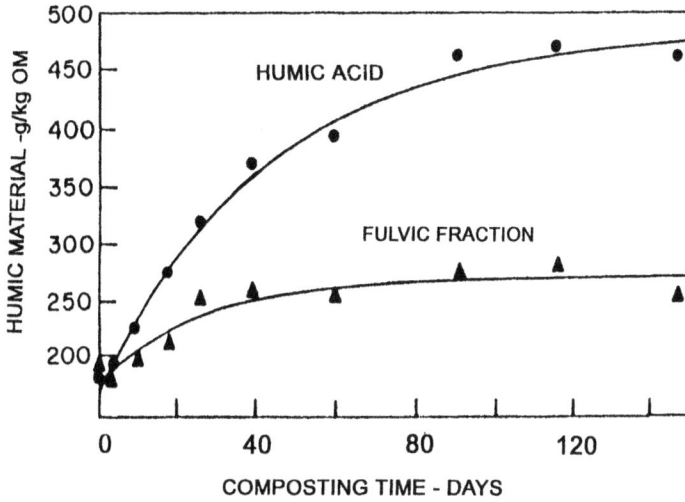

FIGURE 4.15a. Changes in the organic matter content of humic acid and fulvic fraction of separated cattle manure. (Reprinted with permission from Inbar et al., 1990.)

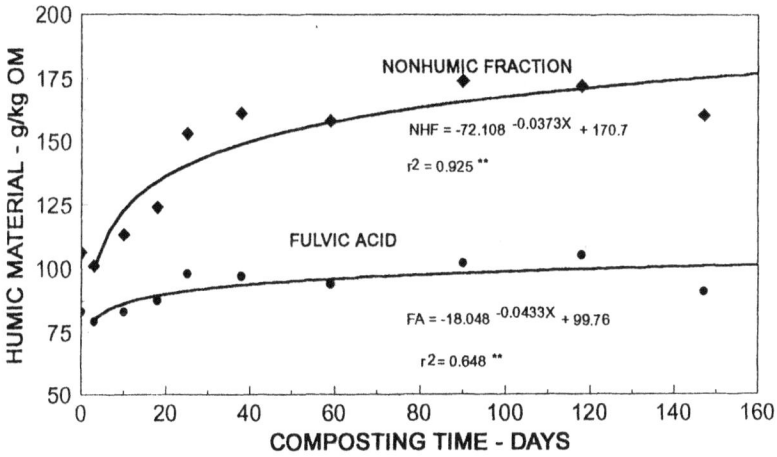

FIGURE 4.15b. Changes in organic matter content of the non-humic fraction and fulvic acid during composting of separated manure. (Reprinted with permission from Inbar et al., 1990.)

TABLE 4.6. *Regression Equations and Correlation between Various Components of Humic Substances and for Humification Indexes and C/N Ratio and Time.*

Component	Regression Equation	R^2 – C/N	R^2 – Time
Humic substances	$1000.4X \exp(-0.887X \text{ C/N})$	0.962[4]	0.976[4]
Humic acid	$808.2X \exp(-0.1002X \text{ C/N})$	0.981[4]	0.989[4]
Fulvic fraction	$226.8X \exp(-0.0661X \text{ C/N})$	0.825[4]	0.885[4]
Non-humic fraction	$178.9X \exp(-0.0560X \text{ C/N})$	0.877[4]	0.925[4]
Fulvic acid	$45.012X \exp(-0.0699X \text{ C/N})$	0.636[4]	0.648[4]
HI[1]	$-0.2527X \exp(-0.965X \text{ C/N})$	0.894[4]	0.992[4]
HR$_1$[2]	$2.115X \exp(-0.10001X \text{ C/N})$	0.912[4]	0.984[4]
HR$_2$[3]	$6.885X \exp(-0.0951X \text{ C/N})$	0.967[4]	0.990[4]

[1]HI = humification index [non-humic fraction/(humic acid + fulvic acid)].
[2]HR$_1$ = humic acid/fulvic fraction.
[3]HR$_2$ = humic acid/fulvic acid.
[4]Denotes that the data is highly statistically significant.
Source: Inbar et al., 1990.

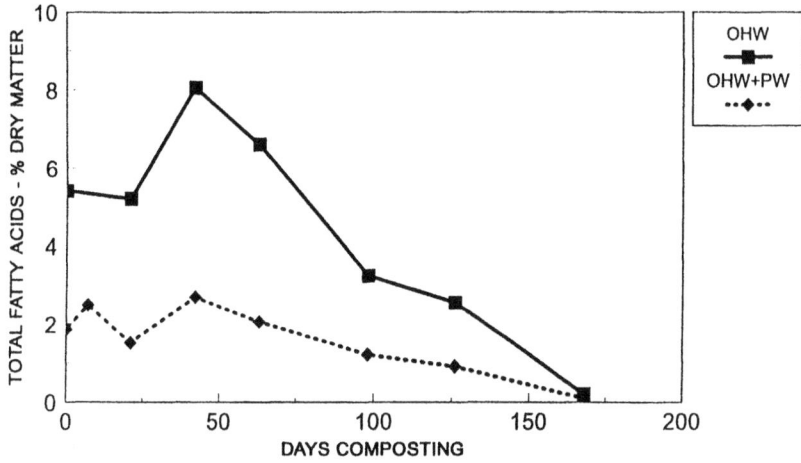

FIGURE 4.16. Changes in total fatty acids during the composting of organic household wastes (OHW) and organic household wastes with park wastes (OHW + PW). (Reprinted with permission from Kirchmann and Widen, 1994.)

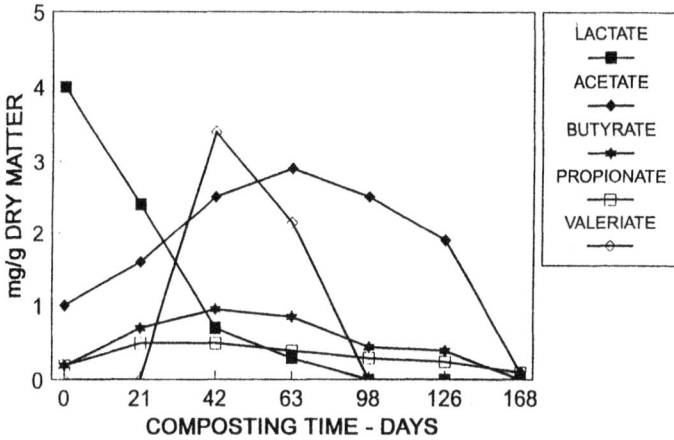

FIGURE 4.17a. Changes in fatty acids concentration during composting of organic household wastes.

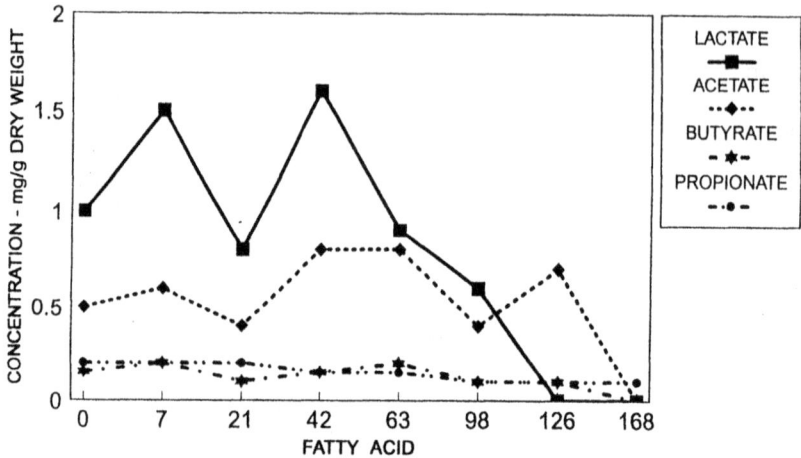

FIGURE 4.17b. Changes in fatty acid concentration during composting of household wastes and park wastes (50/50% by volume). (Reprinted with permission from Kirchmann and Widen, 1994.)

BIOCHEMICAL MANIFESTATIONS OCCURRING WHEN COMPOST IS APPLIED TO SOIL

The chemical composition of a plant changes during its growing period. Thus, the composition of organic residues derived from plants varies with the type of plant and its growing environment. As the plant matures its content of sugars, starches, and other water-soluble constituents decreases. Proteins and nitrogen also decrease whereas hemicellulose, cellulose, and lignin increase. Therefore, fresh residue or organic matter, when added to the soil, decomposes more rapidly than older plant materials.

Tenney and Waksman (1929) found that soluble organic fractions were first to decompose, followed by pentosans (simple sugars) and cellulose. Figure 4.18 shows their data on the chemical composition of mature corn stalks initially and after 205 days of decomposition by soil microorganisms, both with and without added nutrients (Tenney and Waksman, 1929). Sugars and other polysaccharides (water soluble fraction), fats, waxes, oils, and resins (ether- and alcohol-soluble fraction, hemicellulose, and cellulose in cornstalks) decreased with time, whereas, lignins, crude protein and ash increased.

The importance of the nature and quantity of organic matter for the rate of decomposition in the soil has been demonstrated by numerous investigators (Bodily, 1944; Broadbent and Bartholomew, 1948; Miller et al., 1936;

Tenney and Waksman, 1929). The same principles apply to composting, except that during composting the process is accelerated as a result of optimization of conditions.

Numerous attempts have been made to model the manifestations in soil organic carbon and nitrogen (Jenny, 1941; Campbell, 1978; Van Veen and Paul, 1981; Parton et al., 1987). Parton et al. (1987) suggested a compartmentalized soil organic matter formation model, termed the "century model." An adaption of this model to compost is shown in Figure 4.19.

The model indicates three major soil organic matter fractions in terms of rate of decomposition and persistence in the soil: (1) the active fraction, consisting of organic matter with a short turnover time of 1–5 years and live microbes and microbial products; (2) a slow fraction, which is the portion of the organic matter pool that is physically protected and/or in chemical forms relatively resistant to microbial decomposition with a turnover time of 20–40 years; and (3) a passive fraction, which is chemically resistant to microbial decomposition and which also may be physically protected, having the longest turnover time of 200–1500 years.

Compost applied to soil directly adds organic matter to the soil organic matter pool. This pool consists of plant residues and microbial products. Compost enhances plant growth and thus contributes indirectly to the soil organic matter pool through the additional root mass. In time, compost applied to soil is transformed to humus.

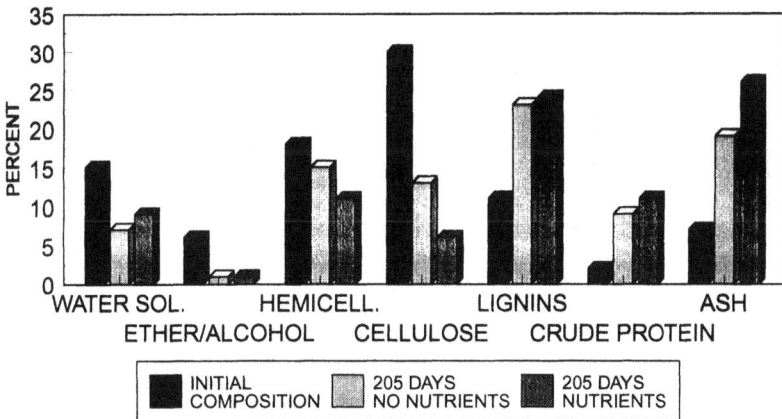

Water soluble includes hot and cold;- sugars, starches amino acids, urea, other soluble N compounds.

FIGURE 4.18. Changes in composition of cornstalks after decomposition with mixed soil microflora with and without added nutrients. (Data from Tenney and Waksman, 1929.)

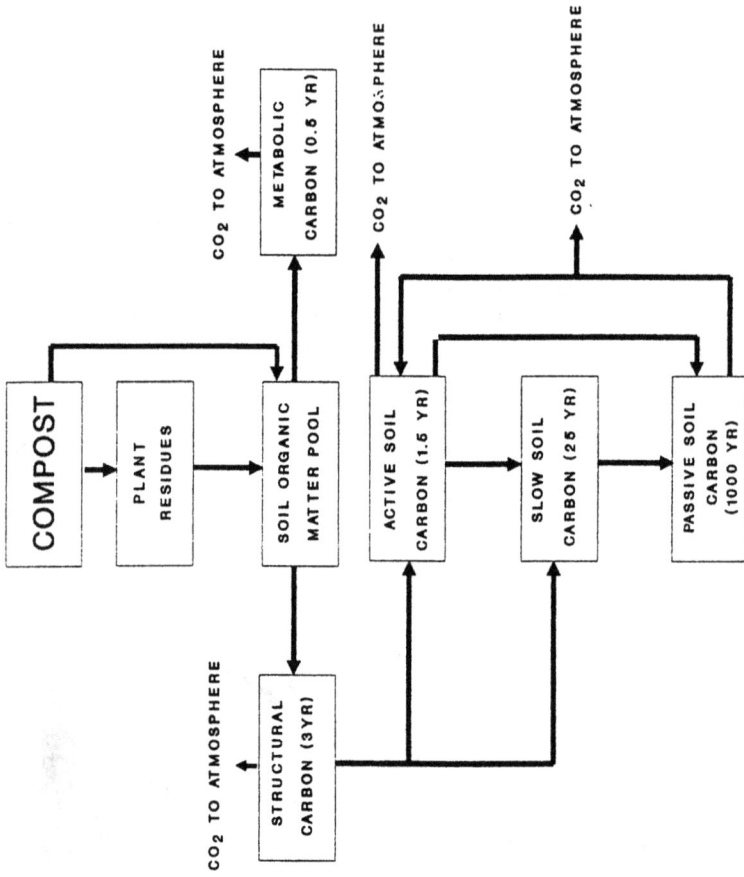

FIGURE 4.19. Flow diagram for carbon from compost based on the century model. (After Parton et al., 1987.)

HUMUS FORMATION

The term "humus" has been used to mean many different things, resulting in confusion. One of the major difficulties associated with humus is that it is not a specific, identifiable compound. Rather, it is comprised of many complex, inseparable mixtures of molecules, which can vary from source to source (MacCarthy et al., 1990).

Often the terms organic matter and humus are used interchangeably. Organic matter is both undecomposed and decomposed plant and animal material. It can be fresh residues incorporated into the soil, compost at different stages of decomposition, animal residue being decomposed, and soil biomass. Therefore, it contains carbohydrates, sugars, amino acids, proteins, polysaccharides, fats, waxes, lipids, organic acids, and many other compounds. Stevenson (1944) and earlier Waksman (1936) equated humus with soil organic matter, excluding the remains of plant residues and partial decomposition products. Much of our knowledge of humus and organic matter comes from the soil science literature.

One of the major objectives of aerobic composting is to obtain a stabilized product. This stabilized product, compost, is often incorrectly termed humus. The process by which organic matter is decomposed during composting or in the soil is humification. Humification involves the decomposition of high and low molecular-weight plant, animal and microbial cell constituents and products synthesized by soil organisms. The process of humification occurs mainly under aerobic conditions (Flaig et al., 1975). Nitrogen plays an important part in humification and the N compounds of humus are the natural reservoir for N nutrition under natural conditions (Flaig et al., 1975).

Kononova (1975) classified organic substances in soil into two major groups. The first group consists of fresh and incompletely decomposed plant and animal residues. These are primary sources of humus. The second group is soil humus. This group is further subdivided into (1) strictly humic substances, which include humic acid, fulvic acid, humins and hymatomelanic acid; and (2) products of advanced decomposition or organic residues and products of microbial resynthesis. These two subgroups are difficult to separate. During the early stages of composting, the material being composted falls primarily into the first group. As the composting process proceeds, however, humus-like material begins to form.

What makes this material called humus so important? Some of the more important attributes are listed below:

- It improves the soil's physical properties.
- It increases the soil buffer capacity.
- It adds plant nutrients to the soil.

- It increases the cation exchange capacity of soils.
- It increases the soil water-holding capacity.
- The dark color provides for soil warming during early spring.
- It reduces the potential for leaching and toxicity of natural and xenobiotics through chelation and organometal complexes.
- It supports and enhances the microbial population.

Humus is a material comprised of many organic substances and representing a state during the decomposition of organic materials. It is a semi-finite state of decomposition (the finite state is ash) and is characterized by a slow decomposition rate.

Chemically, 80% of humus consists of humic acid and polysaccharides (Gascho and Stevenson, 1968). Further, it is represented by the humic substances fulvic acid, humic acid, and humin. Hymatomelanic acid is also often included but is a derivative of humic acid.

Descriptively, humic substances are a series of relatively high-molecular-weight, refractory, yellow-to-black colored substances (Aiken et al., 1985).

Stevenson (1994) indicated that the term humin may be inappropriate since it may consist of several other fractions. However, chemically it consists of the alkali-insoluble fraction of humus (Stevenson, 1994). The chemical definition of humus is based on the chemical extraction methods (Stevenson, 1994).

(1) Humic acid—the fraction of humic substances that can be extracted by dilute alkali and other reagents and that is insoluble in dilute acid

(2) Fulvic acid—the fraction of humic substances that is soluble in both alkali and acid

(3) Humin—the alkali-insoluble fraction of humus

Several different theories have been proposed as to the formation of humus. Stevenson (1994) suggested four principal pathways. Pathway 1 is Waksman's lignin hypothesis. This concept is based on the incomplete microbial degradation of lignin. The residue is transformed in conjunction with nitrogen to form humic acid and fulvic acid. In the second pathway, lignin is still involved but phenolic aldehydes and acids derived from lignin during microbial activity undergo enzymatic conversion to quinones, which polymerize to form humic-like substances. This may take place in the presence or absence of amino compounds. The third pathway is similar to the previous pathway, except that polyphenols are synthesized by microorganisms from non-lignin C sources such as cellulose. The enzyme phenyl oxidase oxidizes the polyphenols into quinones, which are then converted to humic substances.

In the fourth pathway humus is formed from sugars. According to this hypothesis, reducing sugars and amino acids, formed as by-products of microbial synthesis and transformation, undergo non-enzymatic polymerization into brown nitrogenous polymers. These are then condensed to form humic substances. The second and third pathways are the basis of the current polyphenol theory. Stevenson (1994) pointed out that humic substances in the soil may be formed by all the pathways described.

The basic formation of humic substances from the constituents in compost is shown in Figure 4.20. Unlike plant residues, which when added to the soil lose a significant amount of C as CO_2, the readily available C in compost has already been metabolized and CO_2 has evolved. For example, over a 12-week period, 59% of the C in cornstalks, when added to a loamy soil, evolved as CO_2 under aerobic conditions (Stott and Martin, 1990). Under warm, moist conditions in the field even greater amounts have been reported. Most of the C in compost, when added to soils, is in the form of lignin, cellulose, and hemicellulose.

Inbar et al. (1990) showed that during the first 40 to 60 days about 50% of the organic matter cattle manure had been oxidized to CO_2 and H_2O. After 140 days of composting lignin, cellulose and hemicellulose accounted for over 80% of the total organic matter. The lignin and other polyphenolic compounds in compost are attacked by microorganisms into lower molecular

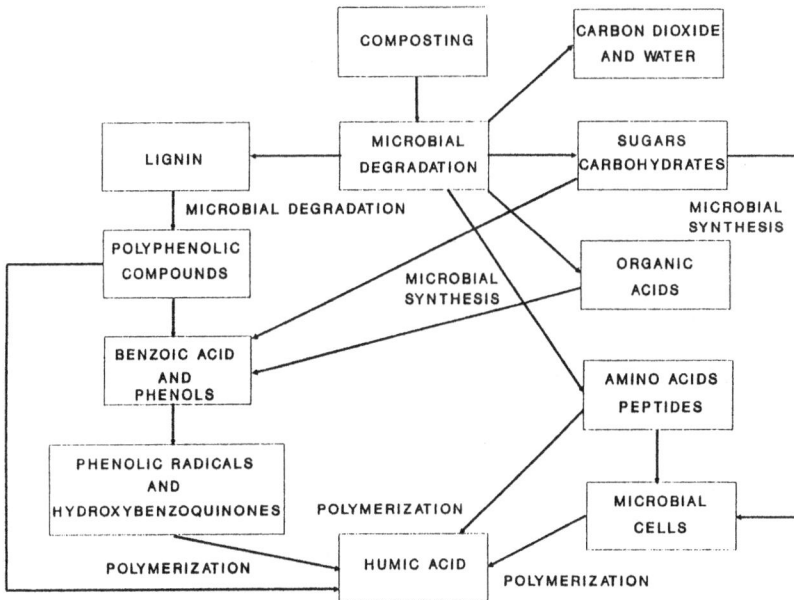

FIGURE 4.20. Potential pathway for the formation of humic acid during composting.

compounds prior to the formation of humus. Quinones of lignin origin, together with those synthesized by microorganisms, are the major building blocks of humic substances (Stevenson, 1994). The formation of humic substances from quinones occurs through the condensation with amino compounds. Carbohydrates and organic and inorganic N are metabolized by microorganisms to form microbial cells and products as well as other intermediate products. These are polymerized to form humic acid.

SUMMARY

Composting does not produce humus. Instead, humus is the result of long-term decomposition of organic matter in the soil. Composting produces a partially decomposed organic matter. If the process is carried out properly, the decomposition rate of the compost is slow and when applied to soils it continues to biodegrade, eventually becoming humus.

Considerable research has been carried out in an effort to identify the characteristics and properties of humus. Today's chemists have at their disposal sophisticated analytical equipment, which is providing insight into the chemistry of organic matter and humus.

One of the major objectives of these and other studies conducted to measure changes in components and the formation of humic substances is to arrive at an assessment of the degree of stabilization and maturity of compost. This aspect is discussed in Chapter 5, Stability, Maturity and Phytotoxicity.

REFERENCES

Aiken, G. R., D. M. McKnight, R. L. Wershaw, and P. MacCarthy. 1985. An introduction to humic substances in soil, sediment, and water, pp. 1–9. In P. MacCarthy, C. E. Clapp, R. L. Malcolm, and P. R. Bloom (eds.). *Humic Substances in Soil, Sediment, and Water: Geochemistry, Isolation and Characterization.* Wiley-Interscience, New York.

Ashworth, M. R. F. 1942. Changes occurring in the organic matter during the decomposition of compost heaps. *J. Agr. Sci.* 32:360–372.

Bodily, H. L. 1944. The activity of microorganisms in the transformation of plant materials in soil under various conditions. *Soil Sci.* 57:341–349.

Broadbent, F. E. and W. V. Bartholomew. 1948. The effect of quantity of plant material added to soil on its rate of decomposition. *Soil Sci. Soc. Amer. Proc.* 13:271–274.

Campbell, C. A. 1978. Soil organic carbon, nitrogen, and fertility, pp. 173–272. In M. Schnitzer and S. U. Khan (eds.). *Soil Organic Matter.* Elsevier Scientific Publ. Co., Amsterdam, The Netherlands.

Chang, Y. 1967. The fungi of wheat straw compost II. Biochemical and physiological studies. *Trans. Br. Mycol. Soc.* 50(4):667–677.

Chen, Y. and Y. Inbar. 1993. Chemical and spectroscopic analyses of organic matter transformations during composting in relation to compost maturity, pp. 551–600. In H. A. J. Hoitink and H. M. Keener (eds.). *Science and Engineering of Composting: Design, Environmental, Microbiological and Utilization Aspects.* Renaissance Pub., Worthington, OH.

Clark, F. E. 1967. Bacteria, pp. 1–49. In A. Burges and F. Raw (eds.). *Soil Biology.* Academic Press, Inc., New York.

Ciavatta, C., M. Govi, and P. Sequi. 1993. Characterization of organic matter in compost produced with municipal solid wastes: An Italian approach. *Compost Sci. & Util. Premier Issue,* pp. 75–81.

DeVleeschauwer, D., O. Verdonck, and P. Van Assche. 1981. Phytotoxicity of refuse compost. *BioCycle.* 27:44-46.

Dindal, D. L. 1978. Soil organisms and stabilizing wastes. *Compost Sci./Land Util.* 19(8):8–11.

Flaig, W., H. Beutelspacher, and E. Rietz. 1975. Chemical composition and physical properties of humic substances, pp. 4–211. In J. E. Gieseking (ed.). *Soil Components. Volume 1: Organic Components.* Springer-Verlag, New York.

Gascho, G. F. and F. J. Stevenson. 1968. An improved method for extracting organic matter from soil. *Soil Sci. Soc. Am. Proc.* 32:117–118.

Gray, K. R., K. Sherman, and A. J. Biddlestone. 1971. A review of composting—Part 1. *Process Biochemistry.* 6(6):32–36.

Greenland, D. J. and J. M. Oades. 1975. Saccharides, pp. 213–261. In J. E. Gieseking (ed.). *Soil Components. Volume 1: Organic Components.* Springer-Verlag, New York.

Gascho, G. F. and F. J. Stevenson. 1968. An improved method for extracting organic matter from soil. *Soil Sci. Soc. Am. Proc.* 32:117–118.

Higuchi, T. 1982. Degradation of lignin: Biochemistry and potential application. *Experientia.* 38:159–166.

Inbar, Y. 1989. Formation of humic substances during the composting of agricultural wastes and characterization of their physicochemical properties. Ph.D. Diss., Hebrew University of Jerusalem.

Inbar, Y., Y. Chen, and Y. Hadar. 1989. Solid state carbon-13 nuclear magnetic resonance and infrared spectroscopy. *Soil Sci. Soc. Am. J.* 53:1695–1701.

Inbar, Y., Y. Chen, and Y. Hadar. 1990. Humic substances formed during the composting of organic matter. *Soil Sci. Soc. Amer. J.* 54:1316–1323.

Jenny, H. 1941. *Factors of Soil Formation.* McGraw-Hill, New York.

Karim, A. and M. U. Chowdhury. 1958. Decomposition of organic wastes. *Soil Sci.* 85:51–54.

Keller, P. 1961. Methods to evaluate maturity of compost. *Compost Sci.* 2(3):20–26.

Kirchmann, H. and P. Widen. 1994. Fatty acid formation during composting of separately collected organic household wastes. *Compost Sci. & Utilization.* 2(1):17–19.

Kononova, M. M. 1975. Humus of virgin and cultivated soils, pp. 475–526. In J. E. Gieseking (ed.). *Soil Components. Volume 1: Organic Components.* Springer-Verlag, New York.

Leszkiewicz, C. G. and N. E. Kinner. 1988. The effect of temperature and oxygen concentration on lignin biodegradation. *Environ. Tech. Letters.* 9:359–368.

Lynch, J. M. 1986. Lignocellulolysis in compost, pp. 178–189. In M. de Bertoldi et al. (eds.). *Compost: Production, Quality and Use.* Elsevier Applied Sci., London.

Lynch, J. M. 1993. Substrate availability in the production of composts. In H. A. J. Hoitink and H. M. Keener (eds.). *Science and Engineering of Composting: Design, Environmental, Microbiological and Utilization Aspects.* Renaissance Pub., Worthington, OH.

MacCarthy, P., R. L. Malcolm, C. E. Clapp, and P. R. Bloom. 1990. An introduction to soil humic substances, pp. 1–12. In P. MacCarthy et al. (eds.). *Humic Substances in Soil and Crop Sciences: Selected Readings.* Amer. Soc. Agron., Inc., Madison, WI.

Martin, J. P. and D. D. Focht. 1977. Biological properties of soils, pp. 115–169. In L. F. Elliott et al. (eds.). *Soils for the Management of Organic Wastes and Waste Waters.* Soil Sci. Soc. Amer., Madison, WI.

Miller, R. C., F. B. Smith, and P. E. Brown. 1936. The rate of decomposition of various plant materials in soils. *J. Amer. Soc. Agron.* 28:914–923.

Parton, W. J., D. S. Schimel, C. V. Cole, and D. S. Ojima. 1987. Analysis of factors controlling soil organic matter levels in Great Plains grasslands. *Soil Sci. Soc. Amer. J.* 51:1173–1179.

Poincelot, R. P. 1975. The biochemistry and methodology of composting. *Conn. Agr. Expt. Sta. Bull. 754,* New Haven, CT.

Roletto, E., R. Barberis, M. Consiglio, and R. Jodice. 1985. Chemical parameters for evaluating compost maturity. *BioCycle.* 26(2):46–47.

Saviozzi, A., R. Levi-Minzi, and R. Riffaldi. 1988. Maturity evaluation of organic wastes. *BioCycle.* 29:54–56.

Stentiford, E. I. 1993. Diversity of composting systems, pp. 95–110. In H. A. J. Hoitink and H. M. Keener (eds.). *Science and Engineering of Composting: Design, Environmental, Microbiological and Utilization Aspects.* Renaissance Pub., Worthington, OH.

Sequi, P., M. De Nobili, L. Leita, and G. Cercignani. 1986. A new index of humification. *Agrochimica.* 30:175–179.

Stevenson, F. J. 1994. *Humus Chemistry. Genesis, Composition, Reactions.* 2nd edition. John Wiley & Sons, Inc., New York.

Stott, D. E. and J. P. Martin. 1990. Synthesis and degradation of natural and synthetic humic material in soils, pp. 37–63. In P. MacCarthy et al. (eds.) *Humic Substances in Soil and Crop Sciences: Selected Readings.* Am. Soc. of Agronomy, Madison, WI.

Tenney, F. G. and S. A. Waksman. 1929. Composition of natural organic materials and their decomposition in the soil: IV. The nature and rapidity of decomposition of the various organic complexes in different plant materials under aerobic conditions. *Soil Sci.* 28:55–84.

Van Veen, J. A. and E. A. Paul. 1981. Organic C dynamics in grassland soils. I. Background information and computer simulation. *Can. J. Soil Sci.* 61:185–201.

Waksman, S. A. 1936. *Humus.* Williams and Wilkins, Baltimore, MD.

Wiley, J. S. and J. T. Spillans. 1961. Refuse-sludge composting in windrows and bins. *J. Sanitary Eng. Div.* 87(SA5):33–52.

CHAPTER 5

Stability, Maturity, and Phytotoxicity

INTRODUCTION

The ultimate goal in composting is to produce a humus-like product that can be used for soil improvement and plant growth. Analytical procedures for determining inorganic elements and organic compounds that can affect plant growth have been well defined. Soil and plant analytical procedures are known and accepted (*Methods of Soil Analysis,* American Society of Agronomy, 1982).

It has been much more difficult to assess and quantify the biological manifestations that occur during composting, however. In lieu of definitive chemical methods, the composting industry for decades relied on bioassays that determined plant growth response. At the same time, there was a constant search for chemical methods that could provide quick results and could be related to the state of decomposition of the compost. Some of the early literature relied on techniques taken from soil science. Others sought to examine chemical changes that occurred during the decomposition of organic matter while being composted.

Soil science has long recognized that from a microbiological view, the measurement of total carbon in a soil was a poor representation of the potential activity. Methods by Schollenberger (1945), Walkley and Black (1934), Walkley (1947) and Mebius (1960) attempted to measure available carbon in soils. The Walkley-Black method is still readily accepted and used in the evaluation of available carbon in soils.

Using these techniques and adapting wet chemistry, Epstein (1956) studied the decomposition of organic matter in soil and its effect on soil aeration. Figure 5.1 depicts carbon dioxide evolution while Figure 5.2 shows the

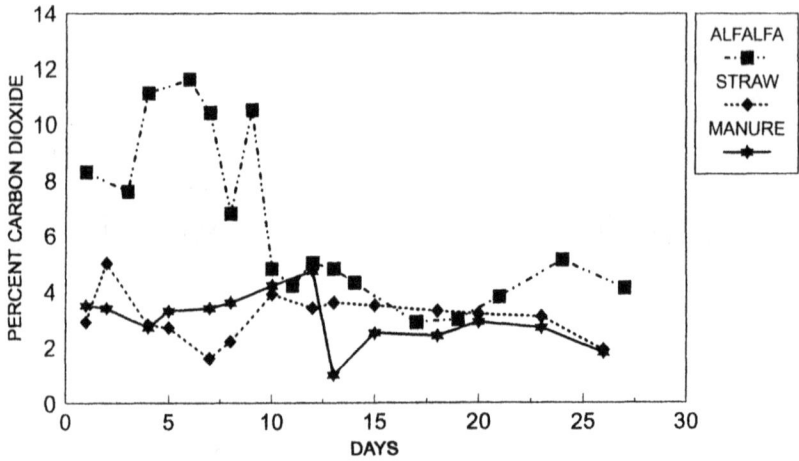

FIGURE 5.1. Percent carbon dioxide at 20 cm soil depth with incorporation of different types of organic matter. (From Epstein, 1956.)

oxygen content in soil as a result of incorporating different types of organic matter into soil. When fresh organic matter, alfalfa or an unstable compost, is incorporated into a soil, the rate of CO_2 evolution is high. Percent CO_2 increases and percent O_2 decreases. As the microbial community consumes the readily available carbon, the rate of CO_2 production decreases; the percent CO_2 decreases and, if the soil pore space is free (water does not

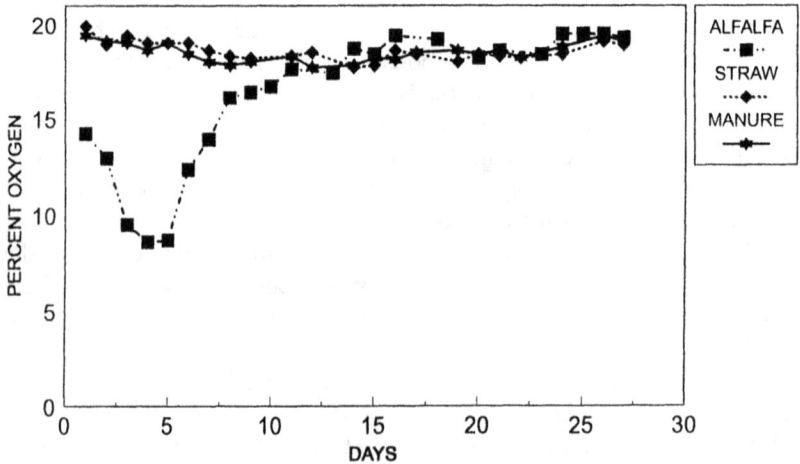

FIGURE 5.2. Percent oxygen at 20 cm soil depth with incorporation of different types of organic matter. (From Epstein, 1956.)

occupy the pore space), oxygen diffuses into the soil and the percent O_2 increases. If the soil pores are blocked, however, anaerobic conditions can set in and plants suffer. Soil aeration can affect plant water relations and nutrient uptake (Chang and Loomis, 1945; Lawton, 1946; Epstein, 1956; Epstein and Konke, 1957).

As early as 1924, Starkey (1924) reported that CO_2 evolution in the soil can serve as an indication of the process of decomposition. This was substantiated by Allen et al. (1934) and Bodily (1944). It was difficult to establish a relationship between microbial counts and CO_2 production (Engle, 1934; Vandecavey and Katznelson, 1938).

Other researchers delved into the chemical makeup of organic matter and the changes that occurred during the decomposition process. Waksman's (1936) proximate analysis of organic matter attempted to evaluate the organic components through fractionation of organic matter. These compounds consisted of sugars, starches, and other simple carbohydrates, fats, fatty acids, cutins, sterols, hemicellulose, cellulose, and lignin. It was argued that if one could follow the changes that occurred in such compounds as sugars, hemicellulose, cellulose, and lignin, or evaluate the formation of humus, it would be possible to identify the path of degradation and determine a stage where the rate of decomposition is very slow. Scientists involved in composting research attempted to use these techniques to identify the state of decomposition during composting. Table 5.1 provides a comprehensive list of the methods used to assess compost stability and maturity.

Two terms subsequently appeared in the literature, "stability" and "maturity," and these were often used interchangeably. The two terms are not synonymous, however. *Stability is a stage in the decomposition of organic matter and is a function of biological activity. Maturity is an organo-chemical condition of the compost which indicates the presence or lack of phytotoxic organic acids.* It must be noted that the term "stabilization" is used in many different contexts. For example, USEPA classes biosolids stabilization as related to pathogen reduction. Biosolids stabilization has also been used to indicate odor and volatile solids reduction (Bruce and Fisher, 1984).

Figure 5.3 illustrates the organic-matter decomposition curve during composting. At the peak of decomposition when microbial activity is greatest, the compost is highly unstable; that is, it is undergoing rapid changes. These changes are principally the result of microbial utilization of the easily decomposable compounds. As these compounds are utilized, the organisms still seek other sources of carbon for energy. They proceed then to attack the more complex-structured carbonaceous compounds. As a result, activity begins to decrease. When the rate of decomposition becomes very low, the compost is "stable." Stable does not mean that no further change will occur, but rather that the rate of decomposition is stable at a very low rate. In fact,

TABLE 5.1. *Methods Used to Assess Compost Stability and Maturity.*

1. Chemical Methods
   a. Carbon/nitrogen ratio
   b. Nitrogen species
   c. pH
   d. Cation exchange capacity
   e. Organic chemical constituents
   f. Acetic acid
   g. Starch-iodine
   h. Reactive carbon
   i. Humification parameters
      —Humification index
      —Relative concentrations of fulvic acid to humic acid
      —Humic substances
      —Functional groups
   j. Optical density

2. Physical Methods
   a. Temperature
   b. Color, odor, specific gravity
   c. Fluorescence

3. Plant Assays
   a. Cress seed
   b. Wheat and rye grass germination
   c. Root color

4. Microbiological tests and activity
   a. Respiration—oxygen depletion
   b. Respiration—carbon dioxide evolution
   c. Microbial changes—content of fungi, actinomycetes, etc.
   d. Enzyme activity

the rate is often so low that it is difficult to detect changes over short periods of time.

Maturity is a function of the organo-chemical properties of the compost as related to phytotoxicity to distinguish between phytotoxic effects due to inorganic chemicals and salinity. The organo-chemical phytotoxic effects are primarily attributed to fatty acid formation. However, other organic compounds are present in immature compost, which could also result in phytotoxicity.

Factors other than organo-chemicals that can cause phytotoxicity or inhibit germination and plant growth include salinity, trace elements, heavy metals, ammonia, carbon dioxide, soil compaction, soil moisture, and soil aeration.

Why should compost be stable and is a stable product always necessary?

An unstable product continues to decompose rapidly. Unless the compost is cured under aeration, anaerobic conditions will occur in the center of piles. Anaerobic conditions generate reduced compounds of C, N, S and P. Amines, sulphamines, mercaptans, skatoles, and H_2S are malodorous (Mather et al., 1993b).

In addition, methane and phosphine can be produced from C and P, which are inflammatory gasses (Mathur et al., 1993a). Fires in immature biosolid compost have occurred at the Baltimore composting facility and others. Similarly, in Riverside, California, fires have occurred in large piles of immature manure compost. Mather et al. (1993b) also indicated that stockpiles of immature compost may release methane and nitrous oxide to the atmosphere and that these gasses are considerably more effective than CO_2 as greenhouse gasses. Mather et al. (1993b) reported that bagged immature compost erupted as a result of gasses.

A stable product, on the other hand, does not produce odors in storage. If the product is bagged at the proper moisture content, opening the bag should not release putrescible odors. Further, when incorporated into the soil, a stable product does not decompose rapidly and utilize nitrogen required for plant growth. When a compost that has a high C/N ratio is added to soil, the microbial population competes with plants for soil nitrogen. In this case, plants typically exhibit chlorosis, yellowing of the leaves, indicating nitrogen deficiency. If organic materials with low C/N ratios are added to soil, ammonia can be released, which can cause phytotoxicity (Dowdy et al., 1976). Ammonia in refuse extracts has been shown to reduce germination and root elongation (Wong, 1985).

The decrease in oxygen in soil can result in reduced oxygen levels or anaerobic conditions, which not only affect plant growth but also soil chemical species, heavy metal solubility and uptake of nutrients.

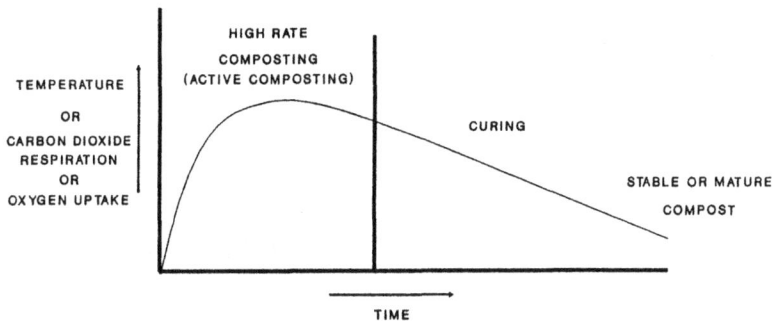

FIGURE 5.3. Phases during composting as related to temperature, and carbon dioxide respiration or oxygen uptake.

Composting to the point of stability often requires considerable time, particularly if cellulolytic feedstocks (e.g., MSW) are used. Stability with high cellulolytic materials can be achieved by proper feedstock preparation such as particle-size reduction and C/N ratio adjustment. Many composting operators want to market material as soon as possible in order to reduce storage space requirements and materials handling.

In certain situations, an unstable product could be used. For example, in remote land-reclamation areas, unstable compost could be applied and the amended soil could be planted at a later date. However, even this practice should be avoided since storage and transportation might release odors. Further, E&A Environmental Consultants, Inc. (1987) found that using uncured compost resulted in depressed growth of jack pine, red pine, white spruce, Siberian pea shrub, and chokecherry seedlings, which were to be used for mine land reclamation. In a subsequent application of cured compost to a taconite mine tailing, the composted plots maintained vegetation for two growing seasons whereas the fertilized plots had no vegetation. Applying uncured compost to land without planting for one to two months would reduce potential phytotoxic effects.

One objective of this chapter is to discuss the various methods that have been used to measure stability and maturity of composts. To be valid, a stability method must relate to physical changes in the compost as a result of microbial degradation. These physical parameters include color, odor, and inability to produce heat if bagged or minimal heat in large piles. Furthermore, a stable compost product should not produce excess CO_2 in the soil or utilize soil nitrogen, which would affect plant growth.

A mature compost should not reduce germination or result in decreased plant growth. Tests for maturity should distinguish between biological impacts on plant growth and chemical effects such as salinity. The stability method should be valid for composts of different feedstocks.

Several recent reviews of the literature focused on stability and maturity; for example, Jimenez and Garcia (1989), Beck & Associates (1990) and Mathur et al. (1993b).

STABILITY AND MATURITY

Many methods have been suggested and evaluated to assess the state of decomposition and the suitability of the compost as a medium for plant growth (Table 5.1). Not all these methods will be discussed here, since several of them have had limited success and were abandoned after initial evaluation.

One of the major early centers of research on composting in the United States was the University of California's Sanitary Engineering Department at Berkeley. A 1953 Sanitary Engineering Research Project Report stated, "A compost is considered finished when it may be stored in large piles indefinitely without becoming anaerobic or generating appreciable heat and may safely be put on agricultural soil because of its low C:N ratio or the poor availability of its carbon." After 42 years this statement is still valid.

An early review of the methods of measuring maturity as related to raw refuse and/or sewage sludge was made by Keller of the Swiss Water Control Federation (Keller, 1961). Many of Keller's observations are currently being revisited. For example, he stated that first of all a raw refuse-biosolids compost must be hygienic. He also questioned whether short-time decomposition can destroy resistant organisms such as spores and cysts as well as weed seeds. Further, Keller (1961) stated that the most important purpose of applying compost is to raise or maintain the humus content of soils and that the production and use of fresh compost does not result in an increase of humus. Furthermore, he indicated that the use of fresh compost can result in crop failures. He does point out, however, that Dutch experiments have shown favorable results with immature compost on clay soils. This may have resulted from the interaction of the organic matter and the clay particles, yielding increased water permeability and improved soil aggregation.

Chemical Methods

More effort and research have gone into chemical methods for determining stability and maturity than into the biological or physical evaluations. In general, chemical methods require less time and are often more rapid.

CARBON/NITROGEN RATIO (C/N)

C and N are the building blocks of plant and animal cells and, therefore, are impacted by microbial activity. Hence they are the most studied parameters during the decomposition process. One of the most difficult aspects of C determinations is the evaluation of what fraction of the total C is readily available C for microorganisms.

The C/N ratio has been used to indicate the stability of compost. Ratios below 20 were assumed to be indicative of a stable compost. Keller (1961) noted that the C/N ratio was not a reliable indicator of stability due to the large variability. This was further substantiated by Hiari et al. (1983), who found that the C/N ratio varied from 5 to 20 for different feedstocks. Mathur (1991) indicated that a mature compost should have a C/N ratio of about 10 as in humus. This is not readily achieved since different compounds decompose at different rates.

Chanyasak et al. (1983) suggested that the ratio of organic C to organic N (org.-C/org.-N) of water extracts from compost was indicative of maturity. Evaluating the ratio of org.-C/org.-N in water extracts, Hiari et al. (1983) found that with the exception of biosolids, this ratio was almost 5 to 6 regardless of material (Figure 5.4). The reason that this parameter was not applicable for biosolids was the low initial org.-C/org.-N ratio. Hiari et al. (1983) also found a good correlation between total-C and the ratio of organic-N to total N in water extracts, but not with solid material.

Morel et al. (1985) proposed that the final C/N to initial C/N ratio was a better indicator of maturity or stability, based on the finding that nitrogen-rich feedstocks often resulted in low C/N ratios. No specific values were given for the ratio, which would indicate stability or maturity.

Jimenez and Garcia (1989) utilized data in the literature to determine if this concept was valid. The data indicated that the final C/N to initial C/N ratio ranged widely (0.49 to 0.85) for different composting times. Some of the lowest and highest values were found for the same duration of composting. As a result, Jimenz and Garcia (1985) concluded that the final C/N ratio to initial C/N ratio could be used as a guide but not as an absolute indicator of the degree of compost maturity.

Inbar et al. (1990) found that the C/N ratio decreased rapidly from 27 to 10 during the first 60 days of composting separated cattle manure (Figure 5.5). After 60 days the C/N ratio decreased slightly, to 8.7. The C/N ratio was highly correlated with composting time ($r^2 = 0.991$). Since the data were not related to plant growth, it is difficult to assess when after the 60-day period the compost was considered stable or mature.

ORG.-C/ORG,-N

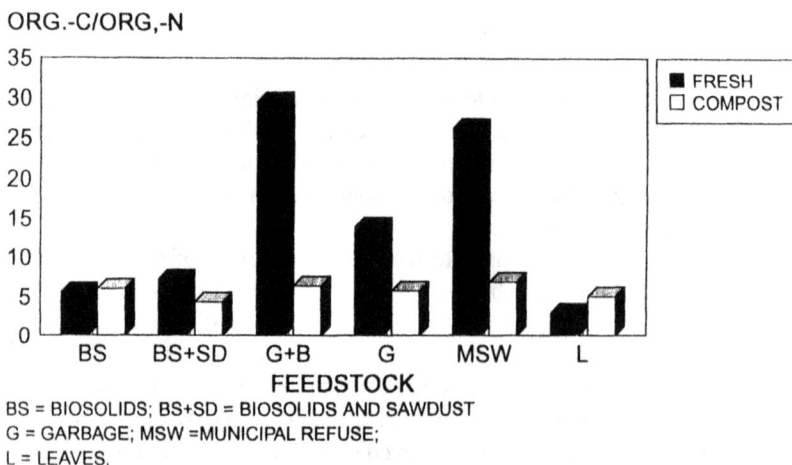

BS = BIOSOLIDS; BS+SD = BIOSOLIDS AND SAWDUST
G = GARBAGE; MSW =MUNICIPAL REFUSE;
L = LEAVES.

FIGURE 5.4. Changes in the ratio of organic carbon to organic nitrogen during composting. (Reprinted with permission from Hirai et al., 1983.)

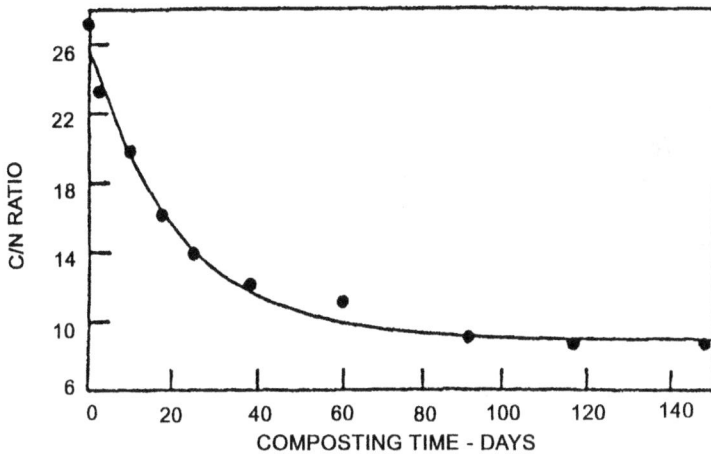

FIGURE 5.5. Changes in C/N ratio during composting of separated cattle manure. (Reprinted with permission from Inbar et al., 1990.)

NITROGEN SPECIES

Changes in nitrite (NO_2), nitrate (NO_3) and ammonia (NH_4) have also been suggested as indicators of maturity.

Spohn (1978) reported that the presence of ammonia indicated that the compost was not cured (ripe), whereas the presence of nitrate indicated that it was well "ripened." Spohn (1978) indicated that N tests alone were not sufficient and that they should be complemented with a sulfide test and a cress seed germination test.

Keller (1961) reported that in some samples of unripe compost, large amounts of NO_2 in relation to NO_3-N were found. However, the ratio of NO_2/NO_3 tests gave only an indication of maturity.

Finstein and Miller (1985) related maturity to the amount of NO_2^- and NO_3^-. The appearance of considerable quantities of these radicals during composting was a sign that the compost was mature. In aerated static pile methods with composting of biosolids, NO_2^- appeared after 86 to 113 days; NO_3^- appeared after 96 to 123 days.

PH

Changes in pH have been noted to occur during the composting period and, therefore, have been considered as a possible indicator of biological activity. Generally, the pH drops during the very early stages of composting and then increases to a range of 6.5 to 7.5. A typical pH curve was shown in Chapter 3, Figure 3.28. However, exceptions to this "typical" curve exist.

Jimenez and Garcia (1989) noted that acid pH values indicate a lack of maturity due to short composting time or the occurrence of anaerobic conditions. Often with very acid food processing wastes (grape pomace, cranberry waste), the initial pH is low and even under aerobic conditions over long periods of time it remains acid.

Jann et al. (1960) found that when an aerobic stable compost is subjected to anaerobic conditions, there is no further change in pH, and odors are not produced. The pH of the material was slightly alkaline at a pH of 7.5. In general, pH is not a good indicator of stability.

CATION EXCHANGE CAPACITY (CEC)

CEC is a measure of an inorganic or organic particle's ability to sorb or retain cations on its surfaces. Since many cations (K, Ca, Mg, etc.) are important plant nutrients, the higher the CEC, the greater the particle's ability to retain these nutrients so that they may be available to plants.

Harada et al. (1981) found that during aerobic composting, the CEC increased from approximately 40 me/100 g to 70 to 80 during five to eight weeks and then remained essentially the same. A highly significant negative correlation ($r = -0.94$) was found between CEC and the C/N ratio.

Inbar et al. (1990) noted that the CEC increased rapidly with composting time and that over a period of 150 days it was threefold that of the original material (Figure 5.6). A correlation coefficient, $r^2 = 0.990$, was obtained. Similar data were obtained in other studies. A minimum value of 60 meq/ 100 g is an acceptable state of maturity. Mathur et al. (1993b) indicated that the CEC is not a reliable indicator of maturity since humic substances may vary in CEC partly as a result of the blocking of their exchange sites by complexing ions such as Cu, Fe, and Al, partly due to interaction with amorphous Fe and Al compounds.

ORGANIC CHEMICAL CONSTITUENTS

During composting, changes take place in the major organic constituents that make up the feedstock. One of the earlier attempts to analyze these constituents and follow their changes during composting was carried out by Waksman (1936) using the proximate analysis to evaluate sugars and simple carbohydrates, fats and waxes, hemicellulose, cellulose, and lignin. This was discussed in Chapter 4.

Cellulose content yields a good index of the degree of maturity of the compost (Keller, 1961). Figure 5.7 shows that cellulose decreased with

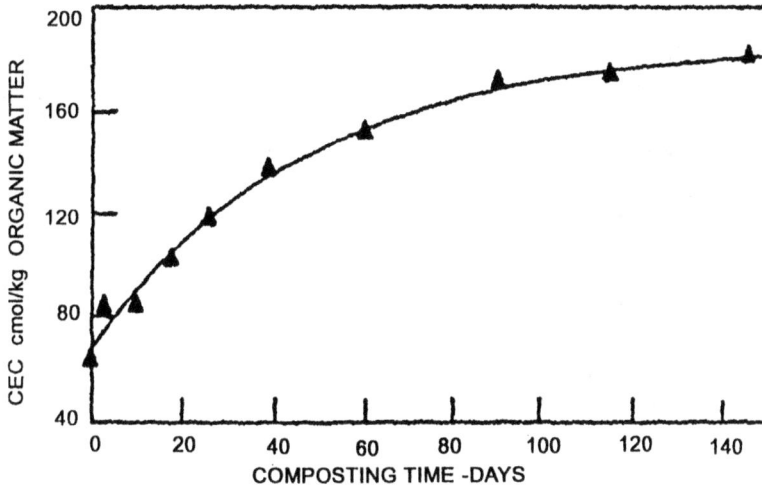

FIGURE 5.6. Changes in cation exchange capacity during composting of separated cattle manure. (Reprinted with permission from Inbar et al., 1990.)

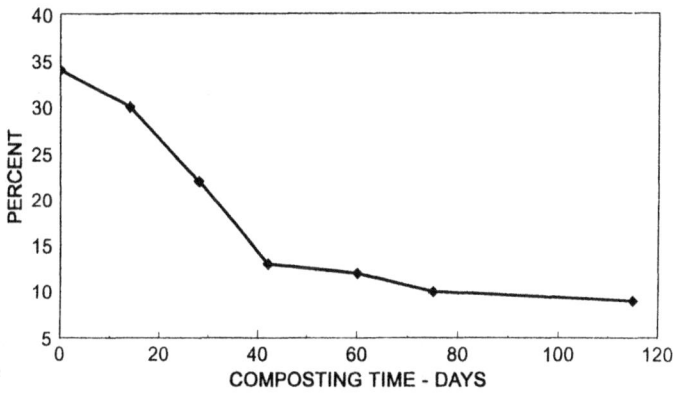

FIGURE 5.7. Changes in percent cellulose during the windrow composting of refuse. (Data from Keller, 1961.)

117

duration of composting. Several methods have been suggested to determine cellulose.

Inoko et al. (1979) studied the changes in hemicellulose and cellulose (reducing sugars) during composting of city refuse. They reported that the content of these polysaccharides decreased from approximately 36% of the total dry weight to about 20% after 60 days. A positive correlation was found between the polysaccharides and the C/N ratio (Jimenez and Garcia, 1989). With nine samples, the regression equation was:

$$\text{Reducing sugars} = (1.749 \ C/N) - 6.424r = 0.909$$

Microbial metabolic activities on organic matter take place in the water film on the surfaces of particles. Hirai et al. (1983), therefore, studied the changes of water soluble components during composting. Amino acids, low fatty acids, and polysaccharides greatly decreased during the composting of refuse and garbage; however, composting of biosolids resulted in a decrease of amino acids and low fatty acids but not polysaccharides or peptides. In fact, polysaccharides and peptides increased. Peptides also increased during composting of garbage. Finally, total organic C decreased during the composting of refuse, garbage, and biosolids (Figure 5.8).

Recently, Zhang et al. (1992) suggested that reactive carbon could be used to assess maturity. Reactive-C measurement is an attempt to quantify the available C, which microorganisms utilize for energy. The approach modi-

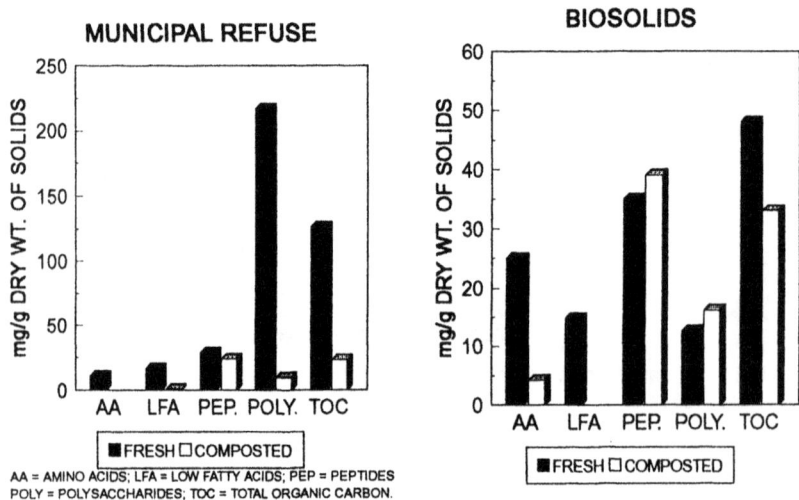

FIGURE 5.8. Changes in several chemical components during composting of refuse and biosolids. (Reprinted with permission from Hirai et al., 1983.)

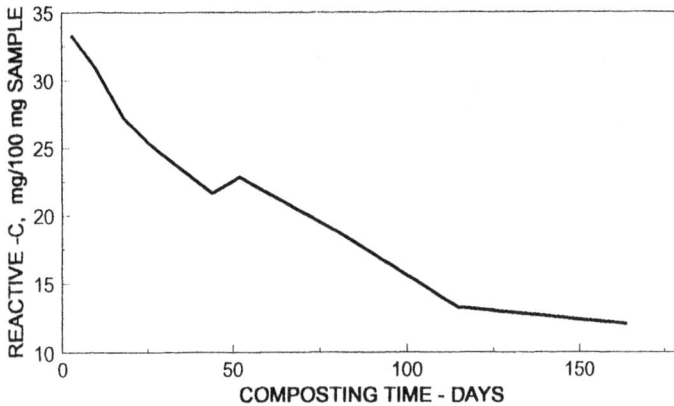

FIGURE 5.9. Changes in reactive-C during composting of MSW. (Data from Zhang et al., 1992.)

fied the Walkley-Black methods used in soils for measuring available C. Changes of reactive C during composting are shown in Figure 5.9.

As illustrated, the initial values of approximately 33 mg reactive-C/100 mg sample and final values of approximately 11 mg/100 mg were similar to 12 to 16 mg/100 mg obtained with yard waste. However, in the case of a biosolids-MSW co-compost, the final value remained near 30 mg/100g. This may indicate that the method does not apply to all organic materials to be composted or that in the case of biosolids or other material, some factor may interfere with the determination.

The authors indicated that since the MSW tested retained some phytotoxic properties, reactive-C is not a good indicator of maturity. One of the difficulties with this method is the small sample size of 200 mg. In fact, the authors stated that the sample size must be less than 200 mg dry weight per assay. Above this size, the results were not linear. This method may require a larger number of samples to ascertain its reproducibility. Data for other materials did show good reproducibility.

HUMIFICATION PARAMETERS

The changes in the constituents of organic matter represent a potential for evaluating the degree of stabilization. Several studies have looked at changes in humic acid, fulvic acid and other products of decomposition.

Humification Index

Roletto et al. (1985) coined the terms "humification ratio" and "humification index." Humification ratio (HR) was the percentage of total extractable humic-C (C-ext) as related to the total organic-C (C-org) (i.e., HR = C-ext × 100/C-org). The humification index (HI) was defined as the percentage humic acid C (CHA) as related to the C-org (i.e., HI = CHA × 100/C-org).

As a result of the extraction procedure, the authors indicated that the method overestimated the extent of humification in the early stages of composting. No comparisons were made at different stages of composting with other parameters.

Relative Concentrations of Humic Acid to Fulvic Acid

Saviozzi et al. (1988) attempted to follow the changes in decomposition (i.e., maturation process) by evaluating various humification parameters. They used a different "humification index," which was based on the ratio between the organic C of the non-humified fraction and that of the humic acid (HA) + fulvic acid (FA). The HI seemed to be appropriate for evaluating stability but was deemed to be complicated and slow. The authors recommended that considerably more analyses with different materials needed to be tested before the methodology could be accepted.

Humic Substances

Another approach has been to evaluate the total content of humic substances (HS). Inbar (1989) reported that the total content of humic material extracted from separated cattle manure compost increased from 377 to 710 g/kg OM (see Figure 4.14, Chapter 4). Humic material increased rapidly during the first 60 days and from 60 to 140 days a very gradual change was noted.

Functional Groups

The advent of sophisticated organic chemical methods provided scientists with tools to evaluate changes in specific C compounds. Two more recent methods include solid-state cross-polarization magnetic angle spinning ^{13}C-nuclear magnetic resonance (CPMAS[13]C-NMR) and infrared spectroscopy. Distribution of C in several organic compounds using CPMAS[13]C-NMR for fresh and composted manure is shown in Figure 4.13, Chapter 4 (Inbar et al., 1990). This method reflected the changes in various C compounds during decomposition of cattle manure.

The humification methods require sophisticated expensive equipment and, therefore, are not suited for routine evaluation of stability for field operations. However, they can be extremely valuable by furthering our knowledge on the transformations that occur during composting. They may also provide validation of simple techniques that could be used for routine stability evaluations.

OPTICAL DENSITY

Mathur et al. (1993a) and Schnitzer et al. (1993) proposed the use of optical density, or colorimetry, to evaluate maturity. The method is based on absorption by water abstracts of composts at 665 nm of visible light. Data presented compared the results to temperature, % O_2, biological oxygen demand (BOD), dissolved organic carbon (DOC), NH_4^+, NO_3^-, and cress germination tests. The optical density data (Mathur et al., 1993a) only showed lower adsorption (E665) at the end of 59 days of composting. Variation between replicates was great. For three of the four treatments, the difference in optical density between 40 and 59 days of composting was significant at a 95% confidence level. Average cress germination tests for four manure treatments at 40 and 59 days were 78% and 102%. The coefficient of determination (R^2) between absorption at various optical densities and DOC was low (< 0.73).

No other statistical data were provided. At this time, limited testing on composted manures using passive aeration in bins cannot justify the use of this method to determine what the authors termed "biomaturity" or "maturity" as a result of biological activity.

Schnitzer et al. (1993) indicated that optical density of water extracts was due to many intermediate products of decomposition. Optical density decreased during composting as a result of the formation of "new" humus. Composting does not produce humus, but rather a state leading to the ultimate production of humus.

Physical Methods

TEMPERATURE AND HEAT OUTPUT

Temperature is a reflection of the microbiological activity during composting. Depending on the feedstock and its physical state, the temperature usually rises within the first few days of composting from ambient to 60°C to 70°C. It remains at this level with minor fluctuations for several days and then gradually decreases to a constant state near ambient. The point at which temperatures approach ambient may be considered the state of stability of the product.

Heat output is a function of microbial activity. Niese (1963) used Dewar flasks to measure heat production as related to respiration. It was found that the temperature increased to a maximum and then declined as the process progressed. Heat loss could not be prevented even with Dewar flasks when using small amounts of organic matter. When heat was not dissipated temperatures increased.

Maximum temperatures of about 75°C occurred. The maximum temperatures were dependent on the ambient temperature. Unless the surrounding temperature of the flask was controlled, inaccurate results were obtained. Dewar's data indicated that temperatures of self-heating, with large temperature differentials between flask and ambient temperature, were not the same as those present during spontaneous self-heating and were, therefore, inaccurate.

Niese (1963) indicated that with a high bacterial count, the temperature rose immediately; but at low bacterial counts, there was a delay in temperature increase. In 8 of 10 experiments, the number of mesophilic bacteria were between 2×10^9 and 15×10^9/g dry weight. Self-heating temperatures increased rapidly, reaching 45°C in 7–15 hours. The maximum temperature ranged from 73.5°C to 77.5°C. Based on these results, it was concluded that self-heating resulted in maximum temperatures if sufficient nutrients were available to the microorganisms.

From these studies, Niese stated that a judgment of decomposition of refuse and refuse compost can be made using several stages of temperature (Table 5.2). However he cautioned that this is only a guide and that comparisons with plant growth need to be made. Spohn (1978) stated that heating in a pile or a well-insulated aerated vessel is indicative that the compost is unstable or "unripe," but that the lack of temperature does not indicate that the compost is "ripe" or stable.

TABLE 5.2. The Relationship of Degree of Decomposition to Temperature as Determined by Self-Heating in a Dewar Flask (Niese, 1963).

Maximum Temperature Achieved	Degree of Decomposition
Above 70°C	0 Raw refuse, very slight decomposition
60–70°C	1 Moderate decomposition
45–60°C	2 Medium decomposition
30–45°C	3 Good decomposition
Under 30°C	4 Decomposition mostly or completely finished

Recently, the Dewar flask-heat output method was reintroduced by German and U.S. scientists (Brinton et al., 1993; Woods End Research Laboratory, Inc., 1993). Currently there is little information on the relation of this method to other stability measurements or plant growth.

Jann et al. (1960) felt that temperature was not a good criterion by which to evaluate maturity because "a low temperature could be the result of loss of aerobic conditions and of loss of moisture and the production of excessive heat during the high plateau could inhibit microbial activity." This has been observed at numerous composting facilities; temperatures drop because of inhospitable conditions for microbial activity, and material is considered stable. Later, rewetting or long-term storage of the material results in reactivation and subsequent anaerobic conditions.

COLOR, ODOR, STRUCTURE AND SPECIFIC GRAVITY

Keller (1961) indicated that the physical properties of color, structure, odor, and specific gravity were not specific enough to assess stability. The following were some of his criteria. A mature compost should be dark brown to black regardless of the feedstock. Mature compost should have "smells like forest soil (typical soil odor is caused by actinomycetes)."

Recently two USDA scientists (Becker, 1995) determined that the smell of soil is primarily the result of two gasses, geosmin and 2-methylisoborneol, which are by-products of fungi and actinomycetes. The scientists are developing an odor-based soil test as a barometer of organic activity. If these two gasses are present in compost, it is possible that they could be used to determine stability.

Odor and color are too subjective to provide accurate assessment of stability, and structure is too difficult to describe and measure accurately. The specific gravity increases during composting. Ranges in a mature compost are from 0.5 to 0.9 g/cc. Because of this wide range, this parameter has not been deemed accurate.

PLANT ASSAYS

Plant assays are used to evaluate maturity but have not been used to determine stability. Keller (1961) used plants to determine phytotoxicity in field trials. However, he felt that these tests were subject to error and recommended they only be used in conjunction with other tests. Keller (1961) studied the relationship between wheat growth and age of compost. With 100% compost, yield was depressed for 45-day-old compost but 60-day-old compost produced a 49% increase over the control.

In 1969 the cress seed germination and elongation tests were described along with two analytical methods (Spohn, 1969). Prior to that time, other plant tests were used. Zucconi et al. (1981a) studied the cress seed germination test and established a germination index, obtained by multiplying the percent germination by the percent root growth as related to control. The presence of phytotoxic compounds indicated that the compost was not stabilized.

Windrow and static piles were compared. Toxicity disappeared much more rapidly in the static piles than in windrows. This is not surprising since within a very short time after turning the windrow it becomes anaerobic, and under these conditions organic acids are produced, which can be phytotoxic.

Studies by E&A Environmental Consultants, Inc., found a correlation between poorly stabilized compost and cress germination, but highly stable compost did not relate to cress germination. As indicated earlier, the confounding effects of other chemical parameters including salinity, C/N ratio, and pH can affect germination and provide erroneous conclusion regarding the maturity of the compost.

Other plants, cucumber, ryegrass and radish have been used. Iannotti et al. (1994) found that these plants differed in their response to MSW compost treatments.

No single plant bioassay has been identified as related to maturity of compost. Several germination and growth tests may be needed using different plants to assure users of the maturity of compost. Alternatively, a method that would remove the confounding effects of salinity and other factors could be valid.

MICROBIOLOGICAL TESTS AND ACTIVITY

Since stability represents the state of microbiological activity, measurements of respiration either through CO_2 evolution or O_2 uptake should provide the best indication of this state. The problem with respiration measurements, as well as most chemical measurements, is the use of small, disturbed samples. If a material is homogeneous and represents the entire mass, then the measurement would indicate the state of stability. However, many compost products are far from uniform. For example, biosolids are often composted with wood chips, where a single wood particle can offset results. MSW often includes inerts, plastics, and other contaminants that affect the results. Yard waste also may contain woody particles that skew the data. Pressel and Bidlingmaier (1981) indicated that sample conditions and preparation could significantly affect measurements of oxygen uptake. Moisture content of the sample was directly related to respiratory activity.

Frost et al. (1992) pointed out that a good assay avoids excessive screening or grinding as it will increase the surface area and provide for an increase in microbiological activity and an indication of instability of the compost. Their data confirmed Pressel and Bidlingmaier's (1981) finding that moisture content of the sample can affect stability determinations. Frost et al. (1992) indicated that stability measurements must be made on samples having a moisture content of 50% to 65% on a weight basis. The major microbial-related tests are:

(1) Respiration—carbon dioxide evolution

(2) Respiration—oxygen depletion

(3) Microbial changes—content of fungi, actinomycetes, etc.

(4) Enzyme activity

Respiration—Carbon Dioxide Evolution

The methodologies for CO_2 have been derived from respiration studies in soils (Epstein, 1957; Bartha and Pramer, 1965; Anderson, 1982). Two principal methods may be used to determine the carbon dioxide respiration rate. One method (Bartha and Pramer, 1965) is based on a single point in time. CO_2 is trapped in an alkaline solution and alkaline solution is titrated to determine the extent of the reaction. CO_2 is determined over a four-day period. Respiration is expressed as mg/CO_2 C per gram of compost per day, or mg CO_2 per kg of biological volatile solids (VS) per hour. The term "biological VS" is used to distinguish from non-biological material such as plastics.

High respiratory activity indicates greater microbial activity as C is being transformed to CO_2. A low respiration activity indicates that the available C has essentially been utilized. E&A has established a compost stability index based on a large number of samples for different feedstocks in relation to their decomposition time. Different stages of stability as related to CO_2 evolution are shown in Table 5.3. Compost is considered relatively stable when the respiration rate is less than 5 mg CO_2-C/g compost C. Rates over this amount reflect different stages of instability. The advantage to this method is its simplicity and low cost.

Another approach that measures continuous CO_2 evolution has been used to measure soil organic matter decomposition (Epstein and Kohnke, 1957). This method allows for continuous monitoring of CO_2 and determining the rate of respiration over time.

Respiration—Oxygen Uptake

Oxygen uptake as a result of microbial activity has been used for many years (Epstein and Kohnke, 1957; Pressel and Bidlingmaier, 1981). Pressel

TABLE 5.3. Compost Stability Index Based on Carbon Dioxide Evolution.

Respiration Rate (mg CO_2-C/g compost C-day)	Rating	Characteristics
<2	very stable	well cure; no malodors; earthy odor
2–5	stable	cured compost; minimal impact on soil dynamics
5–10	moderately stable	uncured compost; some malodor potential; addition to soil may immobilize N; high phytotoxicity potential; not recommended for growing compost from seed
10–20	unstable compost	very immature compost; high malodor and phytotoxicity potential; not recommended for growing plants from seed
>20	very unstable compost	extremely unstable material; very high malodor and phytotoxicity potential; not recommended for use

Source: E&A Environmental Consultants, Inc., 1994.

and Bidlingmaier (1981) stated that the specific biological oxygen consumption or respiratory activity in a sample of compost was indicative of the compost's state of decomposition. Willson and Dalmat (1986) determined oxygen consumption rate by measuring the decrease in partial pressure of oxygen. A 75% decrease in respiration was observed during curing of biosolids composting.

Frost et al. (1992) and Iannotti et al. (1994) evaluated a dissolved O_2 test, which requires only several hours in contrast to much longer times for CO_2 evolution determinations. Again, this method determines a single point in time. The method provides for on-site measurement of respiration and compost stability. As Frost et al. (1992) noted, sample drying and anaerobic samples give false results. Samples must be pre-incubated overnight. The change in % O_2 in air is converted to an O_2 based on VS as follows:

$$\text{Mean } O_2 \text{ uptake} = \frac{-CVS60D}{KW_iVS}$$

where

C = oxygen content by volume in the air, which is usually 20.8% and expressed as a fraction of 0.21

V = volume of air in flask (mL)

S = slope of relative O_2 uptake rate (% saturation per minute)

60 = factor to change minutes to hours

K = constant factor based on the calibration value corrected for elevation above sea level based on the dissolved oxygen recommendations of the instrument manufacturer

W_t = compost dry matter weight used in grams

Figure 5.10 shows data obtained by Iannotti et al. (1994) and compares the O_2 uptake and CO_2 evolution data. A correlation coefficient of -0.80 was obtained between composting time and O_2 respirometry. In comparison, CO_2 respirometry data had a correlation coefficient of -90. Both were highly significant. Oxygen respirometry correlated well with maturity as determined by ryegrass growth (Figure 5.11). The index for O_2 is shown in Table 5.4.

E&A Environmental Consultants, Inc., has used the CO_2 respirometry method for several years applied to a wide variety of feedstocks. Based on the large data base, an index was developed as shown in Table 5.3.

MICROBIAL CHANGES

During composting the microbial populations change (see Chapter 3, Microbiology). Several researchers indicated that these changes can be indicative of the state of stability or maturity (Keller, 1961; Citernesi and De Bertoldi, 1979). Keller (1961) noted that actinomycetes appeared at the later stages of composting. Citernesi and De Bertoldi (1979) found that thermophilic bacteria decrease as the compost approaches maturity and that a total count of microorganisms may be indicative of the state of compost maturity. Microbiological methods are more cumbersome and not suited for routine evaluation of stability.

ENZYME ACTIVITY

Microbial enzymes' activities may indicate changes in carbon substrates and may, therefore, be good indicators of the status of the composting process (Herrmann and Shann, 1993).

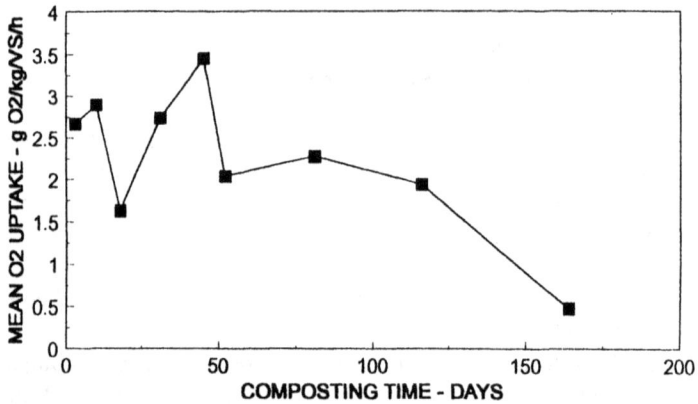

FIGURE 5.10. Changes in rates of respiration based on oxygen uptake per gram of volatile solids during MSW composting. (Reprinted with permission from Iannotti et al., 1993.)

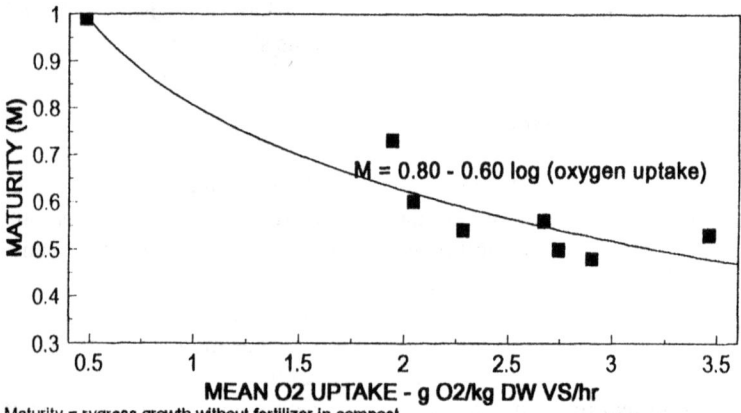

Maturity = rygrass growth without fertilizer in compost divided by growth in fertilized peat.

FIGURE 5.11. Relationship between rate of respiration as measured by oxygen uptake and growth of ryegrass. (Reprinted with permission from Iannotti et al., 1993.)

128

TABLE 5.4. Compost Stability Index Based on O_2 Uptake.

Respiration Rate (mg O_2/g VS × hr)	Rating	Characteristics
0–0.5	very stable	—well cure —no odors —no continued decomposition
0.5–1.0	stable	—cured compost —limited odor potential —minimal impact on soil carbon and nitrogen dynamics
1.0–1.5	moderately stable	—uncured compost —minimal odor production —addition to soil may result in nitrogen immobilization —high phytotoxicity potential —not recommended for growing plants from seed
1.5–2.0	unstable compost	—very immature compost —high odor and phytotoxicity potential —not recommended for growing plants from seed
>2.0	unstabilized material	—extremely unstable —very high odor and phytotoxicity potential —not recommended for use

Source: E&A Environmental Consultants, Inc., based on Ionnatti et al. (1994).

Keller (1961) evaluated the reductase enzyme activity proposed by Bucksteeg and Thiele (1959). It was reported that reductase activity in biosolids was related to oxygen consumption and the number of organisms. However, Keller (1961) did not find that the method was appropriate for compost.

A more recent study evaluated several enzymes including endo-cellulase, glucosidase, lipase, and phosphatase (Herrmann and Shann, 1993). For both endo-cellulase and glucosidase activity, there was relatively little change during the first 80 days of composting. Peak activity occurred during the curing phase at approximately 112 days and then decreased rapidly. This peaking was believed to be the result in a shift towards utilization of the more recalcitrant carbon sources of cellulose, lignocellulose and lignin. The activity of these enzymes does not appear to be a good indicator of changes occurring during composting. With the alkaline and acid phosphatase activities, which make organic phosphorus available, a very rapid decrease was noted during the first few days. Subsequently, for the next 40 to 50 days the activity level remained low. After the compost was removed from the bins to be placed into curing, the activity level rose. No correlations were given between enzymatic activities and other changes occurring during the composting period.

Based on the above reports, it does not appear that enzymatic activity is a good indicator of compost stability.

Phytotoxicity

Phytotoxicity can occur from heavy metals, other inorganic elements, soluble salts and organic compounds. In this chapter only phytotoxicity from organic compounds will be discussed. Phytotoxic compounds during composting may be produced during composting as a result of anaerobic conditions, which arise from insufficient aeration or excessive moisture. A result of anaerobic metabolism is the production of low- molecular-weight organic acids such as acetic, propionic, and butyric acids. Phytotoxicity caused by organic compounds generated during composting can be remedied by increasing the period of aerobic decomposition. The formation of fatty acids during composting was presented in Chapter 4, Biochemistry.

Still et al. (1976) found that composting bark for 30 days reduced or eliminated phytotoxic substances that inhibit growth of cucumber roots (Figure 5.12). Zucconi et al. (1981a) reported that olive tree stunting was a function of the amount of compost. This was the result of partial destruction of the root system. This damage appeared to be transitional, however. Cress seed (*Lepidium sativum*, L) was selected to determine the phytotoxicity of compost. Zucconi et al. (1981a) established a "germination index" by multiplying germination and root elongation.

DeVleeschauwer et al. (1981) found numerous organic acids in fresh compost and only small amounts of acetic acid in five-month-old compost. Cress seed germination was inhibited until the compost had been composted for 120 days. They found little difference in germination between aerated and non-aerated MSW compost.

Wong (1985) found that dry weight of *Brassica parachinensis* Bailey was inhibited for over 75 days. Highest yields occurred after 105 days. A minimum of four months is needed to avoid phytotoxicity of MSW compost. Ammonia and ethylene oxide appeared to inhibit root growth. The author did not investigate the presence of other compounds that have been shown to inhibit germination and root elongation.

Manios et al. (1987) found that phytotoxicity was positively related to

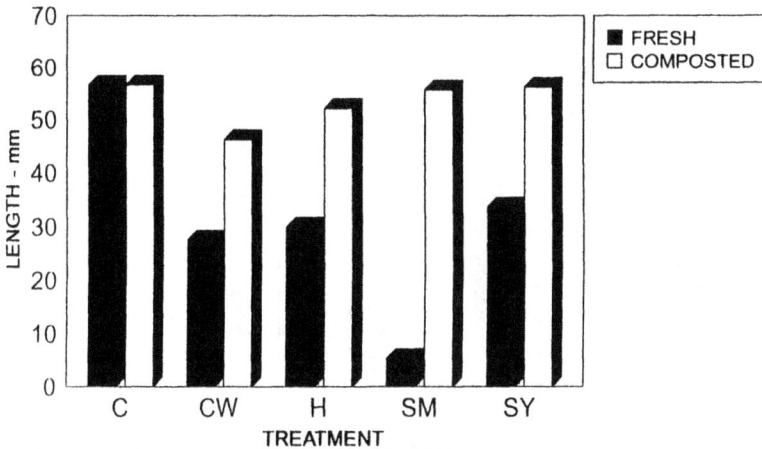

C= Control; CW= Cottonwood bark extract; H= Hackberry bark extract;
SM= Silver Maple bark extract; SY= Sycamore bark extract.

FIGURE 5.12. Effect of composting on phytotoxicity from several bark extracts. (From Still et al., 1976.)

organic acid production during composting of olive tree leaves. Organic acids decreased with composting time and germination of lettuce increased. It took 80 to 180 days for the phytotoxic effect to disappear. Shiralipour and McConnell (1991) found that the presence of a water-soluble substance in the compost inhibited seed germination.

Phenolic acids are the primary phytotoxic compound in eucalyptus. These compounds were degraded after 84 days in the laboratory but required 300 days for complete degradation in the soil (Ratcavek, 1989). E&A Environmental Consultants, Inc. (1990) conducted a study using compost produced from municipal biosolids with eucalyptus or sawdust. Rooted plant cuttings of *Pittosporum tovira* were transplanted into pots containing compost soil mixtures.

The following six treatments were used: eucalyptus compost:soil 2:1, 3:1, 4:1; sawdust compost:soil 2:1, 3:1, and 4:1. In addition to these treatments, a control consisting of a 3:1 mixture of a commercial organic mix and soil was also used. Each treatment was replicated 15 times.

The final vertical growth measurements depicted in Figure 5.13 show that plants grown in the sawdust compost treatments grew to almost twice the height of plants grown in the eucalyptus compost treatment. However, the growth of eucalyptus treatments was similar to that in the control. This indicated that the eucalyptus compost probably did not have a toxic effect on plant growth. Orthogonal contrasts determined that the eucalyptus compost treatments were significantly lower ($p < 0.01$) than the sawdust

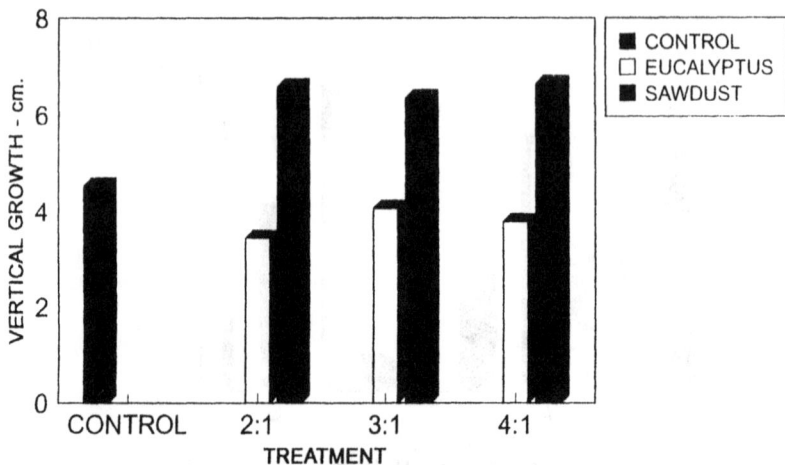

Treatments consist of soil as control and ratios of Eucalptus-
biosolids compost to soil or sawdust - biosolids compost to soil.

FIGURE 5.13. Effect of eucalyptus-biosolids compost and sawdust-biosolids compost on the growth of *Pittosporum tovira*. (From E&A Environmental Consultants, Inc., 1990.)

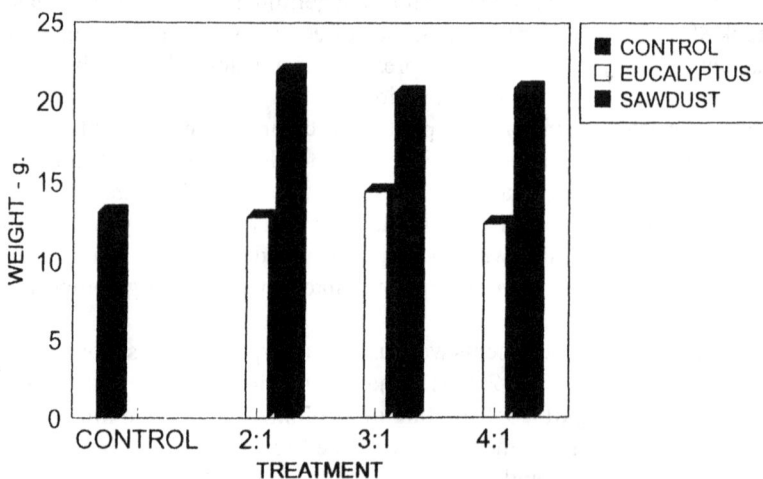

Treatments consist of soil as control and ratios of Eucalyptus-biosolids
compost to soil or sawdust-biosolids compost to soil.

FIGURE 5.14. Effect of eucalyptus-biosolids compost and sawdust-biosolids compost on the weight of *Pittosporum tovira*. (From E&A Environmental Consultants, Inc., 1990.)

compost treatments. No significant differences were found between the eucalyptus compost treatments and the control. Similar results were obtained for wet weight measurements (Figure 5.14).

SUMMARY

Although there is no one single method to evaluate maturity (Inbar et al., 1990), the producer and user of compost has several good methods to assess stability and maturity. For stability, it is recommended that respiration, as measured by CO_2 evolution or O_2 uptake, be used. A seed germination test could be used for maturity. The test should be carried out on the least tolerant plant where the compost is to be used or on a similar plant species. If the compost is to be used for lawn, sod, or turf production from seed, grass seed germination test would be most appropriate. Combination of grass, cucumber and other plants may be the best indicators for general use of the compost. The maturity test should be used in conjunction with a test for soluble salts, since a high salt content is phytotoxic to most plants.

The use of sophisticated chemical/physical methods may provide direction to the changes that occur during composting and relate these changes to simpler stability and maturity methods.

A reliable stability test must be applicable to composts prepared from different feedstocks. It should require minimal preparation so as not to alter the physical characteristics of material, which could affect the respiration rate. This includes grinding or excessive screening (Frost et al., 1992). Sample size should be sufficiently large to represent the matrix tested.

Stability tests should be easy to perform by staff at composting facilities and be inexpensive to allow for frequent testing. The latter is important because the margin of error is reduced by increasing the number of samples. When a material is heterogeneous, more samples are needed to obtain a more realistic average.

Three predominant methods are currently in use, CO_2 and O_2 respirometry and the self-heating Dewar-flask method. The CO_2 method had been used for a considerably longer time and is based on a much larger data base than the other two techniques. It is recommended that at present the CO_2 respirometry method be used unless a facility develops a reliable data base for a specific feedstock using the O_2 or Dewar self-heating method.

Phytotoxicity can result if an immature compost is used. Under anaerobic conditions phytotoxic compounds can be formed, but they can disappear after long periods of curing. Some natural compounds found in bark or wood chips can be phytotoxic in fresh or immature compost. Composting under aerobic conditions reduces the phytotoxic effects.

REFERENCES

Allen, O. N., F. A. E. Abel, and O. C. Magistad. 1934. Decomposition of pineapple trash by bacteria and fungi. *Trop. Agr. (Trinidad).* 11:285–292.

Anderson, J. P. E. 1982. Soil respiration, pp. 831–871. In A. L. Page et al. (eds.). *Methods of Soil Analysis. Part 2.* 2nd ed. *Chemical and Microbiological Analysis.* ASA and SSSA, Madison, WI.

Bartha, R. and D. Pramer. 1965. Features of a flask method for measuring the persistence and biological effects of pesticides in the soil. *Soil Sci.* 100:68–70.

Beck, R. W. & Associates. 1990. *Evaluation of Methods for Determining Compost Maturity. Report to the Solid Waste Composting Council,* Alexandria, VA.

Becker, H. 1995. Good earth. *Agr. Res.* 42(6):16. U.S. Dept. Agr., Agr. Res. Serv.

Bodily, H. L. 1944. The activity of microorganisms in the transformation of plant materials in soil under various conditions. *Soil Sci.* 57:341–349.

Brinton, W. et al. 1993. *Compost Maturity Assessment.* Report at the BioCycle Annual Compost Conf. Albany, NY.

Bruce, A. M. and W. J. Fisher. 1984. Sludge stabilization—methods and measurement. In A. M. Bruce (ed.). *Sewage Sludge Stabilization and Disinfection.* Ellis Harwood Ltd., Chichester, UK.

Bucksteeg, W. and H. Thiele. 1959. *Die Beurteilung von Abwasser und Klarschlamm Mittels TTC* (2,3,5, Triphenyltetrazoliumchlorid). *Gas- und Wasserfach.* 100, 1.

Chang, H. T. and W. E. Loomis. 1945. Effect of carbon dioxide on absorption of water and nutrients of roots. *Plant Phys.* 20:221–232.

Chanyasak, V., A. Katayama, M. F. Hirai, S. Mori, and H. Kubota. 1983. Effects of compost maturity on growth of komatsuna (*Brassica Rapa* var. *pervidis*) in Neubauer's pot. *Soil Sci. Plant Nutr.* 29(3):251–259.

Citernesi, U. and M. De Bertoldi. 1979. Il compostaggio dei fanghi miscelati alla frazione organica dei rifiuti solidi urbani. *Inquinamento.* Anno XXI, 2:3–8.

DeVleeschauwer, D., O. Verdonck, and P. Van Assche. 1981. Phytotoxicity of refuse compost. *BioCycle.* 27(1):44–46.

Dowdy, R. H., R. E. Larson, and E. Epstein. 1976. Sewage sludge and effluent use in agriculture. In *Land Application of Waste Material.* Soil Conservation Soc. of Amer., Ankeny, IA.

E&A Environmental Consultants, Inc. 1987. *The Potential of Composting Combined Munici- pal and Paper Mill Sludge as a Product for Mine Spoil Reclamation.* Report to Iron Range Resources and Rehabilitation Board. Calumet, MN.

E&A Environmental Consultants, Inc. 1990. *Results of a Growth Trial Examining the Phytotoxicity of Eucalyptus-Sewage Sludge Compost and Sawdust-Sewage Sludge.* Re- port to Las Virgenes Municipal Water District. Las Virgenes, CA.

E&A Environmental Consultants, Inc. 1994. In-house data.

Engle, H. 1934. Kritische Bemerkungen zur "Bodenatmung." *Centra. für Bakt. Parasitenk II. Abt.* 90:156–161.

Epstein, E. 1956. The effect of organic matter on soil aeration with particular reference to nutrient uptake. Ph.D. Diss. Purdue Univ.

Epstein, E. and H. Kohnke. 1957. Soil aeration as affected by organic matter application. *Soil Sci. Soc. Amer. Proc.* 21(6):585–588.

Finstein, M. S. and F. C. Miller. 1985. Principles of composting leading to maximization of decomposition rate, odor control, and cost effectiveness, pp.13–26. In J. K. R. Gasser (ed.). *Composting of Agriculture and Other Wastes.* Elsevier Applied Sci. Pub., New York.

Frost, D. I., B. L. Toth, and H. A. Hoitink. 1992. Compost stability. *BioCycle.* 33(11):62–66.

Harada, Y. A., A. Inoko, M. Tadaki, and T. Izawa. 1981. Maturing process of city refuse compost during piling. *Soil Sci. Plant Nutr.* 27(3):357–364.

Herrmann, R. F. and J. R. Shann. 1993. Enzyme activities as indicators of municipal solid waste compost maturity. *Compost Sci. & Util.* 1(4):54–63.

Hirai, M. F., V. Chanyasak, and H. Kubota. 1983. A standard method for measurement of compost maturity. *BioCycle.* 24:54–56.

Iannotti, D. A., M. E. Grebus, B. L. Toth, L. V. Madden, and H. A. J. Hoitink. 1994. Oxygen respirometry to assess stability and maturity of composted municipal solid waste. *J. Environ. Qual.* 23:1177–1183.

Iannotti, D. A., T. Pang, B. L. Toth, D. L. Elwell, H. M. Keener, and H. A. J. Hoitink. 1993. A quantitative respirometric method for monitoring compost stability. *Compost Sci. Util.* 1(3):52–56.

Inbar, Y. 1989. Formation of humic substances during the composting of agricultural wastes and characterization of their physicochemical properties. Ph.D. Diss. Hebrew University of Jerusalem, Israel.

Inbar, Y., Y. Chen, Y. Hadar, and H. A. J. Hoitink. 1990. New approaches to compost maturity. *BioCycle.* 31(12):64–69.

Inoko, A. K. Miyamatsu, K. Sugahara, and Y. Harada. 1979. On some organic constituents of city refuse composts produced in Japan. *Soil Sci. Plant Nutr.* 25:225–234.

Jann, G. J., D. H. Howard, and A. J. Salle. 1960. Method for determining completion of composting. *Compost Sci.* 1(3):31–34.

Jimenez, E. I. and V. P. Garcia. 1989. Evaluation of city refuse compost maturity. *Biol. Wastes.* 27:115–142.

Keller, P. 1961. Methods to evaluate maturity of compost. *Compost Sci.* 2(7):20–26.

Lawton, K. 1946. The influence of soil aeration on the growth and absorption of nutrients by corn plants. *Soil Sci. Soc. Amer. Proc.* 13: 311–317.

Manios, V. I., P. E. Tsikalas, and H. I. Siminis. 1987. Phytotoxicity of olive tree leaf compost. In M. De Bartoldi, M. P. Ferranti, P. L'Hermite, and F. Zucconi (eds.). pp. 296–301. Elsevier, London, New York.

Mathur, S. P. 1991. Composting processes, pp. 147–186. In A. M. Martin (ed.). *Bioconversion of Waste Materials to Industrial Products.* Elsevier, New York.

Mathur, S. P., H. Dinel, G. Owen, M. Schnitzer, and J. Dugan. 1993a. Determination of compost biomaturity. II. Optical density of water extracts of composts as a reflection of their maturity. *Biol. Agric. and Hort.* 10:87–108.

Mathur, S. P., G. Owen, H. Dinel, and M. Schnitzer. 1993b. Determination of compost biomaturity. I. Literature review. *Biol. Agric. and Hort.* 10:65–85.

Mebius, L. J. 1960. A rapid method for the determination of organic carbon in soil. *Anal. Chim. Acta.* 22:120–124.

Methods of Soil Analysis. Part 1 and 2. 1982. 2nd edition. Agronomy No. 9. Amer. Soc. of Agronomy, WS.

Morel, J. L., F. Colin, J. C. Germon, P. Godin, and C. Juste. 1985. Methods for the evaluation of the maturity of municipal refuse compost, pp. 56–72. In J. K. R. Glasser (ed.). *Composting of Agricultural and Other Wastes.* Elsevier Applied Sci. Pub., New York.

Niese, G. 1963. Experiments to determine the degree of decomposition of refuse compost by its self-heating capability. International Research Group on Refuse Disposal. *International Bull. No. 17.* Agric. Microb. Institute, Justice-Liebig-Universität, Giessen, Germany.

Pressel, F. and W. Bidlingmaier. 1981. Analyzing decay rate of compost. *BioCycle.* 22:50–51.

Ratcavek, B. 1989. Composting eucalyptus leaves. *Fine Gardening,* July/August.

Roletto, E., R. Berberis, M. Consiglio, and R. Jodice. 1985. Chemical parameters for evaluating compost maturity. *BioCycle.* 26(2):46–47.

Saviozzi, A., R. Levi-Minzi, and R. Riffaldi. 1988. Maturity evaluation of organic wastes. *BioCycle.* 29:54–56.

Schnitzer, M., H. Dinel, S. P. Mathur, H. R. Schulten, and G. Owen. 1993. Determination of compost biomaturity. III. Evaluation of a colorimetric test by ^{13}C-NMR spectroscopy and pyrolysis-field ionization mass spectrometry. *Biol. Agric. and Hort.* 10:109–123.

Schollenberger, C. J. 1945. Determination of soil organic matter. *Soil Sci.* 59:53–56.

Shiralipour, A. and D. B. McConnell. 1991. Effects of compost heat and phytotoxins on germination of certain Florida weed seeds. *Soil and Crop Sci. Soc. Fla. Proc.* 50:154–157.

Spohn, E. 1969. How ripe is compost? *Compost Sci.* 10(3):24–26.

Spohn, E. 1978. Determination of compost maturity. *Compost Sci./Land Util.* 19(3):26–27.

Starkey, R. L. 1924. Some observations on the decomposition of organic matter in soils. *Soil Sci.* 17:293–314.

Still, S. M., M. A. Dirr, and J. B. Gartner. 1976. Phytotoxic effects of several bark extracts on mung bean and cucumber growth. *J. Amer. Soc. Hort. Sci.* 101(1):34–37.

Vandecavey, S. C. and H. Katznelson. 1938. Microbial activities in soil: V. Microbial activity and organic matter transformation in Palouse and Helmer soils. *Soil Sci.* 46:193–197.

Waksman, S. A. 1936. *Humus.* The Williams & Wilkins Co. Baltimore, MD.

Walkley, A. 1947. A critical examination of a rapid method for determining organic carbon in soils: Effect of variations in digestion conditions and or inorganic constituents. *Soil Sci.* 63:251–263.

Walkley, A. and I. A. Black. 1934. An examination of the Degtareff method for determining soil organic matter and a proposed modification of the chromic acid titration method. *Soil Sci.* 37:29–38.

Willson, G. B. and D. Dalmat. 1986. Measuring compost stability. *BioCycle.* 24:25–27.

Woods End Research Laboratory, Inc. 1993. *Compost Self-Heating Flask.* Woods End Res. Lab. Mt. Vernon, ME.

Wong, M. H. 1985. Phytotoxicity of refuse compost during the process of maturation. *Environ. Poll.* (Series A). 37:159–174.

Zhang, L., F. Chuang, and M. A. Cole. 1992. *A Simple Chemical Assay for Estimating Compost Maturity.* Poster presented in ASA Annual Meeting. Minneapolis, MN.

Zucconi, F. A. Pera, M. Forte, and M. De Bertoldi. 1981a. Evaluating toxicity of immature compost. *BioCycle.* 22(2):54–57.

Zucconi, F., A. Pera, M. Forte, A. Monaco, and M. De Bertoldi. 1981b. Biological evaluation of compost maturity. *BioCycle.* 22(4):27–29.

CHAPTER 6

Trace Elements, Heavy Metals, and Micronutrients

INTRODUCTION

Trace elements are elements found in minute quantities in the earth's crust. The soil is the major source of trace elements for plants, and it is through food and feed crops that these elements are transferred to animals and humans (Welch et al., 1991). Soil parent material and soil physical conditions of water, aeration, drainage and other factors affect the chemical nature of the elements and their uptake by plants. Waste applications, fertilizers and other chemicals applied to soils and deposition by water and air add trace elements to the soil.

Several of these elements are essential to plants, animals, and humans. Others may be toxic or have no effect. Further, many elements are essential at low levels but toxic at high levels. The literature often refers to several of the trace elements as "heavy metals." This term is derived from their position in the periodic table. Heavy metals are elements with a relatively high molecular weight (density >5 g/cu. cm) and are comprised of 40 elements. When several of these elements are taken into the body, they can accumulate in specific body organs. The term includes the metals cadmium (Cd), chromium (Cr), copper (Cu), lead (Pb), mercury (Hg), nickel (Ni), and zinc (Zn). Several of these such as Cd, Hg, and Pb are toxic to humans and animals; others such as barium and antimony are not. Yet others such as Cu, Ni and Zn can be phytotoxic.

As a result, the term heavy metal has become synonymous with an element that is harmful to the environment, plants, animals, and humans. Micronutrients in agriculture are elements that are required by plants in small quantities. Examples are B, Cu, and Zn. It is apparent that these three terms, trace elements, heavy metals, and micronutrients, overlap, which has often led to confusion and misuse. It is particularly confusing to regulators and others not familiar with plants and soils.

A major objective of this chapter is to clarify this confusion and to provide information on trace elements, heavy metals and micronutrients found in compost as related to soils, plants, and human nutrition.

Compost contains trace elements. The concentrations depend on the feedstock. However, soils in their natural state also contain trace elements, and plant material may have some levels that are so low as not to be detected by current analytical methods. The concern regarding levels of elements in compost differs, depending on the potential use for the material. For example, elements in compost to be used for crops destined for animal and human consumption may need to be regulated at one level whereas those to be used in silviculture, mine spoil reclamation, and other non-food-chain crops may be allowable at higher levels.

This chapter is subdivided into the following sections:

- essentiality and toxicity: This section discusses the role of the predominant trace elements found in compost and their importance for growth and development of animals, humans and plants. Information is provided on toxic levels to animals, humans, and plants.
- occurrence in the environment: Data are provided on the background levels in soils, plant material, and wastes along with various compost material.
- environmental consequences: Information on the environmental consequences of composting or the application of compost is presented.
- soil-plant interactions: This section discusses the uptake and accumulation of trace elements in plants and the factors that affect their uptake.

ESSENTIALITY AND TOXICITY

It is important to understand the role of the most important trace elements found in wastes as related to plant, animal and human reactions. The elements to be discussed are either regulated heavy metals or other trace elements that impact plant growth.

Arsenic (As)

Arsenic (As) is not essential for plants but is considered essential to animals. Arsenic has been applied in past years as a pesticide, desiccant, and defoliant to many crops. Arsenic is not readily taken up by plants. The two most common forms of arsenic are arsenate and arsenite. Arsenite is both more soluble and more toxic than arsenate, the normal species in aerobic soils.

Arsenite is formed in flooded soils. Because rice is grown in flooded soil, it is the most sensitive crop to toxicity from soil As. If very high concentrations of As are applied to soils, most crops including peas, potatoes, cotton and soybeans can suffer from As phytotoxicity (Stevens et al., 1972; Deuel and Swoboda, 1972). Jacobs et al. (1970) showed that As residue in soils from potato cultivation where Na-arsenite was used as a defoliant, decreased yields of vegetables. Isaac et al. (1978) reported that arsenic in broiler litter applied to soil did not result in arsenic residue hazard to soil, grasses, and water samples. Generally, the level of As in compost is low; therefore, uptake is very low and does not result in phytotoxicity.

In animals, arsenate and arsenite can be toxic, but natural organic arsenic compounds (e.g., arsenobetaine) are much less toxic (Anke, 1986). Only recently it has been found that As may be essential for animals (Nielsen, 1984).

Boron (B)

Boron (B) is essential for plants at varying levels depending on the species and cultivars. Boron deficiency in plants is well known and has been described in detail by Bradford (1966). It is characterized by a slowdown of root extension, inhibition of cell division, abnormal thickening of cell wall, accumulation of cellulose in conducting tissues, and an increase in indole acetic acid (Gupta, 1980). Jackson and Chapman (1975) indicated that B deficiency may be related to RNA metabolism. Symptoms of B deficiency may occur when plant B content is <15 mg/kg (National Research Council, 1980). Several plants including alfalfa, apples, and sugar beets have high B requirements. Consequently, B is added as a micronutrient.

There is indication that B is essential to animals, and it occurs regularly in animal tissues. For example, in rats fed 0.001 ppm of B, RNA synthesis was stimulated.

Phytotoxic levels in plant tissue vary greatly between plants. Concentrations as low as 1 mg/kg may be toxic to sensitive crops. Richards (1954) classified plants according to their tolerance to B. Boron toxicity is manifested in leaves by marginal and tip chlorosis, which is quickly followed by necrosis. Chauhan and Power (1978) found that 1.5 ppm B in irrigation water was toxic to wheat. Similarly, in pea plants, 1 ppm B in irrigation water was toxic. Cereals also appear to be sensitive to B. Gupta et al. (1976) found that under greenhouse conditions, 0.5 ppm added B reduced yields of wheat and 1.0 ppm B reduced yields of barley.

Boron toxicity is often a concern in arid and semi-arid climates since irrigation water can contain high levels. Organic matter can reduce boron toxicity by binding the element. Soil moisture status is one of the most important factors affecting the availability to plants (Moraghan and Mascagni, 1991).

Phytotoxicity due to B in compost was reported by Purves and Mackenzie (1974), who found severe B toxicity in snap beans. Leaching of the compost to below 3 ppm water extracted B, reduced toxicity.

Toxicity to animals has been shown (Underwood, 1977). Thus, accumulation in the brain may result from extensively high rates and result in toxic effects.

Cadmium (Cd)

Cadmium (Cd) is not essential for plants or animals. Although Cd is phytotoxic when added to acidic soils, it has not been found to be toxic to plants under natural conditions. Chaney and Ryan (1993) indicated that when the ratio of Zn:Cd is >100:1, Zn phytotoxicity occurs before concentrations of Cd in the crop become toxic to humans. Uptake differs among plant species and among cultivars within species, and accumulation varies in different plant organs (leaves > storage roots > fruits and grain).

Cadmium can be toxic to animals and humans. Cases of industrial and environmental contamination affecting human health have been reported. The first health effect of chronic Cd ingestion in humans is renal tubular proteinuria, a mild dysfunction of the kidney cortex. With continued excessive Cd intake, other kidney effects have been reported: decreased ability to concentrate urine, increased excretion of amino acids and glucose, and impaired handling of uric acid, calcium, and phosphorus. Excessive Cd intake increases the body's need for Zn and Cu (National Research Council, 1980). Elinder et al. (1976) indicated that Cd accumulates in the kidney cortex with age up to 50 years and then decreases. Hypertension was noted in sensitive strains of rats fed excessive Cd, but Cd has not been found to be a health effect in humans (even in persons with Cd-induced tubular proteinuria). Lung cancer has been observed in test animals with excessive exposure to aerosol Cd, but cancer has not been found to result from oral exposure. Teratogenesis has been found in animals fed high levels of Cd in diets with limited Cu or Zn, but this is due to Cd interference in Cu and Zn absorption at the intestinal level (Varma and Katz, 1978).

Human exposure to Cd predominantly takes place through the diet. Bioavailability is a key to adsorption and toxicity. Smoking is a major contributor to the cadmium level in the human body. A 20-package of cigarettes can provide 30 μg of Cd. Smokers have a significantly increased body burden of Cd from cigarettes since adsorption through the respiratory system is much greater than through the intestinal system (Sharma et al., 1983). It is estimated that 45% of the inhaled Cd is retained. This is nearly one-third of the Cd intake from all sources. Since the tobacco plants can accumulate Cd, waste materials or fertilizers containing Cd should not be used in tobacco cultivation.

Gastrointestinal absorption of Cd by humans is 2–3% (Shaikh and Smith, 1980). The gastrointestinal uptake of Cd is reduced by normal or increased dietary levels of Ca, Zn and Fe (Levander and Cheng, 1980).

Human Cd disease resulting from crop uptake of soil Cd was found in Japanese farm families who consumed rice grown in paddies that were contaminated by mine wastes or smelter emissions containing Cd (Nogawa, 1978; Kobayashi, 1978; Yamagata and Shigematsu, 1970). The disease manifested in women as Cd-induced osteomalacia resulting in repeated bone fractures. As a result of the pain, it was named "Itai-Itai" or "ouch-ouch." Interestingly, because rice is grown in flooded soils, the grain from these contaminated fields was highly enriched in Cd, but showed no change in Zn even though the soil contained 100 times more Zn than Cd.

In contrast to the Japanese experience, very high Cd intake rates by New Zealand adults who consumed an oyster variety that accumulates Cd did not result in tubular proteinuria (Sharma et al., 1983; McKenzie-Parnell et al., 1988). This contrast in Cd effect is attributed to bioavailability and the composition of different diets containing Fe, Ca and Zn (Chaney and Ryan, 1993). Crops such as lettuce and wheat grown in aerobic soils contaminated with Zn and Cd always have a large increase in crop Zn if crop Cd is increased. As noted by McKenna and Chaney (1991) and McKenna et al. (1992), the presence of Zn along with Cd in a crop markedly alters the potential for Cd risk from crops with the exception of rice.

Copper (Cu)

Copper (Cu) is essential for plants and animals. The amount of copper in plants usually ranges from 2 to 25 μg/g dry weight. Copper concentrations in plants of less than 2–5 μg/g are indicative of Cu deficiency (National Research Council, 1977). Toxicity symptoms may occur at levels above 25–40 μg/g (Hemphill, 1972).

Ruminant animals are the most sensitive class of livestock for Cu deficiency, and high levels of many nutrients interact to worsen Cu deficiency (Mo, S, Zn, Cd, Fe). In ruminants, copper deficiency may result in anemia, depressed growth, bone disorders, cartilage disorders, depigmentation of hair and wool, neonatal ataxia, impaired reproductive performance, heart failure, and cardiovascular defects (Davis and Mertz, 1987).

Prolonged consumption of low Cu, high Mo, high sulfate, high Zn or high Fe diets can induce Cu deficiency. Copper toxicity to animals has been reported for Cu-fertilized pastures where sheep consumed the Cu fertilizer, but no Cu toxicity has been reported for ruminants consuming field-grown Cu-rich forages. Sheep tolerance of ingested Cu-rich biosolids and swine manures has been tested; Cu toxicity did not result even though dietary Cu far exceeded toxic levels of soluble Cu salts (Poole et al., 1983). Bioavailability of diet Cu varies widely for ruminants, and biosolid, manure, or compost Cu has low bioavailability. Copper toxicity to humans and animals occurs very infrequently. Although humans can suffer a genetic Cu toxicity (Wilson's disease), no linkage has been made between Cu levels in normal foods and this disease.

Copper toxicity in plants has been reported near copper deposits, smelters, and where excessive amounts of Cu pesticides and fertilizers were applied to strongly acidic sandy soils (Gough et al., 1979; Fraser, 1961). However, when biosolids and composts with normal Cu concentrations have been land applied, even at very high cumulative loading rates, no evidence of Cu phytotoxicity was observed. Only when biosolids with very high Cu concentrations (>2000 mg/kg) were applied to strongly acidic soils did Cu phytotoxicity occur in sensitive crops (Webber et al., 1981, Marks et al., 1980).

Copper toxicity to animals has been reported for Cu-fertilized pastures where sheep consumed the Cu fertilizer, but no Cu toxicity has been reported for ruminants consuming field-grown Cu-rich forages. Sheep tolerance of ingested Cu-rich biosolids and swine manures has been tested; Cu toxicity did not result even though dietary Cu far exceeded toxic levels of soluble Cu salts (Poole et al., 1983). Copper toxicity to humans and animals occurs very infrequently.

Lead (Pb)

Lead (Pb) is not essential for plants or animals, and it can be toxic to both. Plant tolerance to soil lead is very high because Pb is strongly adsorbed by soils. Very high lead levels, 3200 kg/ha, applied as Pb salts, did not reduce corn yields (Baumhardt and Welch, 1972). Plants with apparent genetic tolerance to Pb have been found on highly infertile Pb mine wastes. Except in highly contaminated soils, plants accumulate little soil Pb when the soil fertility is appropriate for crop production, because phosphate inhibits Pb transport to the plant leaves and other edible plant tissues. Absorbed Pb remains predominantly in the fibrous roots. Lead content in edible portions of crops is low and crops provide very low amounts of Pb for human diets. The potential for risk from Pb in compost is not through plant uptake of compost-applied Pb, but rather through direct soil ingestion of Pb by children or livestock or the result of deposition on edible plant portions.

The potential toxicity of lead to humans has received attention principally as the result of lead poisoning of children in urban environments, largely from chipping or flaking Pb-based paints (Agency for Toxic Substances and Disease Registry, 1988). Lead in gasoline has decreased since 1970 (Arizona still sells leaded gasoline), which has resulted in the decrease of airborne Pb deposits on soil. However, since Pb does not readily move through the soil, levels in urban soils have not decreased. Soil Pb can result in house and atmospheric dust (Wixson and Davies, 1994). The Centers for Disease Control have recently revised their recommendations for management of Pb risk to children in order to prevent neurobehavorial impairment, the most sensitive health effect of excessive blood Pb (Centers for Disease Control, 1991). Soils rich in Pb are now considered to constitute one possible source of excessive Pb in children; some children eat soil and all children ingest some soil by inadvertent hand-to-mouth play. Because Pb stays in the soil, soil ingestion by children may be very important in Pb-rich soils.

Although workers in composting facilities are exposed to dust, which can be a source of lead intake, there is no evidence that workers' lead blood levels have increased. Furthermore, the amount of dust that would have to be inhaled or consumed to cause an increase in blood levels would be very high. Worker protection by dust masks reduces potential exposure.

Absorption of different forms of ingested Pb (food, water, ores, compost) is very complex. Lead is adsorbed by organic matter, Fe, Ca, and P, which reduces its availability to animals. The Pb in soil, biosolids, and compost has low bioavailability to monogastric animals and ruminants until the Pb concentration reaches high levels (Chaney et al., 1989). Decker et al. (1980) found that increase in compost application to a pasture significantly increased soil lead but did not result in an increase in the intestine and bone of grazing animals even though there was an increase in their feces, which indicated that the animals had ingested large quantities of Pb (Table 6.1). This was the result of feeding 215 μgPb/g dry weight for 189 days.

Chaney and Ryan (1993) reported similar data from the feeding of biosolids compost to livestock. Based on this research, the authors concluded that compost products containing <300 μgPb/g dry weight do not significantly result in an increase in blood Pb levels even for the pica child, since lead is greatly adsorbed by the compost.

Mercury (Hg)

Mercury (Hg) is not essential for plants or animals. Uptake of Hg by plants is low, especially in the above-ground portions of plants (Hogg et al., 1978).

*TABLE 6.1. Levels of Lead in Tissues of Cattle Fed
Biosolids Compost in the Diet.*

Dietary Compost (percent)	Lead Concentration (mg Pb/kg dry weight)					
	Diet	Feces	Duodenum	Liver	Kidney	Femur
0	6.0a	14.7a	2.81a	2.36b	3.96ab	3.70a
3.3	11.2b	23.8b	3.18a	2.48b	5.26a	4.74a
10.0	19.9c	46.7c	4.21a	8.44a	2.92b	3.37a

The same letter within each column indicates that mean lead concentrations were not statistically different.
Source: Decker et al., 1980.

Some mushroom species accumulate Hg from their substrate. Consequently, it is not recommended that waste material containing Hg be used for mushroom cultivation. There is little evidence that Hg in compost can cause excessive Hg in either food crops or liver of livestock grazing on compost-amended pastures. Cappon (1987) found that Hg accumulated in vegetables when yard waste compost was applied to soil.

Mercury poisoning of humans from fish and contaminated seed have been reported (Friberg and Vostal, 1972; Bakir et al., 1973).

Molybdenum (Mo)

Molybdenum (Mo) is an essential element in plants and animals. In plants it is an essential mineral for growth (Le Gendie and Runnels, 1975). Mo is part of the enzyme nitrogenase, which is responsible for molecular nitrogen formation. It is also present in the nitrate reductase enzyme, which is responsible for nitrification. Mo does not appear to be phytotoxic even at high concentrations in plants. Thus, phytotoxicity has not been observed in the field but has been created in solution cultures (Adriano, 1986).

Mo can be toxic to animals. Animals' tolerance varies with species and age and on the amount of copper in the soil and animal; the inorganic and organic sulfate and organic sulfur oxidizable to sulfate in the diet; and the intake of other metals such as zinc and iron (United States Environmental Protection Agency, 1978). Excessive Mo in diets of some animals can cause a disease termed molybdenosis, which is related to copper and phosphorus deficiency. All cattle are susceptible to molybdenosis, although sheep are affected to a lesser extent. Forages containing Mo concentrations exceeding 10 to 20 ppm may produce molybdenosis in ruminants (United States

Environmental Protection Agency, 1978). Unlike most trace elements, the availability of Mo increases with an increase in soil pH.

Data on Mo's role in human health are limited. Human exposure to Mo is mainly through ingestion. Approximately 50% of ingested Mo is absorbed in adults and 77% in children. Once absorbed, Mo is distributed in the kidney, liver and bone. Approximately 99% of the absorbed Mo is excreted within 24 hours (Frieberg and Lener, 1986).

Nickel (Ni)

Nickel (Ni) is essential for plants and animals. Deficiency has seldom or ever been observed in normal agriculture (National Research Council, 1980). Nickel can be phytotoxic (Welch et al., 1991). In Ni-rich strongly acidic soils, significant yield reduction occurs in all economic plant species when leaves exceed 25-50 ppm (Walsh et al., 1976). However, nickel toxicity to plants occurs before the levels of nickel in the plants becomes toxic to livestock or humans.

Selenium (Se)

Selenium (Se) is an essential element for animals and humans but is considered non-essential for plants (Gough et al., 1979; Underwood, 1977). It is toxic to animals, and the range between toxicity and deficiency is very narrow. All livestock and humans are susceptible to Se poisoning (James et al., 1990).

Dietary requirements for animals and humans are usually met by levels of 0.1 to 0.3 ppm, although levels of up to 1 ppm sometimes yield animal responses (National Research Council, 1980). Selenium functions as a component of glutathione peroxidase and interacts with a wide range of other nutrients, especially vitamin E. Deficiency is reflected by muscular degeneration in herbivores, which can result in white muscle disease. Selenium deficiencies in livestock are common as the result of low levels in soils and forages.

Selenium poisoning of animals and humans is well documented (Gough et al., 1979; Welch et al., 1991; Van Campen, 1991; Miller et al., 1991). Selenium poisoning of animals is often caused by ingestion of plants that accumulate the element. Selenium is readily taken up by plants. Cattle and sheep ingesting plants containing high amounts of Se develop a clinically acute syndrome characterized by an abnormality known as the blind staggers, which results in anorexia, emaciation, and eventual collapse (James et al., 1990).

Zinc (Zn)

Zinc (Zn) is essential for plants and animals. In a wide variety of plants, a Zn tissue concentration of less than 15–20 ppm dry leaves indicates deficiency, while over 400 ppm indicates possible phytotoxicity. Only when biosolids with high Zn concentrations were applied to strongly acidic soils (or when soil pH was allowed to drop to levels of pH < 5.5) did Zn phytotoxicity occur even in sensitive crops (Marks et al., 1980). Zinc requirements and tolerance in animals are affected by several nutrients, vitamins A and D, and by a variety of elements, including copper, manganese, iron, lead, and cadmium. Several examples of Zn deficiency in humans have been reported (Hambridge et al., 1972; Halstead et al., 1972).

As with Ni, Zn toxicity in plants occurs before reaching concentrations in tissues that could be harmful to humans. Overt toxicosis of Zn is seldom observed in relation to natural-ingredient diets until levels exceeding 1,000 ppm and higher are reached (National Research Council, 1980).

OCCURRENCE IN THE ENVIRONMENT

The presence of trace elements in the environment (i.e., soil, water, plants and animals) is the result of the natural occurrence of elements and human activities. The ubiquitous existence of trace elements and the regulated heavy metals for natural and agricultural soils are shown in Table 6.2. Holmgren et al. (1993) analyzed 3045 samples from 307 different soil series throughout the United States. The evaluation also provided data on regional differences. The wide range in concentration in natural soils for most of the elements is a result of different geological material that produced the soil.

The use of fertilizers, pesticides, herbicides, and the application of waste materials have resulted in increased trace elements in many soils. Some recent data on the analysis of fertilizer material are shown in Table 6.2. Lee and Keeny (1975) estimated that 2150 kg of Cd is added annually to Wisconsin soils through fertilizers and wastewater biosolids applications, with much more coming from fertilizers than biosolids.

Since soils contain various amounts of trace elements, plants growing on different soils will contain different levels of trace elements. This is best illustrated by examining the trace elements in yard waste (Table 6.3). The high level of lead found in some samples is probably due to previous widespread use of leaded gasoline. Since lead does not move readily through soils, it is found in the surface soil along roads and is, therefore, picked up in the yard material. Similar values were reported by Miller et al. (1992). Boron ranged from below the detection limit (<DL) to 193; Cd from <DL

TABLE 6.2. Trace Elements in Natural Soils, Agricultural Soils and Fertilizers in the United States.

Element	Range in Natural Soil[1] (mg/kg)	Range in Agricultural Soils[2] (mg/kg)	Range in Fertilizers[3] (mg/kg)
Arsenic	5–13	NA	0.1–22.5
Cadmium	0.01–7	<0.0010–2.0	0.1–101
Chromium	23–15,000	NA	1.1–15,000
Copper	1–300	<0.6–495	1.0–597
Lead	2.6–25	<1.0–135.0	0.3–2700
Molybdenum	3–8	NA	1–27.1
Nickel	3–300	0.7–269.0	1.1–303.0
Selenium	0.0001–3.4	NA	NA
Zinc	10–2000	<3.0–264.0	1.1–8800

[1]Based on Conner and Shacklett, 1975.
[2]Holmgren et al., 1993.
[3]Data from Massachusetts DEP and Milwaukee Metropolitan Sewage District, Milwaukee, WS.

TABLE 6.3. Trace Metal Content of Yard Waste.

Heavy Metal	n	Mean (mg/kg)	SD	Min. (mg/kg)	Max. (mg/kg)	CV
Arsenic	5	4.8	5.05	1	12.8	106.16
Boron	30	28.7	17.93	0.2	76	62.41
Cadmium	29	0.32	0.20	0.04	0.81	62.22
Chromium	35	39.4	45.41	3.7	236	115.39
Copper	35	64	65.47	8	327	102.15
Lead	35	69.6	54.49	11.4	235	78.34
Mercury	22	0.19	0.11	0.04	0.5	59.57
Molybdenum	17	0.22	0.32	0.05	1.09	143.59
Nickel	33	26.89	28.27	3.27	152	105.13
Selenium	17	0.33	0.10	0.1	0.55	31.89
Zinc	35	153.0	74.13	41.6	295	48.47

to 3.67; Cr from <DL to 58.0; Hg from 0.023 to 1.91 and Pb from <DL to 154 mg/kg.

The concentration of trace elements in compost is a function of several factors. The most important include concentration in the feedstock, type of feedstock, and use of amendments. When composting biosolids, the type and amount of bulking agent or amendment used affects the concentration of heavy metals. Because of the considerable variation in concentration of heavy metals in biosolids, it is impossible to provide a meaningful range.

In recent years, as a result of the Clean Water Act and enforcement of industrial discharges into municipal wastewater facilities, the biosolids produced contain much lower concentrations of heavy metals. Consequently, the compost derived from biosolids also has much lower concentrations of heavy metals. Table 6.4 shows concentration of heavy metals in biosolids and some composts. Table 6.5 shows the decrease in heavy metal content during composting of biosolids with wood chips. Mass is lost during composting, increasing the concentration of heavy metals in the biosolids. The addition of other material such as wood chips or other bulking agents lowers the concentration of heavy metals in the final compost product.

Many different feedstocks can be composted. In addition to manures and food-processing, waste materials such as pharmaceutical wastes, paper mill

TABLE 6.4. Concentration of Trace Elements in Biosolids and Some Biosolid Compost.

Source	Trace Element (mg/kg dry weight)						
	Cd	Cr	Cu	Hg	Ni	Pb	Zn
U.S., Biosolids median	4	39	456	2	18	76	755
Alberta, Canada Biosolids	3.6	59	503	1.9	39	122	534
East Bay Municipal Utilities, CA Compost, 1987	16	111	354	NA	40	155	757
Akron, OH Compost, 1988	7.1	66	180	<0.01	18	14	860
Denver Compost, 1992	7.5	100	375	3.0	42	147	744
New York City, Wards Island Compost, 1994	5.1	110	890	3.5	41	340	990

TABLE 6.5. Effect of Bulking Agent Usage on Heavy Metal
Content of Biosolids Compost.

Element	Digested Biosolids (mg/kg)	Digested Biosolids Compost (mg/kg)	Raw Biosolids (mg/kg)	Raw Biosolids Compost (mg/kg)
Cadmium	19	9	10	8
Copper	723	250	419	300
Lead	577	320	426	290
Zinc	1760	1000	978	770

sludges, waste oil, munitions residues, wax-coated cardboard, and milk cartons have been composted (Bugbee and Frink, 1989). Most of these materials are low in heavy metals and the compost can be beneficially used. Table 6.6 provides information on compost from several feedstocks. The data are only examples of materials composted, since complete databases with adequate ranges to represent many composts are not available.

With the increased interest in composting of MSW, there have been several evaluations of trace element concentrations as influenced by separa-

TABLE 6.6. Heavy Metal Composition of Compost
Produced from Several Feedstocks.

Feedstock	Concentration of Trace Elements (mg/kg dry weight)							
	As	Cd	Cr	Cu	Hg	Ni	Pb	Zn
Pharmaceutical[1]	NA	0.2	NA	0.1	NA	0.1	NA	14
Papermill sludge[1]	NA	<.5	<12	7	NA	5	60	14
Papermill sludge[1]	NA	<.8	30	45	NA	0.1	110	126
Papermill sludge[1]	NA	<10	27	49	ND	20	24	90
Produce and yard material[1]	3	ND	23	20	ND	21	13	95
RDF[2] unders compost	NA	4.7	NA	691	NA	NA	663	655

[1]*Source:* E&A Environmental Consultants, Inc., 1994.
[2]RDF is refuse derived fuel; the unders represent the organic fraction, which is not burned. From Stillwell and Sawhney, 1993.

tion of non-compostables from the organic fraction (Epstein et al., 1992; Richard and Woodbury, 1992). Separation of non-compostables can be achieved either at the source or at a composting facility. At the present, the data indicate that the trace element concentrations are often lower when material is source separated. However, the levels found in mixed MSW (facility separated) or source-separated MSW are lower than the USEPA 40CFR503 levels in biosolids or biosolids compost, which were based on a risk-approach analysis.

Tables 6.7a and 6.7b show a comparison of the trace element contents in the United States' mixed waste composts versus European separated organic waste composts.

Zn, Cu, and Pb were significantly higher ($p < 0.01$) in mixed MSW composts than in separated MSW composts; Cd was also significantly higher ($p < 0.05$). As, Cr, Hg, and Ni were not significantly different.

The separation of MSW into compostable and non-compostable fractions reduces the level of heavy metals. Richard and Woodbury (1992, 1994) evaluated the effect of different separation technologies on the concentration of heavy metals. Some of their data are summarized in Table 6.8. A major issue is the significance of the reduction in heavy metals in terms of human health and impacts to the environment. The U.S. Environmental Protection

TABLE 6.7a. A Comparison of Trace Element Content of Compost Derived from Mixed MSW from U.S. Facilities.

Element	Number of Samples	Mixed MSW Compost (USA)			
		Range (mg/kg)	Mean (mg/kg)	Median (mg/kg)	Standard Deviation
As	8	1–4.8	2.6	3.7	1.45
Cd	46	1–13.2	2.9*	2.9	2.69
Cr	41	8.2–130	34.8	34	29.6
Cu	46	31–623	154**	162	129
Pb	46	22–913	215**	221	170
Hg	17	0.46–3.7	1.27	1.2	0.91
Ni	40	7–101	24.8	28	17.9
Zn	45	152–1363	503**	469	280

*Statistically significant at 5% level.
**Statistically significant at 1% level.
Source: Epstein et al., 1992.

TABLE 6.7b. A Comparison of Trace Element Content of Compost Derived from Source Separated MSW from European Facilities.

| Element | Number of Samples | Separated MSW Compost (Europe) | | | |
		Range (mg/kg)	Mean (mg/kg)	Median (mg/kg)	Standard Deviation
Cd	26	0.4–9.8	1.1	0.9	2.59
Cr	25	0.6–71.4	29.4	28	19.5
Cu	27	24–224	57	47	56
Pb	27	40.7–777	112	86	159
Hg	14	0.17–3.8	0.90	0.5	1.23
Ni	26	8–73	19.9	19.3	17
Zn	27	125–1570	281	248	298

Source: Epstein et al., 1992.

TABLE 6.8. Effect of the Separation of MSW in Europe and North America on Heavy Metal Concentration in Compost.

| Heavy Metal | Europe Sorting Method | | North America Sorting Method | |
	Central (mg/kg)	Source (mg/kg)	Central (mg/kg)	Source (mg/kg)
Cd	3.9 ± 0.4[1]	1.2 ± 0.2	3.7 ± 0.4	1.1 ± 0.2
Cr	117 ± 18	39 ± 9	29 ± 7	15 ± 5
Cu	354 ± 53	53 ± 11	349 ± 67	64 ± 20
Hg	2.6 ± 0.3	0.7 ± 0.5	1.6 ± 0.4	1.0 ± 0
Ni	63 ± 14	25 ± 6	31 ± 6	8 ± 2
Pb	565 ± 46	98 ± 13	324 ± 55	74 ± 28
Zn	864 ± 83	282 ± 53	771 ± 141	292 ±131

[1]Mean and standard error.
Source: Richard and Woodbury, 1992, 1994.

Agency's risk analysis in the biosolids regulations (40CFR503) clearly indicates that the levels found in MSW compost do not present a significant risk. If the levels in MSW compost are unacceptable, then the beneficial use of biosolids and biosolid compost is also unacceptable. There is no scientific basis for this conclusion.

ENVIRONMENTAL CONSEQUENCES

Environmental consequences other than plant uptake, accumulation in the food chain, or direct ingestion of soil or dust include potential leaching to groundwater at composting sites or leaching to groundwater or runoff to surface waters from compost application sites. There could be a direct effect of compost application to soil biota through ingestion. There could also be toxic effects of trace elements in compost on the soil microbial population (Woodbury, 1992).

The data on this subject are very limited. Woodbury (1992) cited examples where MSW compost increased the biomass of soil microbiota and the use of biosolid compost decreased soil biomass in comparison to farmyard manure.

Leachate Characteristics of Compost

Data on characteristics of leachate from different composts are meager. Sawhney et al. (1994) evaluated the leaching of heavy metals from growth media amended with different rates of MSW compost. Figure 6.1 shows that over a two-week period, the concentration of heavy metals increased with application rate. However, the average concentration was below the USEPA drinking water limits. Cd, Pb, Cr, Cu, and Zn concentrations in leachates decreased with time.

Initial concentrations were relatively high, but later concentrations were extremely low. Less than 2% of the total metals were leached, with Ni being the most mobile and Pb the least. The potential for groundwater contamination from MSW leachate is not only negligible as a result of the low concentration, but as the leachate moves through the soil additional removal of the trace elements would occur by soil and organic matter binding. This further reduces the potential for movement of the metal to groundwater.

Helfrich (1992) examined the chemical characteristics of yard waste compost leachate in two studies. The first study measured the leachate generated from decaying grass and leaves with grass under anaerobic conditions. This represented the worst case, essentially resulting in piling or storage of yard waste.

The second study involved the collection of runoff from a 4.9 HA windrow

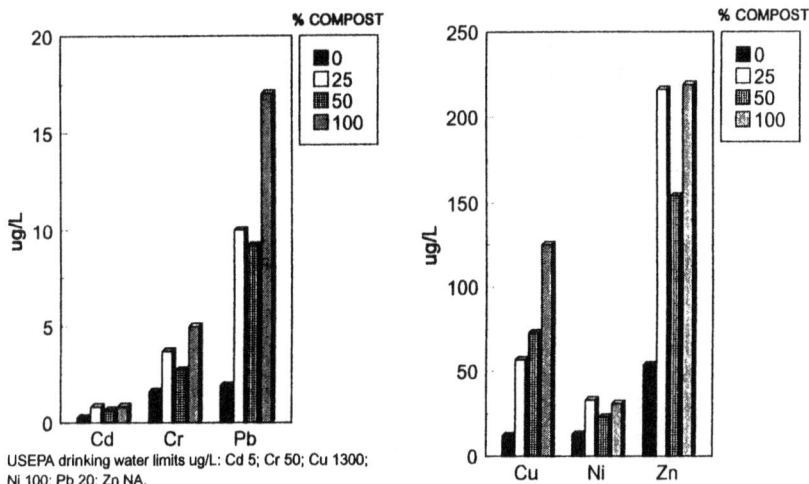

FIGURE 6.1. Average concentration of heavy metals in leachates from plant growth medium containing different percentages of source separated MSW compost. (Data from Sawhney et al., 1994.)

composting site, which represented runoff and leachate from composting. Composting was done on a clay pad sloped towards a retention pond. Collection was done over a six-month period. Leachate generated and collected in a pond was analyzed for 36 parameters, including 14 elements and 9 organic compounds.

Table 6.9 shows the heavy metals in the leachate from the anaerobic decomposition of grass, 50:50 grass:leaves mixture, and the retention pond serving the compost site. The concentration of trace elements in the leachate was generally low. As, Cd, Cr, and Pb were in the ppb range. This was expected since the grass and leaves contain low levels of these trace elements. The difference in B concentration between the grass and mixture of grass and leaves could not be explained. Trace elements in runoff from a well-managed facility were very low and in the ppb range. As, Cd, and Cr were not detected in the retention pond over a six-month period. Pb, Cu, and Zn were detected during the last month of sampling at concentrations of 0.0053, 0.027, and 0.0304 mg/L, respectively.

Richard and Chadsey (1990) studied the chemical changes in the soil and the water infiltrating into the soil beneath a municipal leaf composting site. Soil sample data did not show any increases in heavy metals. Water collected in suction lysimeters also did not show any significant changes.

Using lysimeters, Christensen (1983) studied the potential leaching of several heavy metals from two refuse-biosolids composts. The leaching of Cd, Ni, and Zn decreased rapidly in each successive water application to the

TABLE 6.9. *Trace Elements in Leachate from Anaerobic Piles of Grass and Grass:Leaf Mixture and Runoff from a Yard Waste Composting Site.*

Trace Element	Grass (mg/L)		Grass and Leaves (mg/L)		Runoff (mg/L)	
	Minimum	Maximum	Minimum	Maximum	Minimum	Maximum
Arsenic	ND	ND	ND	ND	ND	ND
Boron	1.06	1.12	435	643	0.277	0.412
Cadmium	ND	ND	ND	ND	ND	ND
Chromium	0.08	0.16	ND	ND	ND	ND
Copper	ND	0.21	ND	0.16	ND	0.027
Lead	ND	0.16	ND	0.07	ND	0.0053
Nickel	ND	0.32	ND	ND	ND	0.0918
Zinc	0.76	0.87	0.31	0.49	ND	0.0304

Source: Helfrich, 1992.

lysimeter. A slow leaching rate was observed for Cd, Cu, Pb, and Zn. Only 0.07% to 0.7% of the compost content of these metals was leached within the first year.

Biosolids and MSW composting operations are either enclosed or operated on asphalt or concrete pads. Therefore, the potential for leaching and contaminating groundwater underneath composting operations is nil. Runoff is usually discharged into sanitary sewers or collected in holding ponds.

The application of compost can increase the trace element content of soils. In order for the trace elements to leach from the compost and move through the soil, they must be in the soil solution. Most of the heavy metals are not soluble unless the soil's pH is low. Metal solubility is low, at a pH above 6.5. Furthermore, the soil will bind the metals on exchange sites of clay minerals or organic matter. Because compost increases the organic content of soil, the addition of compost increases the soil's adsorptive capacity of metals, reducing overall mobility of soil and compost metals. Thus, the potential of leaching to groundwater from compost application is very low.

SOIL-PLANT INTERACTIONS

Trace elements in compost form various compounds or associations when applied to soil, which can affect their uptake by plants and their mobility through soils. They can be complexed by organic compounds, co-precipitated in metal oxides, in a water soluble state, or in an exchangeable form on soil or organic matter colloids. Measuring the total trace element content does not predict the soil-plant interactions.

Several investigators have attempted to fractionate trace elements in compost in order to identify in what form the element exists. This knowledge could possibly predict their potential uptake and bioavailability as well as mobility through soils. The more common extraction procedures are the H_2O soluble, KNO_3 for exchangeable, $Na_4P_2O_7$ for organic matter bound, EDTA for carbonate and sulfide precipitates, and HNO_3 for residual form of the trace element.

In compost the water-soluble fraction that is readily available to plants is small. Leita and De Nobili (1991) observed that during the early stages of composting, the water-soluble fraction of Cd initially increased; however, towards the later stages of composting the water-extractable Cd fraction decreased to non-detectable levels.

Canarutto et al. (1991) indicated that the concentrations of extractable heavy metal are lower at the end of the composting period if the process is carried out properly. Humic and fulvic acids have strong affinities to divalent cations with reactions being strongly pH dependent (Schnitzer and Skinner,

1966). Humic substances are some of the most powerful metal-binding agents among organic substances (Buffle, 1988).

Understanding the soil-plant relationship is extremely important in managing our soils so as to reduce uptake of trace elements or their movement to water resources. One of the most important aspects of beneficial use of compost and other organic materials is the ability to manage and control their application in order to reduce or eliminate potential harmful effects. The use of organic residues in themselves reduces potential uptake by plants of heavy metals. Other factors such as pH and phosphorus also influence plant uptake. Management involves such aspects as crop selection to reduce potential uptake and accumulation of trace elements, soil pH control, and organic matter maintenance.

When compost is applied to soil, there are many potential pathways for the trace elements: uptake by plants, movement with water to ground or surface water sources, volatilization from surface-applied compost, and immobilization in the soil matrix. Immobilization refers to exchange on the soil colloidal system and fixation in forms unavailable to plants. Figure 6.2 shows these potential pathways.

The particular pathway depends on the soil-plant-water relationships of the trace element in question as well as on the amount of the trace element, and interactions among soil-plant-water factors. Trace elements applied to soil may pass through the soil unchanged, react with organic and inorganic compounds to form soluble or insoluble compounds, be adsorbed on the soil colloids, and volatilize from the soil or taken up by plants (Epstein and Chaney, 1978). Various factors influence the potential pathways. These include:

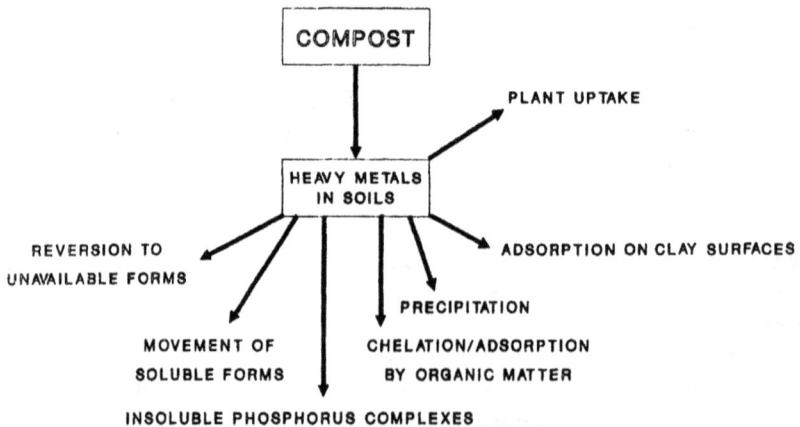

FIGURE 6.2. Major pathways of heavy metals in compost applied to soil.

- type of trace element and chemical state
- soil acidity
- organic matter
- cation exchange capacity
- reversion to unavailable forms

Type of Trace Element and Chemical State

The availability of trace elements to plants and their mobility through the soil often depends on their interaction with other elements and their chemical state. For example, Cu, Ni, Cd, Pb, Zn, and to a lesser extent Mn, behave similarly in soils. In acid soils, these elements could exist as the divalent cations (Cu^{2+}, Zn^{2+}). In alkaline or neutral soils, they may be combined with a hydroxyl ion [$Zn(OH)^+$]. The hydrous oxides of Mn and Fe can control the availability of trace elements by sorption and desorption (Jenne, 1968; Quirk and Posner, 1975).

The importance of the chemical form on plant uptake is well documented. For example, trace elements applied as a salt are more soluble and result in greater uptake than the same element applied in an organic matrix such as biosolids or compost.

Soil Acidity

In general, as the soil acidity increases the solubility of trace elements increases. Further, as the solubility of the metal increases, its potential uptake by plants increases. Tackett et al. (1986) evaluated the effect of pH over a range of 2.5–7.0 on the leaching of Cd, Fe, Pb, and Zn. Zn and Cd were the only metals that leached faster as the pH was lowered. Falahi-Ardakani et al. (1988) observed no increase in Cd uptake as the pH changed from 3.4 to 7.2.

These findings seem to conflict. Since leaching of a metal indicates its greater solubility, it would be expected that uptake would be greater as more metal is soluble. However, many other factors such as compost feedstock, type of soil, and plant species can affect uptake. The compost used by Falahi-Ardakani et al. (1988) was derived from ferric chloride, high-pH, digested biosolids. Both the lime and the Fe could have affected uptake.

Organic Matter

Organic matter has a high cation exchange capacity compared to mineral soil. Thus, it tends to bind or chelate metal ions such as Cu, Ni, Zn, and Cd.

Chelation may be more important in binding of trace elements than cation exchange. Organic matter binds metals more strongly at a soil pH below 7.5 (Schnitzer and Khan; 1972). At lower pHs the organic matter reduces metal availability relative to the same soil without the organic matter. Metal-organic matter complexes are the result of replacement of the water molecules surrounding the metal by other molecules or ions. The organic molecules that combine with the metal are commonly referred to as ligands (Stevenson, 1994). Metal-organic matter complexes play an important role in the micronutrient cycle in the soil (Stevenson, 1991). The important reactions are:

(1) Soluble organic compounds complex metals that otherwise would precipitate at pH values found in most soils.

(2) Under certain conditions, metal ion concentrations may be reduced to non-toxic levels through complexation with organic compounds.

(3) Trace element availability to plants may be enhanced by various organic-metal complexes.

(4) The availability of phosphorus to plants may be enhanced by the complexing of Fe, Al, and Ca, which would form insoluble compounds.

Cation Exchange Capacity (CEC)

CEC is a measure of the soil's ability to bind cations and release them for availability to plants. The CEC of both the mineral portion and the organic matter is important in the availability of micronutrients to plants. Cations are positively charged and are attracted to the negatively charged surfaces of clay micelles and the ionization of COOH groups on the organic fraction (Stevenson, 1994). The effect of organic matter on the CEC of soils is dependent on the pH of the soil.

Reversion to Unavailable Forms

Trace elements can "revert" with time to chemical forms less available to plants. Reversion has been clearly established for Zn and can be quite rapid. The process is not well understood. Factors such as pH and extent of reductive conditions appear to be important. Sims and Kline (1991) showed that although the total levels of Cd, Cu, Cr, Pb, and Zn increased in soils as a result of the application of a co-composted biosolid, the increase was found in the more resistant NaOH, EDTA, and HNO_3 soil fractions. Thus, potential uptake was less.

Other Aspects

Other factors such as the amount of phosphorus, soil temperature, soil moisture, and aeration can affect the solubility and, therefore, the availability of a trace element to plants. For example, phosphate is well known for reducing Zn availability to plants and for decreasing the stunting injury caused by excessive amounts of phytotoxic elements.

The soil's physical factors, temperature, moisture, and aeration also influence plant root growth and proliferation, and thus plant uptake of metals and other elements. Furthermore, manifestations in the soil such as the production of sulfides under reduced conditions can precipitate an element and reduce its availability to plants.

Effect of Compost on Trace Element Uptake

Numerous studies have been conducted on the uptake of trace elements from different types of compost. More has been written in the United States on uptake of trace elements from biosolids or biosolids compost than from solid waste compost. In Europe the literature has primarily been related to solid waste compost.

Uptake differs among plant species and among cultivars within species, and accumulation varies in different plant organs (leaves > storage roots > fruits and grain). Table 6.10 shows uptake of Cd, Zn, Cu, Ni, and Cr by various vegetables (Keefer et al., 1986). Several of these principles are illustrated. Trace elements in edible portions of plants, fruit, and grain were lower than in the non-consumable parts (i.e., the leaves). Sweet corn had lower concentrations of Cd, Cr, Cu, Ni, Pb, and Zn than the corn leaf. Although this has an important implication for human diets, since humans consume only the grain, it indicates that animals fed the leaves would be consuming more of the trace elements than if they consumed the grain. A similar relationship was observed for green beans. For tomatoes, Cd, Cr and Pb were lower in the fruit, whereas Cu, Ni, and Zn were higher.

Figure 6.3 shows the uptake of Cd by corn leaves and grain from biosolids and biosolids-compost-amended soil. Plant availability of Cd and Zn was shown to be lower from composted biosolids or MSW compost than from the uncomposted biosolids (Epstein, 1978; Giordano et al., 1975; Simeoni et al., 1984). The concentration of Cd in leaves and grain grown on biosolids compost was significantly lower than the concentration in leaves and grain from corn grown on biosolid treated soil when equivalent rates of Cd were applied. Two significant findings are shown. First, uptake of a heavy metal varies with the plant organ. Cd concentration was higher in the leaves than

TABLE 6.10. *Concentration of Cadmium, Zinc, Copper,
Nickel, and Chromium in Vegetables.*

Plant and Organ	Concentration of Trace Elements (mg/kg)					
	Cd	Cu	Cr	Ni	Pb	Zn
Radish roots	0.66	4.7	1.34	1.72	0.62	43
Radish tops	0.87	8.4	3.30	4.24	3.74	55
Carrot roots	0.32	6.5	0.40	1.10	0.60	232
Carrot tops	0.73	10.2	0.62	1.20	1.08	36
Green beans	<.10	9.8	0.26	5.04	0.66	35
Bean leaves	0.19	11.7	1.14	5.42	1.94	42
Sweet corn	0.16	3.1	0.52	0.34	<.10	26
Corn leaves	0.50	8.3	1.02	1.42	5.38	32
Tomato fruit early	0.55	11.6	0.65	2.10	0.68	59
Tomato fruit late	0.40	12.7	0.60	2.12	0.40	51
Tomato leaves	0.89	11.1	0.48	1.70	2.08	23

Source: Keffer et al., 1986.

in the grain, which reinforces the importance of crop selection and utilization. Second, the uptake from compost was lower than from the uncomposted biosolids.

Simeoni et al. (1984) showed that for both biosolids and compost, greater uptake of Cd and Zn was found from an acid soil (pH 7.6) (Figure 6.4). The concentrations of Cd and Zn in lettuce were much higher in oats, for example.

Giordano et al. (1975) applied high rates of MSW compost and biosolids, which provided 90, 180, and 360 kgZn/ha. A higher concentration of Cd and Zn was found in the soil treated with biosolids. Cd and Zn were lower in pea vines. Zn in the pods was lower but there was very little difference in Cd concentration. Corn leaves and grain had lower concentration of Zn in soil applied with MSW compost than in soil applied with biosolids.

Logan (1983) stated that uptake from biosolids compost was lower than from plots treated with uncomposted biosolids. This illustrates the fact that different types of organic matter and the stage of decomposition may play roles in plant uptake of heavy metals. Most likely, as a result of decomposition, the compost contained higher amounts of humic substances that chelated the heavy metals and reduced heavy metal uptake.

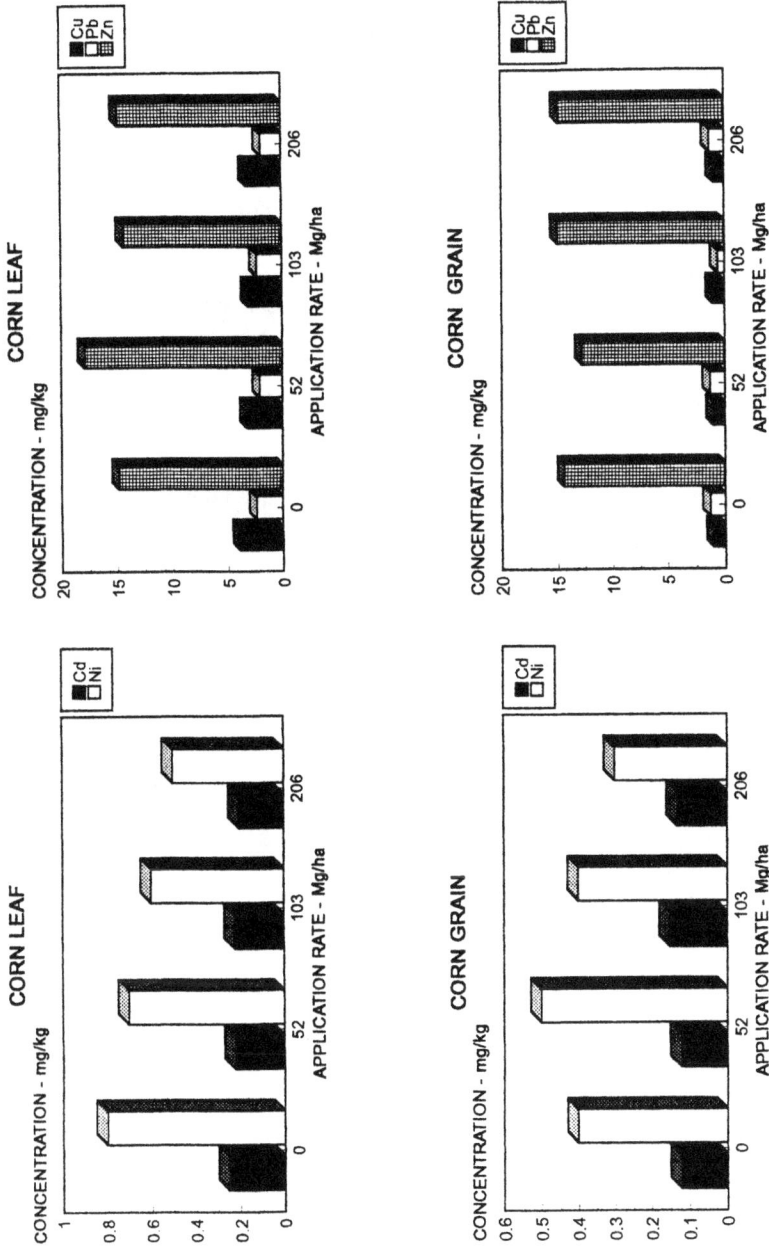

FIGURE 6.3. Effect of biosolids compost application on heavy metal uptake by corn leaves and grain. (Data from Logan, 1983.)

161

Very little data are available on the distribution of heavy metals in different fractions of humic substances. As more data become available, knowledge of the stage of composting in relation to humic substances that bind heavy metals could be important.

The pH of most composts is near neutral (pH 7) and when compost is applied to soils can increase the pH and reduce plant availability of heavy metals. Chaney et al. (1978) found that biosolids compost had a higher soil pH than uncomposted biosolids. In biosolids plots where the soil pH was lowered using elemental sulfur, Zn in soybean leaves was significantly higher than in the soil with the higher pH. There was no difference in Zn concentration in soybean leaves grown on compost-amended soil regardless of soil pH or compost application rate.

Table 6.11 illustrates the effect of pH on Cd uptake in corn grain and leaves for both biosolids and biosolids compost. Unpublished data by Chaney et al. (1978) showed that at the high pH, corn uptake of Cd and Zn was lower in both grain and leaves. In general, there was a reduced uptake of Cd and Zn from compost-amended plots than from biosolids amended plots. Even at a low pH, plant uptake of Cd and Zn was lower. Simeoni et al. (1984) attributed the increase in pH of the soil due to compost application to reduced Cd and Zn availability to plants. The biosolids reduced soil pH, which resulted in higher metal concentration in lettuce and oats.

The differences in uptake can be attributed to the chemical nature of the element in the organic matrix, decomposition of the organic matter in relation to metal solubilization, and formation of hydrous oxides or Cd-organic matter complexes. Interestingly, Sims and Kline (1991) found that Cd behaved very differently than Cu, Ni, Pb, and Zn. The latter four elements

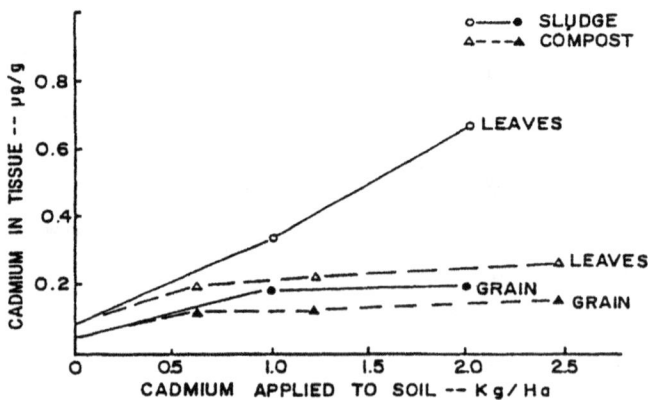

FIGURE 6.4. Uptake of cadmium by corn from biosolids and biosolids-compost applied to soil. (Data from Epstein, 1978.)

TABLE 6.11. Effect of Biosolids and Biosolid Compost
on Zn and Cd Concentration in Corn.[1]

Treatment	Rate (mt/ha)	Soil (pH)	Grain (Cd)	Grain (Cd)	Leaves (Zn)	Leaves (Zn)
Control		5.5	26.5ed	0.004d	35f	0.41e
Biosolids low pH	40	5.3	41.4ab	0.111a–d	180ab	1.11bc
	80	5.7	46.2a	0.214ab	224a	1.74a
	160	6.5	36.4a–c	0.172a–d	168b	1.89a
	240	6.4	45.0a	0.196a–c	143bc	1.69a
Biosolids high pH	80	6.4	35.5a–c	0.154a–d	117cd	1.49ab
	160	6.3	31.8b–d	0.195a–d	113cd	1.82a
	240	6.9	46.3a	0.258a	148bc	1.87a
Compost low pH	40	5.5	30.3b–d	0.183a–d	84def	0.58de
	80	6.0	26.2cd	0.109a–d	64def	0.62de
	160	6.5	27.8cd	0.106bcd	84def	0.81cde
	240	6.8	30.4b–d	0.129a–d	104def	1.10bc
Compost high pH	80	6.9	22.9d	0.104cd	56ef	0.41e
	160	6.9	22.5d	0.129a–d	86def	0.60de
	240	7.0	31.2b–d	0.114a–d	73def	0.93cd

[1]Within each column, values followed by the same letter are not statistically significant at the 5% level according to Duncan's Multiple Range Test.

increased with increased application rates of a biosolids-MSW compost, but Cd decreased slightly.

SUMMARY

Trace elements can be either toxic or beneficial to humans, animals, and plants. The category "heavy metals" includes several elements that have generally been considered toxic. This is a misconception, however, as several of the heavy metals are essential to human and animal nutrition and plant growth. Plant uptake by forage and food crops is the primary pathway for the elements to enter humans and animals. There has been no indication or data to show that the use of compost for growing crops has resulted in concentration of heavy metals in soils or crops that could be toxic to humans or animals.

Different feedstocks contain different amounts of heavy metals, which result in a wide range of heavy metal levels in composts. The Clean Water Act has led to lowering the occurrence of heavy metals in biosolids compost. Separation of MSW, whether at the source or in a facility, leads to lower concentrations of heavy metals in MSW compost. Source separation and utilization of separated organic materials result in lower levels of heavy metals in compost than MSW, which is facility separated.

Numerous soil factors affect the uptake of trace elements by plants and their movement through the soil. For example, organic matter plays a major role in the availability to plants and mobility of trace elements through soils. Humic substances are some of the most effective metal-binding agents. There are no data relating the composting process to the formation of the humic substances most responsible for metal binding.

REFERENCES

Adriano, D. C. 1986. Arsenic, pp. 46–72. In *Trace Elements in the Terrestrial Environment.* Springer-Verlag, New York.

Agency for Toxic Substances and Disease Registry (ATSDR). 1988. *The Nature and Extent of Lead Poisoning in Children in the United States: A Report to Congress.* DHHS Doc. No. 99-2966. U.S. Dept. Health Human Services, Public Health Service, Atlanta, GA.

Anke, M. 1986. Arsenic, pp. 347–372. In W. Mertz (ed.). *Trace Elements in Human and Animal Nutrition. Vol. 2.* Fifth Edition. Academic Press, New York.

Bakir, F., S. F. Damluji, L. Amin-Zaki, M. Murtadha, A. Khalidi, N. Y. Al-Rawi, S. Tikriti, H. I. Dhahir, T. W. Clarkson, J. C. Smith, and R. A. Doherty. 1973. Methylmercy poisoning in Iraq. *Science.* 181:230–241.

Baumhardt, G. R. and L. F. Welch. 1972. Lead uptake and corn growth with soil applied lead. *J. Environ. Qual.* 1:92–94.

Bradford, G. R. (1966). Boron. In H. D. Chapman (ed.). *Diagnostic Criteria for Plants and Soils,* Univ. of California.

Buffle, J. 1988. *Complexation Reactions in Aquatic Systems: An Analytical Approach.* Ellis Horwood Ltd., Chichester, England.

Bugbee, G. J. and C. R. Frink. 1989. Composted waste as a peat substitute in peat-lite media. *HortSci.* 24:625–627.

Canarutto, S., G. Petruzzelli, L. Lubrano, and G. V. Guidi. 1991. How composting affects heavy metal content. *BioCycle.* 32(6):48–50.

Cappon, C. J. 1987. Uptake and speciation of mercury and selenium in vegetable crops grown on compost-treated soil. *Water, Air, Soil Pollut.* 34:353–361.

Centers for Disease Control. 1991. *Preventing Lead Poisoning in Young Children. A Statement by the Centers for Disease Control,* October 1991, US-DHHS, PHS, CDC, Atlanta, GA.

Chaney, R. L., P. T. Hundemann, W. T. Palmer, R. J. Small, M. C. White, and A. M. Decker. 1978. Plant accumulation of heavy metals and phytotoxicity resulting from utilization of sewage sludge and sludge composts on cropland, pp. 86–97. In *Proc. Nat. Conf. Composting of Municipal Residues and Sludges.*

Chaney, R. L., H. W. Mielke, and S. B. Sterrett. 1989. Speciation, mobility, and bioavailability of soil lead. In B. E. Davis and B. G. Wixson (eds.). *Proc. International Conf. Lead in Soils: Issues and Guidelines. Environ. Geochem. Health 11* (Supplement):105–129.

Chaney, R. L. and J. A. Ryan. 1993. Heavy metals and toxic organic pollutants in MSW-compost: Research results on phytoavailability, fate, etc. In H. A. J. Hoitink and H. M. Keener (eds.). *Science and Engineering of Composting; Design, Environmental, Microbiological and Utilization Aspects.* Renaissance Publ., Worthington, OH.

Chauhan, R. P. S. and S. L. Power. 1978. Tolerance of wheat and pea to boron in irrigation water. *Plant and Soil.* 50:145–149.

Christensen, T. H. 1983. Leaching of pollutants from land-disposed municipal compost. In E. I. Stentiford (ed.). *Proceedings of the International Conference on Composting Solid Waste and Slurries.* The Univ. of Leeds, Leeds, England.

Conner, J. J. and H. T. Shacklette. 1975. Background geochemistry of some rock, soils, plants, and vegetables in the conterminous United States. *U.S. Geological Survey Prof. Paper* 574-F.

Davis, G. K. and W. Mertz. 1987. Copper, pp. 301–364. In W. Mertz (ed.). *Trace Elements in Human and Animal Nutrition. Vol. 1.* Fifth Edition. Academic Press, New York.

Decker, A. M., R. L. Chaney, J. P. Davidson, T. S. Rumsey, S. B. Mohanty, and R. C. Hammond. 1980. Animal performance on pastures topdressed with liquid sewage sludge and sludge compost, pp. 37–41. In *Proc. Nat. Conf. Municipal and Industrial Sludge Utilization and Disposal.* Information Transfer, Inc. Silver Spring, MD.

Deuel, L. E. and A. P. Swoboda. 1972. Arsenic toxicity to cotton and soybeans. *J. Environ. Qual.* 1:317–320.

E&A Environmental Consultants, Inc. 1994. In-house data.

Elinder, C., T. Kjellstrom, L. Friberg, B. Lind, and L. Linnman. 1976. Cadmium in kidney cortex, liver, and pancreas from Swedish autopsies. *Archives of Environ. Health.* 31:292–302.

Epstein, E. 1978. Impact and possibilities in the use of sludge and sludge compost in agriculture. In A. Banin and U. Kafaki (eds.). *Agrochemicals in Soils.* Pergamon Press, New York.

Epstein, E. and R. L. Chaney. 1978. Land disposal of toxic substances and water-related problems. *J. Water Pollution Control Federation.* 50:2037–2042.

Epstein, E., R. L. Chaney, C. Henry, and T. J. Logan. 1992. Trace elements in municipal solid waste compost. *Biomass and Bioenergy.* 3(3–4):227–238.

Falahi-Ardakani, A., J. C. Bouwkamp, F. R. Corey, and F. R. Gouin. 1988. Influence of pH on cadmium and zinc concentrations of cucumber grown in sewage sludge. *HortScience.* 23:1015–1017.

Fraser, D. C. 1961. A syngenetic copper deposit of recent age. *Econ. Geology.* 56:951–962.

Frieberg, L. and J. Lener. 1986. In L. Frieberg, G. G. Norberg, and V. B. Vouk (eds.). *Handbook of Toxicology of Metals.* Elsevier Applied Sciences, New York.

Friberg, L. and J. Vostal. 1972. *Mercury in the Environment.* Chemical Rubber Co., Cleveland, OH.

Giordano, P. M., J. J. Mortvedt, and D. A. Mays. 1975. Effect of municipal wastes on crop yields and uptake of heavy metals. *J. Environ. Qual.* 4:394–399.

Gough, L. P., H. T. Shacklette, and A. A. Case. 1979. Element concentrations toxic to plants, animals and man. *Geol. Survey Bull. 1466.* U.S. Govt. Printing Office. Washington, D.C.

Gupta, U. C. 1980. Boron nutrition of crops. *Advances in Agronomy.* 31:273–307. Academic Press, New York.

Gupta, U. C., J. A. MacLeod, and J. D. Sterling. 1976. Effects of boron and nitrogen on grain yield and boron and nitrogen concentrations of barley and wheat. *Soil Sci. Amer. J.* 40:723–726.

Halstead, J. A., H. A. Ronaghy, P. Abadi, G. H. Haghshenass, G. H. Amirhakemi, R. M. Barakat, and J. G. Reinhold. 1972. Zinc deficiency in man: The Shiraz experiment. *Amer. Jour. of Med.* 53:277–276.

Hambridge, K. M., C. Hambridge, M. Jacobs and J. D. Baum. 1972. Low levels of zinc in hair, anorexia, poor growth and hypogeusia in children. *Pediat. Res.* 6:868–874.

Hemphill, D. D. 1972. Availability of trace elements to plants with respect to soil-plant interaction, pp. 46–61. In H. C. Hopps and H. L. Conner (eds.). *Geochemical Environment in Relation to Health and Diseases. ANN. New York Acad. Sci.* Vol. 199, New York.

Helfrich, J. 1992. *An Examination of the Chemical Characteristics of Yard Waste Compost Leachate.* Project No. YWLH 519. Waste Management of North America, Oak Brook, IL.

Hogg, T. J., J. R. Bettany, and J. W. B. Stewart. 1978. The uptake of [203]Hg-labeled mercury compounds by bromegrass from irrigated undisturbed soil columns. *J. Environ. Qual.* 7:445–450.

Holmgren, G. G. S., M. W. Meyer, R. L. Chaney, and R. B. Daniels. 1993. Cadmium, lead, zinc, copper, and nickel in agricultural soils of the United States of America. *J. Environ. Qual.* 22:335–348.

Isaac, R. A., S. R. Wilkinson, and J. A. Stuedemann. 1978. Analysis and fate of arsenic in broiler litter applied to coastal Bermuda grass and Kentucky-31 tall fescue, pp. 207–220. In D. C. Adriano and I. L. Brisbin, Jr. (eds.). *Environmental Chemistry and Cycling Processes.* U.S. Dept. Energy Rpt. Conf.-/60429.

Jacobs, L. W., D. R. Keeney, and L. M. Walsh. 1970. Arsenic residue toxicity to vegetable crops grown on Plainfield sand. *Agron. J.* 62:588–591.

Jackson, J. F. and K. S. R. Chapman. (1975). The role of boron in plants, pp. 213–225. In D. J. D Nicholas and A. R. Egan (eds.). *Trace Elements in Soil-Plant-Animal Systems.* Academic Press, New York.

James, L., K. Panter, H. Maryland, M. Miller, and D. Baker. 1990. Selenium poisoning in livestock: A review and progress. In L. W. Jacobs (ed.). *Selenium in Agriculture and the Environment.* Soil Sci. Soc. Amer. Special Pub. No. 33, Madison, WI.

Jenne, E. A. 1968. Controls on Mn, Fe, Co, Ni, Cu, and Zn concentrations in soils and water. The significant role of hydrous Mn and Fe oxides. *Advan. Chem. Ser.* 73:337, Amer. Chem. Soc., Washington, D.C.

Keefer, R. F., R. N. Singh, and D. J. Horvath. 1986. Chemical composition of vegetables grown on an agricultural soil amended with sewage sludge. *J. Environ. Qual.* 15:146–152.

Kobayashi, J. 1978. Pollution by cadmium and the *itai-itai* disease in Japan, pp. 199–260. In F. E. Oehme (ed.). *Toxicity of Heavy Metals in the Environment.* Marcel Dekker, Inc., New York.

Lee. K. W. and D. R. Keeney. 1975. Cadmium and zinc additions to Wisconsin soils by commercial fertilizers and wastewater sludge applications. *Water, Air, Soil Pollut.* 5:109–112.

Le Gendie, G. R. and D. D. Runnels. 1975. Removal of dissolved molybdenum from wastewaters by precipitates of ferric iron. *Environ. Sci. and Tech.* 9:744.

Leita, L. and M. De Nobili. 1991. Water soluble fractions of heavy metals during composting of municipal solid waste. *J. Environ. Qual.* 20:73–78.

Levander, O. A. and L. Cheng. 1980. Micronutrient interaction: Vitamins, minerals and hazardous elements. *Ann. N.Y. Acad. Sci.* Vol. 355, New York.

Logan, T. J. 1983. Agronomic practices for nutrient availability and heavy metal accumulations in compost applied to cropland. *Proc. of the Natl. Conf. on Municipal and Industrial Sludge Utilization and Disposal.* Hazardous Material Control Research Institute, Silver Spring, MD.

Marks, M. J., J. H. Williams, and C. G. Chumbley. 1980. Field experiments testing the effects of metal-contaminated sewage sludges on some vegetable crops, pp. 235–251. In *Inorganic Pollution and Agriculture.* Min. Agr. Fish. Food Reference Book 326. HMSO, London.

McKenna, I. M. and R. L. Chaney. 1991. Cadmium transfer to humans from food crops grown in sites contaminated with cadmium and zinc, pp. 65–70. In L. D. Fechter (ed.). *Proc. 4th Intern. Conf. Combined Effects of Environmental Factors,* Oct. 1–3, 1990, Baltimore, MD, Johns Hopkins University School of Hygiene and Public Health.

McKenna, I. M., R. L. Chaney, S. H. Tao, R. M. Leach, and F. M. Williams. 1992. Interactions of plant zinc and plant species on the bioavailability of plant cadmium to Japanese quail fed lettuce and spinach. *Environ. Res.* 57:73–87.

McKenzie-Parnell, J. A., T. E. Kjellstrom, R. P. Sharma, and M. F. Robinson. 1988. Unusually high intake and fecal output of cadmium, and fecal output of other trace elements in New Zealand adults consuming dredge oysters. *Environmental Res.* 46:1–14.

Miller, E. R., X. Lei, and D. E. Ullrey. 1991. Trace elements in animal nutrition. In *Micronutrients in Agriculture.* 2nd ed. Soil Sci. Soc. Amer., Madison, WI.

Miller, T. L., R. R. Swager, S. G. Wood, and A. D. Adkins. 1992. *Selected Metal and Pesticide Content of Raw and Mature Compost Samples from Eleven Illinois Facilities. Results of Illinois' Statewide Compost Study.* ILENR/RR-92/09, Illinois Department of Energy and Natural Resources.

Moraghan, J. T. and H. J. Mascagni, Jr. 1991. Environmental and soil factors affecting micronutrient deficiencies and toxicities. In *Micronutrients in Agriculture.* Soil Sci. Soc. Amer., Inc., Madison, WI.

National Research Council. 1977. *Copper.* National Academy of Sciences, Washington, D.C.

National Research Council. 1980. *Mineral Tolerance of Domestic Animals. Committee on Animal Nutrition.* National Academy of Sciences, Washington, D.C.

Nielsen, F. H. 1984. Ultratrace elements in nutrition. *Annu. Rev. Nutr.* 4:21–41.

Nogawa, K. 1978. Studies on *itai-itai* disease and dose-response relationship of cadmium, pp. 213–221. *Proc. First International Cadmium Conf. Metals Bull.,* London.

Poole, D. B. R., D. McGrath, G. A. Fleming, and J. Sinnott. 1983. Effects of applying copper-rich pig slurry to grassland. 3. Grazing trials: Stocking rate and slurry treatment. *J. Agric. Res.* 22:1–10.

Purves, D. and E. J. Mackenzie. 1974. Phytotoxicity due to boron in municipal compost. *Plant Soil.* 40(1):231–235.

Quirk, J. P. and A. M. Posner. 1975. Trace element adsorption by soil minerals, p. 95. In *Trace Elements in Soil-Plant-Animal Systems.* Academic Press, Washington, D.C.

Richard, T. and M. Chadsey. 1990. Environmental impact of yard waste composting. *BioCycle.* 31(4):42–46.

Richard, T. L. and P. B. Woodbury. 1992. The impact of separation on heavy metal contaminants in municipal solid waste composts. *Biomass and Bioenergy.* 3:191–211.

Richard, T. L. and P. B. Woodbury. 1994. What materials should be composted? *BioCycle.* 35:63–68.

Richards, L. A. 1954. *Diagnosis and Improvement of Saline and Alkaline Soils.* USDA Agr. Handbook 60. U.S. Govt. Printing Office, Washington, DC.

Sawhney, B. L., G. J. Bugbee, and D. E. Stillwell. 1994. Leachability of heavy metals from growth media containing source-separated municipal solid waste compost. *J. Environ. Qual.* 223:718–722.

Schnitzer, M. and S. U. Khan. 1972. Reactions of humic substances with metal ions and hydrous oxides, p. 203. In *Humic Substances in the Environment.* Marcel Dekker, Inc. New York.

Schnitzer, M. and S. I. M. Skinner. 1966. Organo-metallic interactions in soils. 5. Stability constants of Cu^{2+}, Fe^{3+}, and Zn^{2+}-fluvic acid complexes. *Soil Sci.* 102:361–365.

Shaikh, Z. A. and J. C. Smith. 1980. Metabolism of orally ingested cadmium in humans, pp. 569–574. In B. Holmstedt et al. (eds.). *Mechanisms of Toxicity and Hazard Evaluation.* Elsevier/North-Holland Biomed. Press.

Sharma, R. P., T. E. Kjellstrom, and J. M. McKenzie. 1983. Cadmium in blood and urine among smokers and non-smokers with high cadmium intake via food. *Toxicology.* 29:163–171.

Simeoni, L. A., K. A. Barbarick, and B. R. Sabey. 1984. Effect of small-scale composting of sewage sludge on heavy metal availability to plants. *J. Environ. Qual.* 13(2):264–268.

Sims, J. T. and J. S. Kline. 1991. Chemical fractionation and plant uptake of heavy metals in soils amended with co-composted sewage sludge. *J. Environ. Qual.* 20:387–395.

Stevens, D. R., L. M. Walsh, and D. R. Keeney. 1972. Arsenic phytotoxicity on Plainfield soil as affected by ferric sulfate or aluminum sulfate. *J. Environ. Qual.* 1(3):301–303.

Stevenson, F. J. 1994. *Humus Chemistry.* 2nd ed. John Wiley & Sons, Inc., New York.

Stevenson, F. J. 1991. Organic matter-micronutrient reactions in soil. In *Micronutrients in Agriculture.* 2nd ed. Soil Sci. Soc. of Amer., Inc., Madison, WI.

Stillwell, D. E. and B. J. Sawhney. 1993. Elemental analysis of municipal solid waste compost II. Refuse derived fuel unders. *Compost Sci. & Util.* 1(4):49–53.

Tackett, S. L., E. R. Winters, and M. J. Puz. 1986. Leaching of heavy metals from composted sewage sludge as a function of pH. *Can. J. Soil Sci.* 66:763–765.

Underwood, E. J. 1977. *Trace Elements in Human Nutrition.* 4th ed. Academic Press, New York.

United States Environmental Protection Agency. 1978. *Guidelines Support Document.* ERT Document p-3249, Washington, D.C.

Van Campen, D. R. 1991. Trace elements in human nutrition, pp. 663–701. In *Micronutrients in Agriculture.* 2nd. ed. Soil Sci. Soc. Amer., Madison, WI.

Varma, M. M. and H. M. Katz. 1978. Environmental impact of cadmium. *J. Environ. Health.* 40:324–329.

Walsh, L. M., M. E. Sumner, and R. B. Corey. 1976. Considerations of soils for accepting plant nutrients and potentially toxic non essential elements, pp. 22–47. In *Land Application of Waste Materials.* Soil Cons. Soc. of Amer., Ankeny, IA.

Webber, M. D., Y. K. Soon, T. E. Bates, and A. U. Haq. 1981. Copper toxicity to crops resulting from land application of sewage sludge, pp. 117–135. In P. L'Hermite and J. Dehandschutter (eds.). *Copper in Animal Wastes and Sewage Sludge*. Reidel Publ., Dordrecht.

Welch, R. M., W. H. Allaway, W. A. House, and J. Kubota. 1991. Geographical distribution of trace element problems. In *Micronutrients in Agriculture*. 2nd Ed. Soil Sci. Soc. of Amer., Inc. Madison, WI.

Wixson, B. G. and B. E. Davies. 1994. Guidelines for lead in soil. *Environ. Sci. Technol.* 28(1):26A–31A.

Woodbury, P. B. 1992. Trace elements in municipal solid waste composts: A review of potential detrimental effects on plants, soil biota, and water quality. *Biomass and Bioenergy*. 3(3–4):239–259.

Yamagata, T. and I. Shigematsu. 1970. Cadmium pollution in perspective. *Bull. Inst. Publ. Health*. 19:1–27.

CHAPTER 7

Organic Compounds

INTRODUCTION

Both natural and xenobiotic (man-made) organic compounds abound in the universe. Many of these compounds are toxic to humans and animals. Industrial and manufacturing enterprises produce a myriad of organic chemicals, and it has been estimated that over 5 million distinct organic compounds are registered. Following World War II, pesticide and herbicide usage in agriculture increased dramatically and many of the compounds used were very persistent in the environment. Kuhn and Suflita (1989) indicate that in recent years 17 pesticides have been found in groundwater in 23 states. In the 1960s there was an increased awareness of the potential harmful effects of many of these organic compounds on humans, fish, and wildlife.

Many bacteria, fungi and other organisms have been found that can degrade organic compounds under aerobic conditions. For example, Saber and Crawford (1985) isolated strains of *Flavobacterium,* which degrade pentachlorophenol (PCP). The white rot fungi have been found to degrade a wide host of organic compounds. Barr and Aust (1994) listed the environmental pollutants degraded by the white rot fungus *Phanerochaete chrysosporium.* These include

(1) Chlorinated aromatic compounds: pentachlorophenol (PCP), 2,4,5-trichlorophenoxyacetic acid, polychlorinated biphenyls (PCB), and dioxin
(2) Polycyclic aromatic compounds: benzo(a)pyrene, pyrene, anthracene, and chrysene

TABLE 7.1. *Examples of Organic Compounds and the Organisms That Degrade Them under Aerobic Conditions.*

Organic Compound	Organism	Environment
2-Chlorobenzoic acid	*Pseudomonas*	Aerobic, soil
4-Chlorobenzoic acid	*Arthrobacter*	Aerobic
3,5-Dichlorobenzoic acid	*Pseudomonas*	Aerobic
3-Chlorobenzene	*Pseudomonas*	Aerobic
1,4-Dichlorobenzene	*Pseudomonas*	Aerobic
1,4-Dichlorobenzene chlorobenzene	*Alcaligenes*	Aerobic
3-Chlorophenol	*Nocardia*	Aerobic
4-Chlorophenol	*Mycobacterium* *Alcaligenes* *Flavobacter*	Aerobic
Pentachlorophenol (PCP)	*Arthrobacter* *Flavobacter* *Pseudomonas* *Coryneform*	Soil Aerobic Aerobic Aerobic
2,4-Dichlorophenoxyacetic acid (2,4-D)	*Pseudomonas* *Azobacter*	Soil Aerobic
2,4,5-Trichlorophenoxyacetic acid (2,4,5-T)	*Pseudomonas*	Soil
1,1,1-Trichloro-2,2-bis(*p*-chloro-phenol)ethane (DDT)	*Escherichia* *Pseudomonas* *Aerobacter* *Clostridium* *Proteus* *Fusarium* *Mucor* *Cylindrotheca* *Nocardia* *Streptomycetes* *Phanerochaete*	Aerobic
Polychlorobiphenyl (PCB)	*Alcaligenes* *Acinetobacter* *Pseudomonas*	Aerobic, soil Aerobic Soil

Source: Boyle, 1989.

(3) Pesticides: 1,1,1-trichloro-2,2-bis(4-chlorophenyl)ethane (DDT), lindane, chlordane, and toxaphene

(4) Nitrogen compounds

Table 7.1 presents examples of organic compounds and organisms that degrade them under aerobic conditions and in the soil (Boyle, 1989).

The potential impact of toxic organic compounds on human health and the environment is a function of levels of concentration, uptake by crops and plants, and bioavailability.

Composting can be an effective way to reduce the levels of toxic organics in organic wastes that would normally be disposed of by other means and consequently remain in the environment.

ORGANIC COMPOUNDS IN VARIOUS COMPOST MATERIALS AND FEEDSTOCKS

Data on toxic organics in compost and their feedstock are limited. This is because, historically, there has been greater concern about the regulated heavy metals, whereas toxic organics, except for PCBs, are not regulated. The cost for their analysis is high, which also discourages their identification.

As a result, data are often conflicting; some investigators report numbers, while others indicate that toxic organics are not detected. Most likely this is the result of different efficiencies in analytical procedures as well as the efficiency in composting. Duration of composting, temperature, and aeration affect the decomposition of an organic compound. Some of the current confusion may also be due to analytical methods and levels of detection. Several different examples are given in the text.

The presence of toxic organics in compost depends on the type of feedstock involved. For example, biosolids can contain organic compounds as a result of the disposal of industrial, commercial, and household wastes. Pesticides can be found in yard wastes and food wastes. Pthalates are found in plastics along with other organic dyes and compounds. Household wastes discharged into the MSW stream contain oils, solvents, pesticides and many other toxic organic compounds. Paper products may contain toxic organics as a result of printing inks and ash discharged from incinerators or boilers may contain dioxins.

Polycyclic or polynuclear aromatic hydrocarbons (PAHs) are compounds containing two or more aromatic rings. Some of these such as benzo(a)anthracene, chrysene, benzo(b)fluoranthene, benzo(b)fluoranthene, benzo(k)-fluoranthene, benzo(a)pyrene, indo(1,2,3-c,d)pyrene, dibenzo(a,h)anthracene, and benzo(g,h,i)perylene are considered probable human carcinogens

(U.S. Environmental Protection Agency, 1985). These compounds are introduced into the environment both through natural and combustion processes such as volcanic eruptions and forest fires and through human activities such as the use of fossil fuels, automobile exhausts, asphalt production, and outdoor broiling. The level of these compounds is usually very low. Because of their ubiquitousness, however, humans are constantly exposed to them.

A Minnesota study (E&A Environmental Consultants, Inc., 1988) analyzed MSW for the 129 EPA priority pollutants. Pesticides, phthalate, dioxins, and PCBs were detected. These data are shown in Table 7.2. As illustrated, the levels detected were very low, and with the exception of chlordane, endrin, and toxaphene, were found in parts per billion levels. PCBs were less than 1.1 ppm and phthalate ranged from 20 to 40 ppm. Dioxin (2,3,7,8-TCDD) was at less than 1 part/trillion. The levels in the compost were lower than in the MSW feedstock.

In a report prepared for the Minnesota Pollution Control Agency, Malcome Pirnie (1990) indicated that many of the pesticides and other toxic organics were not found in the unprocessed MSW, but were found in very low concentrations in the compostable fraction. The compostable fraction is the more concentrated organic portion of the waste stream. Table 7.3 shows some of the data. Many compounds detected in the compostable fraction were not detected in the compost. The authors indicated that this could be due to biodegradation, volatilization or photolysis.

Fricke et al. (1989) reported on the level of several pesticides in composted MSW and biosolids from five different plants in Germany (Table 7.4). The levels found were in the ppb range. Their data also reported on the levels found in organic household waste (Table 7.4). Separation of organic kitchen and garden wastes greatly reduced the level of contamination of pesticides in composts. Although these pesticides are no longer manufactured, they are very persistent in the environment.

Martens (1982) investigated the concentration of 4- and 6-ring polycyclic aromatic hydrocarbons (PHAs) in 12 composts prepared from sorted and ground household garbage. The compounds studied were benzo(a)anthracene/chrysene, 2-benzo(b,j,k)fluoranthene, 3-benzo(e)pyrene, 4-benzo(a)-pyrene, 5-perylene, 6-dibenzo(a,h/a,c)anthracene, 7-indeno(1,2,2-c,d)pyrene, 8-dibenzo(ah/ac)anthracene, and 9-benzo(g,h,i)-perylene.

Most of the compounds were lower in the composted material than in the fresh feedstock, indicating biodegradation. The compound of 2-benzo-(b,j,k)fluoranthene did not change in two composts and benzo(e)pyrene was higher in the Stuttgart compost. In addition Martens (1982) investigated the degradation of four C^{14} labeled PHAs (Figure 7.1). The loss of labeled C^{14} radioactive carbon indicates that the organic compound was degraded.

Recently a study evaluating organic compounds in a "wet bag" separation

TABLE 7.2. Concentration of Toxic Organic Compounds in MSW, Biosolids, and Their Respective Compost.

Organic Compound	MSW (μg/kg)	Biosolids (μg/kg)	MSW Compost (μg/g)	MSW and Biosolids Compost (μg/g)
Aldrin	<0.8–23	<0.13–12	<0.13	<0.13
α-BHC	7.8–51	<1.1	<1.1	<1.1
β-BHC	<4.8	<0.8–1.3	<0.8	<0.8
Lindane	<6.0	<1.0	<1.0	7.8
Chlordane	<1500	<240	<240	<240
p,p-DDD	87–130	<7.0	<7.0	<7.0
p,p-DDE	<14	2.3	4.3	<2.3
p,p-DDT	<49	<8.2	<8.2	<8.2
Dieldrin	57–110	2.9–3.3	<2.7	<2.7
Endosulfan I	<9.4–21	<1.6	<1.6	<1.6
Endosulfan II	34–40	<2.1–3.4	<2.1	<2.1
Endosulfan sulfate	<42	<7.0	<7.0	<7.0
Endrin	160–1400	<2.2	<2.2	<2.2
Endrin aldehyde	91–100	<5.9	<5.9	<5.9
Heptachlor	8.4–26	<0.23	<0.23	<0.23
Heptachlor epoxide	<3.8	<0.63	<0.63	<0.63
Toxaphene	<1000	<170	<170	<170
Methoxychlor	150–260	21–98	12.3	<8.3
Total PCBs	<1.1 μg/g	<0.17 μg/g	<0.33 μg/g	<0.33 μg/g
bis-(2-Ethylhexyl) phthalate	20–40	<3.3	17	19
2,3,7,8-TCDD Dioxin	<1 pg/g	<1 pg/g	<0.73 pg/g	<1.3 pg/g

Source: E&A Environmental Consultants, Inc., 1988.

175

TABLE 7.3. Toxic Organic Chemicals in the Compostable Feedstock and in Composted Municipal Solid Waste.

Organic Compound	Range in Feedstock (ppm)	Range in Compost MSW (ppm)
Aldrin	0.03–0.08	ND
BHC	0.09–0.21	ND
Chlordane	0.01–0.06	ND
Cyanide	ND	0.49–0.49
DDE	0.01–0.11	0.20–0.23
DDT	0.01–0.21	ND
Dieldrin	0.01–0.22	0.04–0.08
Endrin	0.01–0.02	ND
Heptachlor	0.02–0.02	0.00–0.04
Kapone	0.01–0.03	ND
Lindane	ND	0.00–0.08
Methoxychlor	0.01–0.03	0.00–0.07
Mirex	0.01–0.03	ND
PCBs	0.02–0.08	0.32–2.53
PCPs	0.00–0.00	0.02–0.02
Toxaphene	0.08–0.32	ND
Diazinon	0.18–0.35	ND
Ethyl parathion	0.18–1.10	ND
Malathion	0.21–1.70	ND
Methyl parathion	1.81–4.50	ND
2,4-D	0.40–0.51	ND
2,4,5-T	0.02–0.02	ND
2,4,5-TP	0.01–0.01	ND
Polyaromatic hydrocarbons	45.80–45.80	20.80–20.80

ND = not detected.
Source: Malcome Pernie, 1990.

TABLE 7.4. Concentration of Several Chlorinated Pesticides and PCB in MSW and Biosolids Compost as Compared to Levels in Separated Organic Household Waste.

Organic Compound	Concentration in MSW and Biosolids (ppb)	Concentration in Household Organic Wastes (ppb)
Aldrin	114	0.1
Dieldrin	592	0.3
Heptachlor	462	0.2
Sum of DDT/DDE/DDD	2681	33.5
Sum PCB	1493	140

Source: Fricke et al., 1989.

system (household organic material, non-recyclable paper, etc.) determined that the level of organic compounds was very low (National Audubon Society, 1993). As shown in Table 7.5, the compounds were detected in the ppb range. PCBs were undetected at the 2 ppb level.

Several studies have been conducted on organic compounds found in yard wastes. A report to Seattle Solid Waste Utility by Herrera et al. (1991)

PERCENTAGE OF APPLIED 14C MINERALIZED

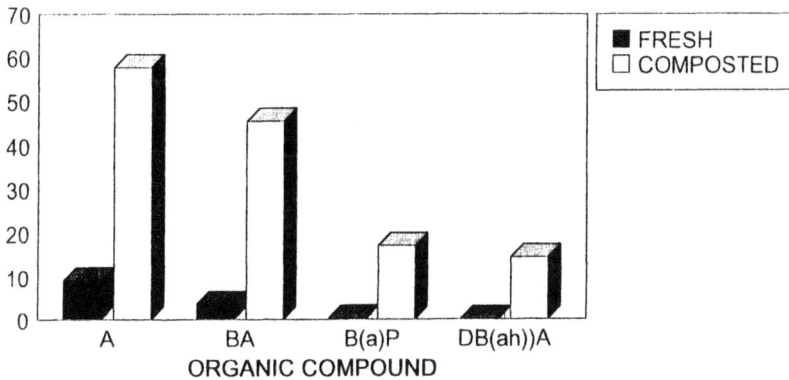

A= 9-14C-anthracene; BA = 12-14C-benz(a)anthracene;
B(a)P = 7,10-14C-benzo(a)pyrene; DB(ah)A = 7-14C-dibenz(ah)anthracene.

FIGURE 7.1. Percentage of applied C^{14}-radioactives mineralized to CO_2^{14} within ten weeks after addition of C^{14}-labeled polycyclic aromatic hydrocarbon to fresh and composted materials. (Data from Martens, 1982.)

TABLE 7.5. *Organic Compounds in Compost Produced from a "Wet Bag" Collection Program.*

Organic Compound	Concentration (mg/kg)
Naphthalene	0.006[1]
Acenaphthalene	0.024
Acenaphthene	<.012[2]
Fluorene	<0.012[2]
Phenanthrene	0.140
Anthracene	0.016
Fluoranthene	0.510
Pyrene	0.380
Benzo(a)anthracene	0.260
Chrysene	0.480
Benzo(b)fluoranthene	0.420
Benzo(k)fluoranthene	0.430
Benzo(a)pyrene	0.380
Indeno(1,2,3-c,d)pyrene	0.360
Dibenzo(a,h)anthracene	0.088
Benzo(g,h,i)perylene	0.340
PCBs	<0.002[2]
Pesticides	<0.002[2]

[1]Estimated value.
[2]Not detected at the concentration indicated.
Source: National Audubon Society, 1993.

reported data on semi-volatiles, volatile organic compounds, and pesticides before and after composting (Tables 7.6, 7.7 and 7.8). The analysis was very limited and involved one facility with analysis of two to three samples.

Although many compounds decreased upon composting, several showed an increase. The authors indicated that changes in detection limits may have accounted for this phenomenon. However, since the data represent the mean of three samples, the heterogeneity of the samples could account for the differences. Furthermore, the initial material was stockpiled resulting in anaerobic conditions.

TABLE 7.6. Concentration of Semivolatile Organic Compounds in Yard Waste Compost at Day 0 and Day 90.

Organic Compound	Initial Concentration (mg/kg dry weight)	Concentration after 90 Days (mg/kg dry weight)
Acenaphthene	0.24	ND
Anthracene	0.13	0.12
Benzoic alcohol	41.0	1.30
Benzyl alcohol	41.0	1.30
Benzo(a)anthracene	0.10	0.17
Benzo(b)fluoranthene	0.20	0.26
Benzo(a)pyrene	ND	0.16
bis(2-Ethylhexyl)phthalate	2.10	5.70
Chrysene	0.21	0.27
Dibenzofuran	0.18	ND
Fluorene	0.27	ND
Fluoranthene	0.41	0.70
2-Methylnaphthalene	0.32	ND
4-Methylphenol	ND	1.90
Phenanthrene	0.78	0.73
Phenol	1.60	2.20
Pyrene	0.21	0.38

ND = not detected.
Source: Herrera et al., 1991.

TABLE 7.7. Concentration of Volatile Organic Compounds
in Yard Waste at Day 0 and Day 90.

Organic Compound	Initial Concentration (mg/kg dry weight)	Concentration after 90 Days (mg/kg dry weight)
Benzene	ND	0.0019
2-Butanone	1.50	0.15
Chlorobenzene	ND	0.0018
2-Hexanone	ND	0.120
4-Methyl-2-pentanone	ND	0.021
Styrene	0.027	ND
1,1,1-Trichloroethane	ND	0.0038
Toluene	0.026	0.031

ND = not detected.
Source: Herrera et al., 1991.

TABLE 7.8. Concentration of Pesticides in Yard Waste
Compost at Day 0 and Day 90.

Pesticide	Initial Concentration (mg/kg dry weight)	Concentration after 90 Days (mg/kg dry weight)
Alpha-BHC	0.03	ND
Alpha-chlordane	0.11	0.10
Gamma-chlordane	0.08	0.09
Gamma-BHC (Lindane)	0.06	0.01
Dieldrin	0.13	0.21
4,4'-DDD	ND	0.07
4,4'-DDE	0.20	0.19
4,4'-DDT	0.65	0.28
Endosulfan II	0.07	ND
Methoxychlor	0.13	0.11

ND = not detected.
Source: Herrera et al., 1991.

The method of composting has considerable effect on the degradation of organic compounds. For example, windrow aerobic conditions are very different from static pile and aerated windrow or other systems that provide aeration. Measurements taken from windrows have shown that oxygen levels at the bottom of the windrows are at 0% within 15 minutes after turning. Unless the windrow has considerable porosity, anaerobic conditions could exist in a major part of it.

The pesticides detected are not currently used or distributed in the United States because of their toxicity and longevity in the environment. Therefore, detection of these compounds is due to residual material.

In 1990, results of an analysis for organic compounds in leaf compost from several communities in New York State showed that only 12 of the 139 compounds analyzed were detected at the parts-per-billion level. The compounds, phenanthrene, anthracene, fluoranthene, pyrene, benzo(a)anthracene, bis(2-ethylhexyl)phthalate, chrysene, benzo(b)fluoranthene, benzo(k)fluoranthene, benzo(a)pyrene, indeno(1,2,2-cd)pyrene and dibenzo(a,h)anthracene, were detected at very low parts-per-billion dry weight. The discrepancy between Herrera et al. (1991) and the New York State data could be the result of the extent of composting, in terms of time, temperature and aeration.

A comprehensive study by Miller et al. (1992) reported on an evaluation of 11 landscape composting facilities throughout Illinois that were analyzed for pesticides. Table 7.9 shows the results of analysis of the uncomposted and composted landscape waste. The samples were analyzed for PCBs but none were detected. In 4 of 22 compounds, the average levels in the compost were lower than in the raw waste. Furthermore, the number of compost samples containing a given pesticide was lower than the number of raw landscape waste samples. The levels detected were below the maximum allowable tolerance values for raw agricultural commodities as established by USEPA 40 CFR Chapter 1, 1991 edition. Richard and Chadsey (1989) analyzed a yard waste compost from Westchester County, New York, and found that four pesticides, captan, chlordane, lindane, and 2,4,-D, averaged 0.0052, 0.0932, 0.1810, and 0.0025 ppm, respectively. In all these studies, the levels of toxic organics found in organic wastes were extremely small.

Dioxins, polychlorinated dibenzo-*p*-dioxins (PCDD) and polychlorinated dibenzofurans (PCDF) are a class of compounds considered highly toxic to humans and animals. This group has been shown to be carcinogenic, mutagenic, and teratogenic in laboratory animals. However, epidemiological studies of residents of Times Beach, Missouri and Seveso, Italy, who were exposed to high levels of dioxin have not shown significant negative health effects.

The compounds are very persistent in the environment. Dioxin is a

TABLE 7.9. Average Concentration of Pesticides in Uncomposted and Composted Landscape Waste.

Pesticide	Average Concentration in Raw Waste (ppm)	Average Concentration in Compost (ppm)
Alachlor	0.749 (91)	0.304 (89)
Atrazine	4.61 (80)	3.03 (57)
Carbaryl	22.5 (86)	11.0 (93)
Chlordane	0.526 (100)	0.400 (98)
Chlorpyrifos	0.0096 (89)	0.00770 (91)
2,4-D	1.04 (100)	0.268 (100)
DDD	0.0641 (98)	0.0505 (98)
DDE	0.0516 (77)	0.807 (73)
Diazinon	0.991 (57)	0.587 (39)
Dichlobenil	0.0144 (93)	0.0133 (84)
Dieldrin	0.00992 (100)	0.00834 (100)
Fonofos	0.0112 (45)	0.00538 (20)
Heptachlor	0.00942 (2)	0
Heptachlor epoxide	0.0216 (100)	0.0152 (100)
Lindane	0.495 (95)	0.314 (95)
Malathion	0.313 (82)	0.169 (61)
Methoxychlor	0.314 (66)	0.507 (43)
Metolachlor	1.06 (84)	0.972 (77)
Parathion	0.235 (68)	0.104 (59)
2,4,5-T	0.788 (91)	1.15 (86)
Trifluralin	0.142 (91)	0.156 (68)
PCBs	0	0

Source: Miller et al., 1992.

by-product of the manufacturing of 2,4,5 trichlorophenol (TCP) from tetrachlorobenzene and 2,4,5-trichlorophenoxyacetate (2,4,5-T). In the United States a mixture of the two herbicides, 2,4,-dichlorophenoxyacetate (2,4-D) and 2,4,5-T, was used extensively (Arthur and Frea, 1989). Commoner et al. (1987) stated that the predominant source of dioxin today is from incineration of municipal waste. A series of papers by Lahl et al. (1990), Mueller et al. (1990), Wilken et al. (1990), Borgas et al. (1990), and Zeschmar-Lahl et al. (1990) provided data on dioxins and furans for four types of waste management, incineration, landfill, composting, and recycling. All four methods were found to contribute to dioxin generation.

Dioxins and furans represent a class of 210 chemical compounds, not all of which are toxic or have the same level of toxicity. The most toxic are the tetra isomers, especially the 2,3,7,8 tetrachlorinated compounds. The tetrachlorinated furans are 10% to 50% as toxic as the tetrachlorinated dioxins. A concept of toxicological equivalents (TE) was established to help evaluate the total or combined toxicological effects. Yet considerable question regarding the toxicity of dioxin remains. For example, Ames and Gold (1991) indicated that in comparison alcohol is a much more potent teratogenic (birth defects) compound. Several scientists and solid waste associations manipulated the data and wrongly tried to conclude that composting is worse than incineration. Goldfarb (1991) reexamined the German data (Fricke et al., 1989), and using data from the United States, showed that incineration stack emissions alone (not considering the ash) would result in a 15-fold contamination of soil compared to compost application to soil. The German scientists concluded that "it is urgently necessary to reduce the content of dioxins, furans and *precursor* substances in wastes, in the INPUT of all these techniques" (Zeschmar-Lahl, 1990).

Lahl et al. (1991) evaluated PCDD and PCDF in screened compost from four different feedstocks and several from unidentified sources (Table 7.10). PCDD ranged from 0.276 to 21.28 and PCDF from <0.01 to 1.76. The lowest values, 1.62 to 3.10 μg/kg, were found in plant waste compost. For vegetable waste compost, the values ranged from 0.394 to 21.67 μg/kg. Mixed MSW compost of two samples was 12.573 and 19.14 μg/kg. In all of the studies on dioxins in compost, the most toxic 2,3,7,8-tetrachlorodibenzo-*p*-dioxin (TCDD) was not found. The toxic equivalents (TE) for all samples ranged from 0.8 to 35.7 ng (TE)/kg. The greatest variation occurred with the vegetable wastes.

It would have been helpful if the authors had provided background on the samples. For example, information on the area of collection (urban versus rural), the proximity to other generators such as incinerators, and whether or not the mixed waste contained incinerator ash or other ash, would have further defined the waste material. The detailed data were presented in an

TABLE 7.10. Dioxin and Furan (PCDD/PCDF) Contents of Different Compost.

Compost Type	PCDD (µg/kg)	PCDF (µg/kg)	PCDD + PCDF (µg/kg)	PCDD + PCDF [ng(TE)/kg]
Mixed waste	12.53	0.04	12.57	22.6
Mixed waste	19.10	0.04	19.14	32.1
Plant material	1.62	<0.01	1.62	1.8
Plant material	2.94	0.16	3.10	5.2
Plant material	1.88	<0.01	1.88	2.2
Bark	2.15	0.05	2.15	3.6
Vegetable waste	15.91	0.05	15.96	19.4
Vegetable waste ($n = 7$)	4.68–21.28	0.14–1.76	5.39–21.67	7.1–35.7

TABLE 7.10. (continued).

Compost Type	PCDD (µg/kg)	PCDF (µg/kg)	PCDD + PCDF (µg/kg)	PCDD + PCDF [ng(TE)/kg]
Vegetable waste	11.06	0.04	11.10	13.4
Vegetable waste	17.92	0.11	18.03	21.8
Vegetable waste	0.276	0.285	0.562	7.7
Vegetable waste	0.734	0.443	1.177	15.4
Vegetable waste	0.338	0.056	0.394	0.8
Unspecified	8.543	0.884	9.427	30.8
Unspecified	4.782	0.523	5.305	19.7
Unspecified	5.394	0.534	5.928	24.2

Source: Lahl et al., 1991.

earlier paper (Wilken et al., 1990) and showed that the principal isomers were the less toxic HeptaCDD and OctaCDD. The authors indicated that the HexaCDD burden was probably from contamination from other sources. The principal sources of dioxin are combustion or incineration. Vegetable and other plant waste could easily be contaminated from atmospheric deposits.

Harrad et al. (1991) analyzed compost from yard waste for PCDDs, PCDFs, chlorophenols, and chlorobenzenes and found these compounds at levels significantly higher than in the local soils. It is interesting to note that the soil contained the more toxic forms of dioxin (T4CDD), which was not detected in the compost (Table 7.11). All occurrences were at the picogram/gram level. Table 7.12 provides data on chlorophenols and chlorobenzenes. Six compounds were analyzed: T3CP, T4CP, P5CP, T4CBz, P5CBz and HxCBz. With the exception of T4CBz, the other compounds were all found in the soil, yard waste, and compost at very low levels. This indicated that the levels were probably not due to atmospheric deposition from combustion processes, but rather from previous and present use of

TABLE 7.11. Dioxin Compound Levels and Toxic Equivalents
in Yard Waste Compost.

PCDD/PCDF Compounds	Soil Range (in pg/g)	Compost Range (in pg/g)
T4CDD	ND–7.3	ND
P5CDD	ND	ND
HxCDD	ND–14	ND–250
HpCDD	67–240	0–4300
OCDD	660–6200	4000–23,000
T4CDF	ND–9.7	ND–74
P5CDF	ND–30	ND–220
HxCDF	5–19	ND–500
HpCDF	ND–22	1–1200
OCDF	ND–14	170–1400
Toxic equivalent (TEQ)	1–7.4	7.7–41

ND = not detected.
Source: Harrad et al., 1991.

TABLE 7.12. Range of Concentration of Chlorophenols and
Chlorobenzene Compounds in Soil, Yard Waste,
and Yard Waste Compost (ng/g).

Compound	Compost	Yard Waste	Soil
T3CP	ND–5.8 (4/11)[1]	ND–8.9 (5/7)	ND–2.6 (2/4)
T4CP	ND–23 (6/11)	ND–2.6 (6/7)	ND–1.3 (2/4)
P5CP	7–190 (0.11)	9.2–210 (0/7)	1.5–2.5 (04)
T4CBz	ND–12 (5/11)	ND–0.74 (4/7)	ND
P5CBz	0.55–11	0.4–1.1	ND–0.14 (2/4)
HxCBz	1.9–97	0.69–20	0.09–0.97

[1]Ratios in parentheses are the number of nondetected values to the total of numbers analyzed.
ND = not detected.
Source: Harrad et al., 1991.

pentachlorophenol (PCP)-based biocides. An analysis of a PCP-based herbicide showed much higher levels than in the soil or compost.

E&A Environmental Consultants, Inc. (1993) conducted a food waste composting study using three feedstocks: yard waste, wood waste and paper waste. The two composting methods employed were static pile and windrow. Data were collected on the following compounds or classes of compounds:

- dioxin
- low molecular weight polycyclic aromatic hydrocarbons
- high molecular weight polycyclic aromatic hydrocarbons
- phenolics
- chlorinated hydrocarbons
- phthalates
- organonitrogen compounds
- miscellaneous oxygenated compounds
- pesticides
- organophosphorus pesticides
- herbicides
- polychlorinated biphenyls

Table 7.13 lists the compounds analyzed for the yard wastes. The low values substantiate previous studies indicating that the levels of toxic organics in source-separated waste are very low and do not represent a risk. Similar low values were found for wood and paper waste. Most of the compounds were not detected in the μg (parts-per-billion range).

Data from a wet bag demonstration project (National Audubon Society,

Compound	Static Pile (μg/kg)	Turned (μg/kg)
Dioxin	0.0012	<0.001[1]
Low Molecular Weight Polycyclic Aromatic Hydrocarbons		
Acenaphthene	<120	<120
Acenaphthylene	<120	<120
Anthracene	<120	<120
Fluorene	<120	<120
2-Methyl-naphthalene	59[2]	<120
Naphthalene	<120	<120
Phenanthrene	<120	<120
Total LPAHs	779	840
High Molecular Weight Polycyclic Aromatic Hydrocarbons		
Benzo(a)anthracene	120	110[2]
Benzo(a)pyrene	260	130
Benzo(b)fluoranthene + Benzo(k)fluoranthene	510	170
Benzo(g,h,i)perylene	190	140
Chrysene	210	210
Dibenzo(a,h)anthracene	<120	110
Fluoranthene	270	240
Indeno(1,2,3-c,d)pyrene	220	110[2]
Pyrene	140	150
Total HPAHs	2040	1470
Phenolics		
4-Chloro-3-methylphenol	<240	<240
2-Chlorophenol	<120	<120

TABLE 7.13. (continued).

Compound	Static Pile (μg/kg)	Turned (μg/kg)
Phenolics *(continued)*		
2,4-Dichlorophenol	<350	<360
2,4-Dimethylphenol	<240	<240
4,6-Dinitro-2-methylphenol	<1200	<1200
2,4-Dinitrophenol	<1200	<1200
4-Methylphenol	<120	<120
2-Methylphenol	<120	<120
2-Nitrophenol	<590	<600
4-Nitrophenol	<590	<600
Pentachlorophenol	<590	<590
Phenol	<240	<240
2,4,6-Trichlorophenol	<590	<600
2,4,5-Trichlorophenol	<590	<600
Chlorinated Hydrocarbons		
2-Chloro-naphthalene	<120	<120
1,3-Dichlorobenzene	<120	<120
1,2-Dichlorobenzene	<120	<120
1,4-Dichlorobenzene	<120	<120
Hexachlorobenzene	<120	<120
Hexachlorobutadiene	<240	<240
Hexachlorocyclopentadiene	<590	<600
Hexachloroethane	<590	<600
1,2,4-Trichlorobenzene	<120	<120
Phthalates		
Butylbenzylphthalates	<120	<120
Di-*n*-butylphthalate	<120	<120

(continued)

189

TABLE 7.13. (continued).

Compound	Static Pile (μg/kg)	Turned (μg/kg)
Phthalates *(continued)*		
Diethylphthalate	<120	<120
Dimethylphthalate	<120	<120
bis(2-Ethylhexyl phthalate)	240	240
Di-*n*-octylphthalate	<120	<120
Organonitrogen		
Carbazole	<120	<120
4-Chloroanaline	<350	<350
3,3'-Dichloro benzidine	<590	<590
2,4-Dinitrotoluene	<590	<600
2,6-Dinitrotoluene	<590	<600
2-Nitroaniline	<590	<600
3-Nitroaniline	<590	<600
4-Nitroaniline	<590	<600
Nitrobenzene	<120	<120
n-Nitrosodiphenyl amine(1)	<120	<120
n-Nitroso-di-*n*-propylamine	<120	<120
Miscellaneous Oxygenated		
Benzoic acid	200[2]	<1200
Benzyl alcohol	<590	<600
4-Bromophenylphenylether	<120	<120
bis(2-Chloroethoxy) methane	<120	<120
bis(2-Chloroethyl) ether	<120	<120
4-Chlorophenylphenylether	<120	<120
Dibenzofuran	<120	<120
Isophorone	<120	<120

190

TABLE 7.13. (continued).

Compound	Static Pile (μg/kg)	Turned (μg/kg)
Miscellaneous Oxygenated (continued)		
2,2'-Oxybis(1-chloropropane)	<120	<120
Pesticides		
Aldrin	<6.8	<6.8
Delta-BHC	<6.8	<6.8
Alpha-BHC	<6.8	<6.8
Beta-BHC	<7.5	<7.8
Gamma-BHC (Lindane)	5.5	9.4
Alpha-chlordane	7.8	8.7
4,4'-DDD	11[2]	21
4,4'-DDE	66	36
4,4'-DDT	14	22
Dieldrin	19	19
Endosulfan I	<6.8	<6.8
Endosulfan II	<7.0	<14
Endosulfan sulfate	<14	<14
Endrin	<14	<14
Endrin aldehyde	<14	<14
Endrin ketone	<14	<14
Gamma-chlordane	12	15
Heptachlor	<6.8	<6.8
Heptachlor epoxide	<6.8	<6.8
Methoxychlor	25[2]	45[2]
Toxaphene	<140	<140

(continued)

TABLE 7.13. (continued).

Compound	Static Pile (μg/kg)	Turned (μg/kg)
Organophosphorus Pesticides		
Atrazine	<120	<120
Bolstar	<20	<20
Chlorpyrifos	<8.0	<8.0
Coumaphos	<17	<17
Demeton	<110	<110
Diazinon	<17	<17
Dichlovos	<13	<13
Dimethoate	<120	<120
Disulfoton	<14	<14
EPN	<28	<28
Ethroprop	<4.0	<4.0
Ethyl parathion	<14	<14
Fensulfothion	<50	<50
Fenthion	<8.0	<8.0
Malathion	<16	<16
Methyl parathion	<19	<19
Mevinphos	<170	<170
Phorate	<10	<10
Ronnel	<7.0	<7.0
Simazine	<8.0	<8.0
Tetrachlorovinphos	<40	<40
Tokuthion	<20	<20
Herbicides		
2,4-D	<85	<85
2,4-DB	<200	<200

TABLE 7.13. (continued).

Compound	Static Pile (µg/kg)	Turned (µg/kg)
Herbicides *(continued)*		
Dalapon	<50	<50
Dicamba	<20	<20
Dichloroprop	<75	<75
Dinseb	<20	<20
MCPA	<25,000	<25,000
Silvex	<20	<20
2,4,5-T	<20	<20
Polychlorinated biphenyls (PCBs)		
Arochlor 1221/1016	<280	<280
Arochlor 1232/1016	<140	<140
Arochlor 1242/1016	<140	<140
Arochlor 1248/1016	<140	<140
Arochlor 1254/1016	68[2]	66[2]
Arochlor 1260/1016	<280	<140
Total PCBs	1048	906

[1]< Symbol = not detected at specified detection limit.
[2]Estimated value.
Source: Herrera Environmental Consultants Inc. 1994 Commercial food waste collection and composting demonstration project. King County Solid Waste Division and Seattle Waste Utility.

1993) showed low levels of organic constituents in separated organic wastes (see Table 7.5). Air samples were collected during the study. The volatile organic compounds (VOC) emitted during composting and curing were greatest during the initial composting, but dropped off dramatically before the curing phase. At all times, VOCs were well below OSHA permissible exposure levels. Some PAHs were detected in the mg/kg (ppm) range. Benzo(a)pyrene was found at 0.380 mg/kg and the total of 16 PAH compounds was 3.900 mg/kg.

Biosolids can contain toxic organic chemicals, principally as discharged from industrial sources, but also from atmospheric deposition (Webber and Lesage, 1989; USEPA, 1990). USEPA (1990) conducted a nationwide

TABLE 7.14. Some Toxic Organics in Biosolids.

Analyte	Percent Detected[1]	Mean (μg/kg)	Standard Deviation	CV[2]
Benzene	2	9.74	2.08	0.21
Dieldrin	3	2.70	238.6	88.22
Heptachlor	2	2.81	4.67	1.66
Lindane	2	3.88	454.00	116.98
PCB 1260	9	112.43	1629.78	14.50
Trichloroethane	5	59.42	6947.10	116.92
4,4'-DDT	3	22.82	13.78	0.60

[1]Percent detected of the samples analyzed.
[2]Coefficient of variation.
Source: USEPA, 1990.

survey of biosolids at publicly owned wastewater treatment plants. The data showed extremely low levels; therefore, toxic organics were excluded from the recently issued 40CFR503 biosolid regulations. Some of the data reported by EPA are shown in Table 7.14. The chlorinated pesticides are no longer manufactured or distributed in the United States. Therefore, compost from biosolids should contain very low levels of toxic organics.

Lisk et al. (1992) reported that compost from biosolids had PCB levels from not detected to 0.20. In biosolids with yard wastes, PCB levels ranged from ND to 0.41 ppm, and in MSW and biosolids the values ranged from 0.15 to 0.38 ppm. Composting significantly biodegraded many of the compounds.

FATE OF ORGANIC COMPOUNDS DURING COMPOSTING

The potential of biodegrading through composting of toxic organics has been investigated by several researchers (Snell Environmental Group Inc., 1982; Deever and White, 1978; Rose and Mercer, 1968). Earlier in the section on the presence of toxic organics in MSW, biosolids, and yard waste, several of the researchers were reported as indicating that compost often contained lower amounts of toxic organics than in the original feedstock and that this could be due to biodegradation, volatilization, or photolysis.

One of the most comprehensive laboratory studies on the biological decompositions of organic compounds through composting was conducted by Snell Environmental Group Inc. (1982). Table 7.15 shows the decompo-

TABLE 7.15. *The Rate of Biodegradation of Toxic Organic Compounds during Composting.*

Toxic Organic Compound	Percent Reduced by Composting	
	7 Days in Compost	30 Days in Compost
Analine	100	100
2-Naphthylamine	100	100
Benzidine	86	0
Benzidine hydrochloride	89	84
3,3'-Dichlorobenzidine	57	64
3,3'-Dichlorobenzidine hydrochloride	100	0
Naphthalene	100	100
Pyrene	0	31
Benzo(a)pyrene	27	0
Benzo(a)anthracene	25	42
2,4,5-Trichlorophenol	71	69
Pentachlorophenol	36	51
Dieldrin	0	11
4,4'-DDT	0	7
Chlordane	0	44
Lindane	66	73
Endrin	47	52
Toxaphene	0	4
Methoxychlor	34	74
Chlorpyrifos	0	31
Hexachloro-1, 3-Butadiene	100	100

(continued)

TABLE 7.15. (continued).

Toxic Organic Compound	Percent Reduced by Composting	
	7 Days in Compost	30 Days in Compost
Hexachlorocyclopentadine	98	100
Hexachlorobenzene	31	62
Hexabromobiphenyl	0	8
Firemaster	0	4
Octachlorodibenzo-*p*-dioxin	32	50
2,7-Dichlorodibenzo-*p*-dioxin	0	21
Curene	4	31
1,2-Dichloroethane	85	85
Tetrachloroethane	85	85
1,1,2-Trichloroethane	21	21
1,1,1-Trichloroethane	95	95
Trichloroethylene	72	99
Methylene chloride	50	50
1,2-Dichlorobenzene	100	100
1,3-Dichlorobenzene	100	100
1,2-Dibromoethane	94	94
Carbon tetrachloride	76	76
Chloroform	64	95
Benzene	50	71
Toluene	95	95
Styrene	79	79
O-xylene	79	79
M-xylene	81	81

TABLE 7.15. (continued).

Toxic Organic Compound	Percent Reduced by Composting	
	7 Days in Compost	30 Days in Compost
Ethyl benzene	74	74
Bis(2-ethyl)hexaylphthalate	25	41
Di-*n*-butyl phthalate	72	79
Benzyl butyl phthalate	75	65
PCB 1221	77	75
PCB 1016	21	72
PCB 1242	29	39
PCB 1254	95	60
PCB 1260	2	7

Source: Snell Environmental Group Inc., 1982.

sition of various organics compounds. Several compounds such as dichlorobenzene, hexachloro-1,3-butadiene, aniline and trichloroethylene were completely biodegraded in 30 days. Others like benzidine hydrochloride, tetrachloroethylene, 1,1-trichloromethane, chloroform, toluene, styrene, PCB 1221 were reduced by over 75% during composting. Many others were decomposed to a lesser extent; PCB 1260, hexachlorobiphenyl were less than 10% decomposed.

Deever and White (1978) found significant reduction in the amount of toluene-hexane extractable grease and oil after composting. Epstein and Alpert (1980) reported that aliphatic and aromatic compounds were easily decomposed during composting of waste oil. Rose and Mercer (1968) showed that the insecticides diazinon, parathion, and dieldrin were significantly degraded in a short time whereas DDT was more resistant. Diazinon was reduced from 3.3 ppm to 0.002 ppm in 42 days. The concentration of parathion was reduced by 50% in 12 days of composting. DDT was biodegraded from 2.2 ppm to 0.8 ppm in 50 days.

Considerable research attention has focused on the use of composting of soil and sediment contaminated with explosives. For example, TNT HMX and RDX were greatly degraded by composting (Williams et al., 1988; Griest et al., 1990). In a field composting study, the concentration of TNT, a nitro-amino-toluene compound, was reduced by 99.9% in 153 days. As a result of these studies, the military is using composting to remediate a contaminated site in Oregon.

Martens (1982) evaluated the contents of four to six polycyclic aromatic hydrocarbons (PAHs) in composted municipal solid waste. Concentration of PAHs ranged from 0.17 perylene to 56.75 benzo(a)anthracene/thrysene μg/g of compost. Data were provided for six composting plants. Three plants showed marked reduction in all PAHs; but three plants showed an increase of some compounds during composting. No data were provided on the type of composting or conditions. The efficiency of the composting process (i.e., temperature, aeration, and moisture control) would have a significant effect on the decomposition of organic compounds.

Davie (1993) reported on the composting of petroleum wax-based coatings on corrugated containers. During 92 days of composting, between 90 and 100% of the wax degraded. Barley was used in germination tests to evaluate the compost produced. The bioassay test showed good germination rates of barley in all composts with no significant differences between them. Mixtures of garden soil and composts were prepared. In one study, barley growth was higher when 10%, 30%, and 100% of compost was used in the mix. In another study, the high wax mixture showed the poorest growth. This was attributed to the high levels of boron.

Gross (1994) evaluated the fate of several organic compounds during laboratory composting of primary and waste-activated biosolids. The pathways considered were volatilization, biodegradation or chemical transformation, and retention in the biosolids compost. Biodegradation or chemical transformation represented the major pathway for phenol, 4-nitrophenol, and nitrobenzene. This transformation pathway ranged from 73–89% for phenol, 90–93% for 4-nitrophenol, and 93–99% for nitrobenzene. Relatively small amounts were lost by volatilization. Volatilization ranged from 8–15% for phenol, 0.3–2% for 4-nitrophenol, and 0.7–5% for nitrobenzene. Volatilization may have been enhanced since the biosolids that were spiked with compounds did not have a chance to become attached to some of the organic matter in the biosolids. Relatively small amounts were found in the compost at the end of the study. Specifically, amounts in the composted biosolids ranged from 3–13% for phenol, 6–10% for 4-nitrophenol, and 0.1–3% for nitrobenzene.

REACTIONS AND MOVEMENT OF TOXIC ORGANICS IN SOIL

If compost contains toxic organic chemicals, what happens when it is applied to a soil? There is a concern that these toxic organics can move through the soil into groundwater; move into surface water through runoff and erosion; and be taken up by plants and soil biota. Finally, through water and plants they can enter the human food chain. Therefore, it is important to understand the potential pathways for an organic chemical in compost when applied to soils.

Toxic organic compounds can undergo numerous chemical and biological reactions in the soil (Figure 7.2). The principal pathways in the soil include:

- adsorption onto the soil and its components
- volatilization
- degradation: microbial, chemical, and photochemical
- leaching into groundwater
- runoff and erosion to surface waters
- uptake by macro- and microorganisms
- uptake by plants

There has been considerable research on the fate of toxic organics in soils, particularly pesticides (Sawhney and Brown, 1989; Chaney, 1984; Jacobs et al., 1987; Kaufman, 1983).

Sorption is considered the most important mechanism affecting the behavior of organic chemicals in soils. Thus, Chiou (1989) indicated that sorption of organic chemicals in soils is more strongly correlated with soil organic matter than with any other soil factor. For toxic organic compounds to leach into groundwater, they must enter the soil solution. To enter the soil solution, desorption from colloidal surfaces or dissolution of the chemical in particulate form must take place (Kaufman, 1983).

This does not occur very readily, however. Hence, leaching and plant uptake are very minimal. Furthermore, there is evidence that toxic organic compounds can undergo degradation in the soil, which further minimizes their potential for leaching into groundwater (Kaufman, 1983; Jacobs et al., 1987). Degradation in the soil can occur by microbial degradation, chemical

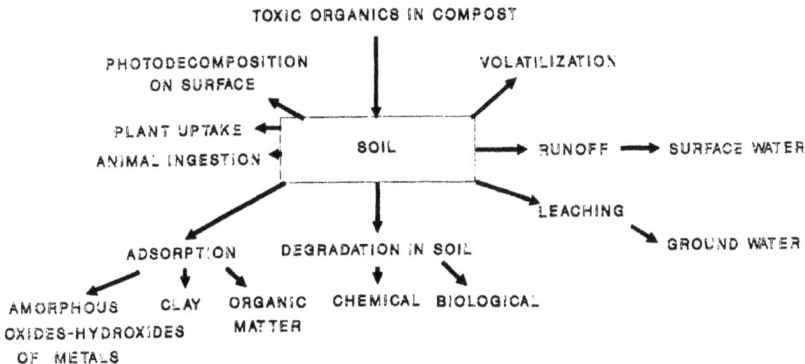

FIGURE 7.2. Fate of toxic organic compounds in compost applied to soil.

degradation, or photochemical degradation. Overcash (1983) noted that very few organic compounds can be said to be non-degradable.

Many microbes can use organic compounds in waste as an energy source. Significant biodegradation has been shown for phenolic compounds, phthalate esters, naphthalenes, and nitrogenous organic compounds. Moderate degradation can occur for aromatics, PCB, halogenated ethers, and halogenated aromatics. Organochlor in compounds is slow to biodegrade (Kaufman, 1983).

Volatilization is another important mechanism in the removal of toxic organic compounds from soils. Most of the research in this area has been done on pesticides applied to agricultural soils. Chang and Page (1984) discussed the volatilization of organics when applied to soils through wastewater. Since volatilization is a surface phenomenon once the compound is incorporated into the soil, the rate of volatilization depends on the movement of the organic compound to the soil surface. Therefore, volatilization may be an important pathway for reduction of organics when compost is applied as a mulch. No data were found on volatilization of toxic organics from compost applied to the soil surface. Currently, however, several studies are being conducted to determine volatilization of organic compounds from compost piles.

Degradation can occur by biological, chemical, or photochemical processes. There is considerable literature on the biodegradation of pesticides in soils (Kaufman, 1983; Jacobs et al., 1987; Sawhney and Brown, 1989).

Ahmed and Focht (1972) reported that PCBs were biodegraded by two species of *Achromobacter*. The white rot fungus (*Phanerochaete chrysosporium*) was shown to have the ability to degrade many diverse organic compounds (Bumpus and Aust, 1986). Bushnell and Haas (1941) showed that *Pseudomonas* spp. was one of the predominant metabolizers of crude oil. Reviewing the data on the biodegradation of petroleum wastes, Hornick et al (1983) concluded that composting could result in accelerated degradation. Boyle (1989) discussed the microbial decomposition of chlorinated aromatic hydrocarbons. He provided a list of 24 compounds and the organisms responsible for their biodegradation. Arthur and Frea (1989) reviewed the properties of dioxin and its potential biodegradation in soils, noting that extensive microbial transformation of TCDD had not been shown. The microbial degradation of several organic compounds in soil is shown in Table 7.16.

Leaching into groundwater is related to the solubility of the organic compound and the sorption and desorption of the compound from colloidal surfaces such as organic matter and clay. Therefore, the chemical structure of the compound is very important as it affects biodegradability, volatilization, sorption, and other reactions.

TABLE 7.16. Examples of Degradation of Several Toxic Organic Compounds in Soil.

Compound	Organism	Reference
3-Chlorobenzoic acid	*Pseudomonas*	Pertsova et al., 1984
1,2,3-Trichlorobenzene	Consortium	Marinucci and Bartha, 1979
Pentachlorophenol (PCP)	*Arthrobacter*	Edgehill and Finn, 1983
2,4,5-Trichlorophenoxyacetic acid (2,4,5-T)	*Pseudomonas*	Kilbane et al., 1983
1,1,1-Trichloro-2,2-bis(p-chloro-phenol)ethane (DDT)	Consortium	Gains and Beard, 1967
Polychlorobiphenyl (PCB)	*Pseudomonas* *Alcaligenes* *Acinetobacter*	Unterman et al., 1988; Brunner et al., 1985
Benzo(a)pyrene		Shabad et al., 1971

Excellent reviews have been done by Leonard et al. (1976), Weber and Miller (1989), and Sawhney and Brown (1989). As pointed out earlier, organic matter has a high affinity for organic compounds and, therefore, reduces the potential for leaching; this provides more opportunity for microbial degradation.

Sawhney and Kozloski (1984) reported on toxic organics in landfill leachates. Although it is difficult and wrong to extrapolate their data to potential leachate from compost, their study does provide a perspective. Five landfills in Connecticut were studied. One town had considerably higher levels of volatile organic compounds than the others. The volatile organic compounds detected included acetone, isopropyl alcohol, methyl ethyl ketone, 2-butyl alcohol, benzene, toluene, and methyl isobutyl ketone. Non-volatile compounds detected were toluene, phenol, *m*-cresol, *p*-cresol, *m*-ethyl phenol, *p*-ethyl phenol, and C$_x$-acids. The authors indicated that the presence of 0.1 to 1.5 mg/L of the substituted phenols in the leachate was probably the result of leaching through the soil under anaerobic conditions. In a previous paper, Isaacson and Sawhney (1983) found that under aerobic conditions, sorption by clays and soil of organic compounds was irreversible.

Runoff is a major non-point source of pollution of water courses; it is primarily a result of agricultural application of pesticides. Pesticide transport in the runoff is the result of the solubility of the compound or its affinity to colloidal particles of soil and organic material (Leonard et al., 1976).

TABLE 7.17. Concentration of Pesticide in Runoff and Sediment.

Date of Storm	DDT (ppb)		Endrin (ppb)		Endosulfan (ppb)	
	Sediment	Runoff	Sediment	Runoff	Sediment	Runoff
5/29/66	8.0	13.9	1.3	1.4	—	—
7/7/66	4.9	18.0	0.7	2.0	—	—
7/13/66	3.7	31.9	3.3	36.7	1.0	18.9
Total amounts in runoff (g/ha)	—	13.29	—	8.62	—	3.69
Percent applied	—	1.83	—	1.00	—	0.35

Source: Epstein and Grant, 1968.

Epstein and Grant (1968) measured the concentration of three chlorinated insecticides in runoff as the result of potato, oats, and sod cropping systems. Table 7.17 shows the concentration of DDT, endrin and endosulfan in runoff (soil-water suspension) and sediment from field plots in continuous potato cropping. The amounts of pesticides in runoff were very small compared to the amount applied. A crop rotation system reduced the amount of pesticides in runoff compared to continuous cropping of potatoes. This could have been the result of the added organic matter provided under a rotation management system. The organic matter increases soil porosity and increases infiltration, thus reducing runoff. Furthermore, the organic matter in the soil bound the pesticide and reduced its potential for leaching while making it more available for microbial degradation. Data on the potential contribution of applied compost to toxic organics in runoff were not found in the literature.

UPTAKE BY PLANTS AND POTENTIAL ENTRY INTO THE FOOD CHAIN

There are several pathways for toxic organic chemicals in compost to enter the food chain. These include:

(1) Uptake by plants
(2) Deposition of the compost on edible portions and direct consumption by humans. Deposition may result from direct application of the compost to edible portions or from dust during farming activities
(3) Direct ingestion of compost or soil-compost mixtures by children
(4) Consumption of animal products that contained the toxic organic as a result of the animal ingesting plants grown on compost-amended soil or ingesting compost deposited on plants

The most important pathway is the uptake by plants. Although the other pathways represent possible scenarios, they are less likely to occur and involve lower levels of intake overall.

Organic compound molecules are much larger than elements; hence the potential absorption and translocation from the root system to the edible portions of a plant are less likely.

Very little data are available on uptake of toxic organic compounds from compost. Much more information is available on uptake of toxic organics from biosolids. Although this information might provide insight into the potential uptake from compost, caution must be used in extrapolating the data from biosolids.

Studies were conducted on a Lakeland sandy loam to evaluate the uptake of two dioxins, DCDD and TCDD, and 2,4-dichlorophenol (DCP) by oats and soybeans. The Lakeland sandy loam has a very low adsorptive capacity. Uptake was low. A maximum of 0.15% of DCDD and TCDD and 0.21% of the DCP present in the soil was translocated to the foliage of soybeans and oats. The compounds were not detected (>1 ppb) in oat grain; no TCDD and only trace amounts of DCP and DCDD were found in bean of soybean (Helling et al., 1973).

Iwata and Gunther (1976) reported that PCBs were absorbed by carrot roots. The carrot peel contained 97% of the PCBs. Very little was translocated to the inner plant tissue, however, Fries and Marrow (1981) used C^{14} labeled tri-, tetra- and pentachlorobiphenyls to measure uptake by soybeans. They separated the aerial portion from the soil to avoid contamination in the plants by volatilization. No C^{14} was found on top of the plants. Phthalate esters are found in municipal biosolids and MSW, since the major sources of phthalates are plasticizers in polyvinyl chloride plastic products. These have been found in soils and compost.

Aranda et al. (1989) studied the uptake of di-(2-ethylhexyl)phthalate (DEPH) in lettuce, carrots, and chile peppers grown on soil amended with biosolids. There was little uptake, and intact DEPH was not detected in any plants.

O'Connor et al. (1990) evaluated the plant uptake of sludge-borne PCBs. The plants studied were fescue, carrot, and lettuce. PCB content was below detection limits (20 μg/kg) in all plant species and parts except carrot peels. These authors also evaluated the potential from volatilization and did not detect PCB contamination. These data supported previous work on carrots and other plants.

Wild and Jones (1992) found that biosolids application to agricultural soils increased the soil PAH burden but did not affect the PAH levels in the carrot foliage. PAHs' level increased in the carrot peel with increasing concentration up to 450 μ/kg. After that, the increase was slight (Figure 7.3).

Uptake of chlorbenzenes (CBs) by carrots from biosolids was studied by Wang and Jones (1994) (Figure 7.4). Both carrot shoots and roots took up CBs. Only 0.03–0.72% of the soil CBs were taken up by carrots. There was a greater loss probably from a soil spiked with CBs than from biosolids, which was the result of retention of CBs on organic matter sites in biosolids.

There was very little uptake of pentachlorophenol (PCP) by tall fescue, lettuce, carrot, and chile pepper. Detection of intact PCP was minimal in fescue and lettuce, indicating very little uptake. Intact PCP was not detected in carrot peel, foliage or pulp or in chile foliage and pods. Soil analysis indicated that PCP was readily degraded in the soil, thereby reducing potential uptake by plants.

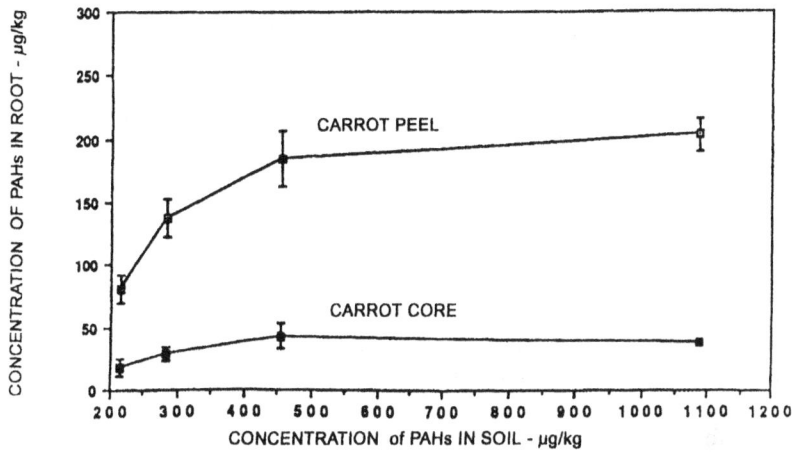

FIGURE 7.3. Concentrations of PAHs in carrot peels and cores. (Data from Wild and Jones, 1992.)

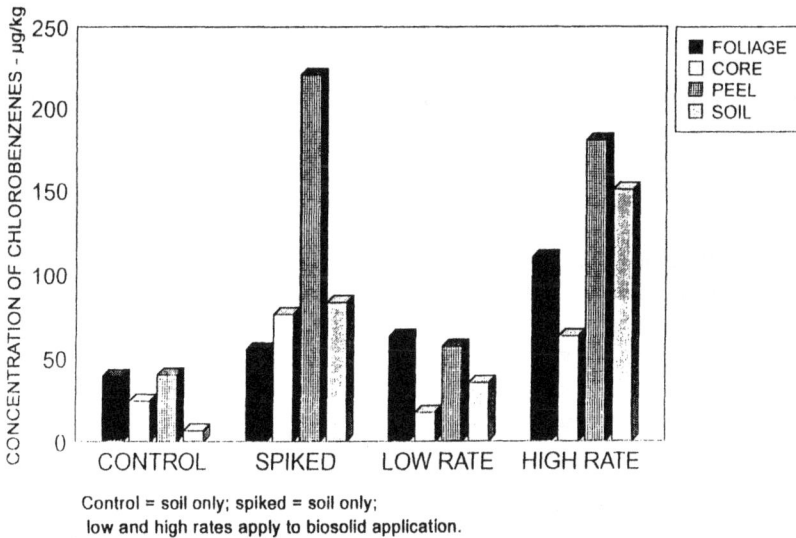

Control = soil only; spiked = soil only;
low and high rates apply to biosolid application.

FIGURE 7.4. Chlorobenzene concentration in different parts of carrots taken up from soil and biosolids-amended soil. (Data from Wang and Jones, 1994.)

205

Hulster et al. (1994) demonstrated that polychlorinated dibenzo-*p*-dioxins (PCDD) and dibenzofurans (PCDF) were taken up by zucchini and pumpkin from soil. All other plant species including cucumber were mainly contaminated from atmospheric deposition. Several hypotheses were provided to explain why the two plants took up PCDD/PCDF. One explanation that is being investigated is the release of root exudates, which may desorb PCDD/PCDF from the soil.

Muller (1976) investigated the uptake of 3,4-benzopyrene by carrots, red radish, and spinach from compost-amended soil. The amount of 3,4-benzopyrene in roots and leaves of carrots increased with increasing amounts of refuse compost incorporated into the soil. When the substrate contained 2 ppm 3,4-benzopyrene, 3.5 ppb was detected in red radish tubers and 20 ppb in the leaves. The authors concluded that the use of composted MSW may slightly increase the 3,4-benzopyrene content in food plants but the levels found did not represent any risk to humans.

CONCLUSION

Any composting feedstock may contain some level of toxic organic compounds. For example, pesticides and herbicide residues may be present in yard and food waste. These wastes may also be contaminated as a result of atmospheric deposits from incinerators or other combustion systems. MSW may contain toxic organics as a result of household disposal or components of plastic and other materials. Biosolids often contain toxic organics as a result of domestic and industrial disposal of various organic compounds. Many of the toxic organic compounds used in the past are very persistent in the environment with long half-lives; consequently, their residues may be detectable for a long time even though they are no longer manufactured or used.

Compost has been shown to be an effective method of microbial degradation of many toxic organics. Thus, many organisms are capable of decomposing toxic organics and use the C as an energy source. Many of the studies reviewed here were conducted under laboratory conditions, which may not have optimized the system for microbial degradation of organic compounds. One of the best examples of large-scale composting of toxic organics is the biodegradation of military wastes from World War II. Research is needed to develop techniques for cost-effective composting of recalcitrant organic compounds. This may include inoculation of specific organisms and providing the environment that sustains them.

The research data on plant growth on soil and biosolids indicate that there is little or no uptake of toxic organics by plants; however, little or no information is available on uptake of toxic organics from compost. The use of spiked samples can lead to erroneous results since the binding organics in the media studied affect the potential uptake. Data on uptake of toxic organics from compost are needed.

REFERENCES

Ahmed, M. and D. D. Focht. 1972. Degradation of polychlorinated biphenyls by two species of *Achromobacter. Can. J. Microbiol.* 19:47–52.

Ames, B. N. and L. S. Gold. 1991. Cancer prevention strategies greatly exaggerate risks. *Chemical and Eng. News.* January 7, pp. 27–32.

Aranda, J. M., G. A. O'Conner, and G. A. Eiceman. 1989. Effects of sewage sludge on di-(2-ethylhexyl)phthalate uptake by plants. *J. Environ. Qual.* 18:45–50.

Arthur, M. F. and J. I. Frea. 1989. 2,3,7,8-Tetrachlorodibenzeno-*o*-dioxin: Aspects of its important properties and its potential biodegradation in soils. *J. Environ. Qual.* 18:1–11.

Barr, D. P. and S. D. Aust. 1994. Mechanisms white rot fungi use to degrade pollutants. *Environ. Sci. Technol.* 28(2):78a–87A.

Borgas, M., M. Wilken, B. Zeschmar-Lahl, and J. Jager. 1990. PCDD/PCDF balance of different municipal waste management methods IV. Recycling. Paper presented at *Dioxin 90 Conference.* Bayreuth, Germany.

Boyle, M. 1989. The environmental microbiology of chlorinated aromatic decomposition. *J. Environ. Qual.* 18:395–402.

Brunner, W., F. H. Sutherland, and D. D. Focht. 1985. Enhanced biodegradation of polychlorinated biphenyls in soil by analog enrichment and bacterial inoculation. *J. Environ. Qual.* 14:324–328.

Bumpus, J. A. and S. D. Aust. 1986. Biodegradation of environmental pollutants by the white rot fungus *Phanerochaete chrysosporium*: Involvement of the lignin degrading system. *BioEssays.* 6:166–170.

Bushnell, L. E. and H. F. Haas. 1941. The utilization of certain hydrocarbons by microorganisms. *J. Bact.* 41:1082–1089.

Chaney, R. L. 1984. Potential effects of sludge-borne heavy metals and toxic organics on soils, plants, and animals, and related regulatory guidelines. *Proc. Workshop on the International Transportation, and Utilization or Disposal of Sewage Sludge, Including Recommendations.* Pan American Health Organization, Washington, D.C.

Chang, A. C. and A. L. Page. 1984. *Fate of Wastewater Constituents in Soil and Groundwater: Trace Organics.* California State Water Resources Control Board, Sacramento, CA.

Chiou, A. C. 1989. Theoretical considerations of the partition uptake of nonionic organic compounds by soil organic matter. In B. L. Sawhney and K. Brown (eds.). *Reactions and Movement of Organic Chemicals in Soils.* Soil Sci. Soc. Amer., Madison, WI.

Commoner, B., T. Webster, and K. Shapiro. 1987. Environmental levels and health effects of PCDDs and PCDFs. *5th International Symposium on Chlorinated Dioxins and Related Compounds.* Bayreuth, West Germany.

Davie, I. N. 1993. Compostability of petroleum wax-based coating. *Tappi J.* 76:167–170.

Deever, W. R. and R. C. White. 1978. *Composting Petroleum Refinery Sludges.* Texaco, Inc., Port Arthur, TX.

E&A Environmental Consultants, Inc. 1988. *Crow Wing Pilot Study.* Crow Wing Planning and Zoning Board, Brainard, MN.

Edgehill, R. U. and R. K. Finn. 1983. Microbial treatment of soil to remove pentachlorophenol. *Appl. Environ. Microbiol.* 45:1122–1125.

Epstein, E. and J. E. Alpert. 1980. Composting of industrial wastes, pp. 243–252. In R. Pojasek (ed.). *Toxic and Hazardous Waste Disposal.* Ann Arbor Science, Ann Arbor, MI.

Epstein, E. and W. J. Grant. 1968. Chlorinated insecticides in runoff water as affected by crop rotation. *Soil Sci. Amer. Proc.* 32:423–426.

Fricke, G. F., W. Pertle, and H. Vogtmann. 1989. Technology and undesirable components on compost of separately collected organic wastes. *Agric. Ecosystems and Environ.* 27:463–469.

Fries, G. F. and G. S. Marrow. 1981. Chlorobiphenyl movement from soil to soybean plants. *J. Agric. Food Chem.* 29:757–759.

Goldfarb, T. D. 1991. Comparing the risk of MSW composting and incineration—A rebuttal to Dr. Jones. Letter sent to Dr. Joel Alpert of E&A Environmental Consultants, Inc.

Greenland, D. J. and J. M. Oades. 1975. Saccharides, pp. 213–261. In J. E. Gieseking (ed.). *Soil Components. Volume I. Organic Components.* Springer-Verlag, New York.

Griest, W. H., R. L. Tyndall, A. J. Stewart, C. H. Ho, D. E. Katon, and W. M. Caldwell. 1990. *Characterization of Explosives Processing Waste Decomposition Due to Composting.* U.S. Army Medical Research and Development Command, Fort Detrick, MD.

Gross M. D. 1994. Fate and toxic effects of selected organic compounds during composting of primary and waste activated sludge. M.S. Thesis, Tulane U., New Orleans, LA.

Guenzi, W. D. and W. E. Beard. 1967. Anaerobic biodegradation of DDT and DDD in soil. *Soil Sci.* 156:1116–1117.

Guenzi, W. D. and W. E. Beard. 1968. Anaerobic conversion of DDT to DDD and aerobic stability of DDT in soil. *Soil Sci. Soc. Amer. Proc.* 32:522–524.

Harrad, S. J., T. A. Malloy, T. A. Khan, and T. D. Goldfarb. 1991. Levels and sources of PCDDs, PCDFs, chlorophenols (CPs) and chlorobenzenes (CBs) in composts from a municipal yard waste composting facility. *Chemosphere.* 23(2):181–191.

Helling, C. S., A. R. Isensee, E. A. Woolson, P. D. J. Ensor, G. E. Jones, J. R. Plimmer, and P. C. Kearney. 1973. Chlorodioxins in pesticides, soils, and plants. *J. Environ. Qual.* 2:171–178.

Herrera Environmental Consultants Inc., Sound Resource Management Group and Woods End Research Laboratory. 1991. *Compost Testing and Analysis: Final Report.* Seattle Solid Waste Utility.

Hornick, S. B., R. H. Fisher, and P. A. Paolin. 1983. Petroleum wastes. In J. F. Parr, P. D. Marsh, and J. M. Kla (eds.). *Land Treatment of Hazardous Wastes.* Noyes Data Corporation, Park Ridge, NJ.

Hulster, A., J. F. Muller, and H. Marschner. 1994. Soil-plant transfer of polychlorinated dibenzo-*p*-dioxins and dibenzofurans in vegetables of the cucumber family. *Environ. Sci. Technol.* 28:1110–1115.

Isaacson, P. J. and B. L. Sawhney. 1983. Sorption and transformation of phenols on clay surfaces: effect of exchangeable cations. *Clay Miner.* 18:253–265.

Iwata, Y. and F. A. Gunther. 1976. Translocation of the polychlorinated biphenyl aroclor 1254 from soil into carrots under field conditions. *Archives Environ. Contamination and Toxicol.* 4:44–59.

Jacobs, L. W., G. A. O'Conner, M. R. Overcash, M. J. Zabik, and P. Rygiewicz. 1987. Effects of trace organics in sewage sludge on soil-plant systems assessing their risk to humans, pp. 101–143. In A. L. Page, T. J. Logan and J. A. Ryan (eds.). *Land Application of Sludge, Food Chain Implications.* Lewis Publishers, Inc., Chelsea, MI.

Kaufman, D. D. 1983. Fate of organic compounds in land applied wastes, pp. 77–151. In J. F. Parr, P. D. Marsh, and J. M. Kla (eds.). *Land Treatment of Hazardous Wastes.* Noyes Data Corp., Park Ridge, NJ.

Kilbane, J. J., D. K. Chatterjee, and A. M. Chakrabarty. 1983. Detoxification of 2,4,5-trichlo-rophenoxyacetic acid from contaminated soil by *Pseudomonas cepacia. Appl. Environ. Microbiol.* 45:1697–1700.

Kuhn, E. P. and J. M. Suflita. 1989. Dehalogenation of pesticides by anaerobic microorganisms in soils and groundwater—A review. In B. L. Sawhney and K. Brown (eds.). *Reactions and Movement of Organic Chemicals in Soils.* SSSA Special Pub. No. 22. Soil Sciences Society of America, Madison, WI.

Lahl, U., M. Wilken, B. Zeschmar-Lahl, and J. Jager. 1990. PCDD/PCDF balance of different municipal waste management methods I. Municipal waste incinerators (MWI). Paper presented at *Dioxin 90 Conference.* Bayreuth, Germany.

Lahl, U., M. Wilkin, B. Zeschmar-Lahl, and J. Jager. 1991. PCDD/PCDF balance of different municipal waste management methods. *Chemosphere.* 23(8–10):1481–1489.

Leonard, R. A., G. W. Bailey and R. R. Swank. 1976. Transport, detoxification, fate and effects of pesticides in soil and water environments, pp. 48–78. In *Land Application of Waste Materials.* Soil Conservation Soc. Amer., Ankeny, IA.

Lisk, D. J., W. H. Gutemann, M. Rutzke, H. T. Kuntz, and G. Chu. 1992. Survey of toxicants and nutrients in composted waste materials. *Arch. Environ. Contam. Toxicol.* 22:190–194.

Malcome Pirnie. 1990. *Compost Health Risk Assessment Model. Report to the Minnesota Pollution Control Agency.* St. Paul, MN.

Marinucci, A. C. and R. Bartha. 1979. Biodegradation of 1,2,3- and 1,2,4,-trichlorobenzene in soil and liquid enrichment culture. *Appl. Environ. Microbiol.* 38:811–817.

Martens, R. 1982. Concentration and microbial mineralization of four to six ring polycyclic aromatic hydrocarbons in composted municipal wastes. *Chemosphere.* 11:761–770.

Miller, T. L., R. R. Swager, S. G. Wood, and A. D Adkins. 1992. *Selected Metal and Pesticide Content of Raw and Mature Compost Samples from Eleven Illinois Facilities.* ILENR/RR-92-09. Illinois Department of Energy and Natural Resources, Springfield, IL.

Mueller, U., B. Hoer, B. Zeschmar-Lahl, M. Wilken, and J. Jager. 1990. PCDD/PCDF balance of different municipal waste management methods. II. Waste disposal and disposal gas incineration. Paper presented at *Dioxin 90 Conference.* Bayreuth, Germany.

Muller V. H. 1976. Aufnahme von 3,4-benzopyrene durch Nahrungspflanzen aus kunstlich angereicherten Substraten. *Z. Pflanzenernaehr. Bodenkd.* 6:686–695.

National Audubon Society. 1993. *Wet Bag Composting Demonstration Project.* National Audubon Society, Procter & Gamble, International Process Systems, Waste Management, McDonalds, Greenwich Audubon Society, Town of Fairfield, CT, Connecticut Agricultural Experiment Station.

O'Conner, G. A., D. Kiehl, G. A. Eiceman, and J. A. Ryan. 1990. Plant uptake of sludge-borne PCBs. *J. Environ. Qual.* 19:113–118.

Overcash, M. R. 1983. Land treatment of municipal effluent and sludges: Specific organic compounds. *Workshop on Utilization of Municipal Wastewater and Sludges on Land.* Univ. of California, Riverside, CA.

Pertsova, R. A., F. Kunc, and L. A. Golovleva. 1984. Degradation of 3-chlorobenzoate in soil by pseudomonad carrying biodegradative plasmids. *Folia Microbiol.* 29:242–247.

Richard, T. and M. Chadsey. 1989. Environmental impact of yard waste composting. *BioCycle.* 31:42–46.

Rose, W. W. and W. A. Mercer. 1968. *Fate of Insecticides in Composted Agricultural Wastes.* National Canners Association, Washington, D.C.

Saber, D. L. and R. L. Crawford. 1985. Isolation and characterization of *Flavobacterium* strains that degrade pentachlorophenol. *Appl. Environ. Microb.* 50:1512–1518.

Sawhney, B. L. and K. Brown. 1989. *Reactions and Movement of Organic Chemicals in Soils.* Pub. No. 22. Soil Sci. Soc., Amer. Madison, WI.

Sawhney, B. L. and R. P. Kozloski. 1984. Organic pollutants in leachate from landfill sites. *J. Environ. Qual.* 13:349–352.

Shabad. L. M., Y. L. Cohan, A. P. Ilnitsly, Y. Khesina, N. P. Shcherbak, and G. A. Smirnov. 1971. The carcinogenic hydrocarbon benzo(a)pyrene in the soil. *J. Natl. Cancer Inst.* 47:1179–1191.

Snell Environmental Group Inc. 1982. *Rate of Biodegradation of Toxic Organic Compounds While in Contact with Organics Which Are Actively Composting.* National Science Foundation, Washington, DC.

Unterman, R., D. L. Bedard, M. J. Brennan, L. H. Bopp, F. J. Mondello, R. E. Brooks, D. P. Mobley, J. B. McDermott, C. C. Schwartz and D. K. Dietrich. 1988. Biological approaches for polychlorinated biphenyl degradation, pp. 253–269. In G. S. Omenn (ed.). *Environmental Biotechnology: Reducing Risks from Environmental Chemicals through Biotechnology.* Plenum Press, New York.

U.S. Environmental Protection Agency. 1990. National sewage sludge survey: Availability of information and data, and anticipated impacts on proposed regulations. *Federal Register.* 55:47210–47283.

U.S. Environmental Protection Agency. 1985. *Evaluation and Estimation of Potential Risks of Polynuclear Aromatic Hydrocarbons.* Carcinogenic Assessment Group, Office of Health and Environmental Assessment, Washington, DC.

Wang, M. and K. C. Jones. 1994. Uptake of chlorobenzenes by carrots from spiked and sewage sludge-amended soil. *Environ. Sci. Technol.* 28:1260–1267.

Webber, M. D. and S. Lesage. 1989. Organic contaminants in Canadian municipal sludges. *Waste Mgt. & Res.* 7:63–82.

Weber, J. W. and C. T. Miller. 1989. Organic chemical movement over and through soil. In B. L. Sawhney and K. Brown (eds.). *Reactions and Movement of Organic Chemicals in Soils.* Pub. No. 22. Soil Science Soc. America, Inc., Madison, WI.

Wild, S. R. and K. C. Jones. 1992. Polynuclear aromatic hydrocarbon uptake by carrots grown in sludge-amended soil. *J. Environ. Qual.* 21:217–225.

Wilken, M. F. Neugebauer, B. Zeschmar-Lahl, and J. Jager. 1990. PCDD/PCDF balance of different municipal management methods III: Composting. Paper presented at the *Dioxin 90 Conference.* Bayreuth, Germany.

Williams, R. T., P. S. Ziegenfuss, and P. J. Marks. 1988. Field Demonstration—Composting of Explosives Contaminated Sediments at the Louisiana Ammunition Plant (LAAP). U.S. Army Toxic and Hazardous Materials Agency. Report No. AMXTH-IR-TE-88242. Aberdeen, MD.

Zeschmar-Lahl, B., M. Wilken, and J. Jager. 1990. PCDD/PCDF balance of different municipal waste management V: Comparison and results. Paper presented at the *Dioxin 90 Conference.* Bayreuth, Germany.

CHAPTER 8

Pathogens

INTRODUCTION

A human pathogen is any virus, microorganism or substance capable of causing disease (Stedman's Medical Dictionary, 1976). There are two broad categories (1) primary pathogens and (2) secondary or opportunistic pathogens. The primary pathogens can invade and infect healthy persons, whereas a secondary pathogen invades and infects a debilitated person or an individual on immunosuppressant medication or drugs.

Exposure to pathogens can occur during the composting process or through use of the product if the composting process was not carried out properly and the product, therefore, was not disinfected. The potential modes of infection for workers are inhalation of aerosols containing airborne microorganisms, dermal contact, or orally through inadvertent ingestion of dust or contaminated food or through other hand-to-mouth contact such as cigarette smoking.

Ingestion of a contaminated product or contamination by the product to cigarettes or food is the greatest potential source of pathogen invasion to workers or users. However, the risk to these individuals is low since normal behavior provides much greater risk from pathogenic organisms in the environment. Clark et al. (1980) evaluated the health hazards to 270 workers in four composting plants and did not find any health problems associated with primary pathogens.

Only the primary human pathogens will be addressed in this chapter. The secondary pathogens in compost are discussed in Chapter 9, Bioaerosols. Plant pathogens will be discussed in the chapter on compost use.

The USEPA 40CFR503 regulations dealing with biosolids and biosolids products include primary pathogens. Only the product is regulated. Thus, there are no regulations for worker health or health to populations living in areas surrounding composting facilities, nor are there regulations for MSW or yard waste composts that do not contain biosolids. Process monitoring requirements in the 503 regulations are designed to ensure pathogen reduction and vector control. This is done through time-temperature criteria for broad classes of composting technologies. Specific pathogen analysis must be done if the time-temperature criteria are not met. The following is an excerpt from the USEPA regulations (1994):

> Processes to Further Reduce Pathogens (PFRPs) Listed in Appendix B of 40CFR Part 503
>
> Composting
>
> Using either the within-vessel composting method or the static aerated pile composting method, the temperature of the biosolids is maintained at 55°C or higher for three days.
>
> Using the windrow composting method, the temperature of the biosolids is maintained at 55°C or higher for 15 days or longer. During the period when the compost is maintained at 55°C or higher, the windrow is turned a minimum of five times.

In addition, to satisfy Class A requirements, the following must be met (USEPA, 1994):

> Pathogen Requirements for all Class A Alternatives
>
> The following requirements must be met for *all* Class A pathogen alternatives.
>
> Either:
>
> - the density of fecal coliform in the biosolids must be less than 1,000 most probable number (MPN) per gram total solids (dry weight basis),
>
> or
>
> - the density of *Salmonella* sp. bacteria in the biosolids must be less than 3 MPN per 4 grams of total solids (dry weight basis).
>
> Either of these requirements must be met at one of the following times:
>
> > —when the biosolids are used or disposed;
> > —when the biosolids are prepared for sale or give-away in a bag or other container for land application; or
> > —when biosolids or derived materials are prepared to meet the requirements for EQ biosolids [see Chapter 2].
>
> Pathogen reduction must take place before or at the same time as vector attraction reduction, except when pH adjustment, percent solids, vector attraction, injection, or incorporation options are met.

Most states in the United States are adopting the USEPA regulations, with some states specifying more monitoring requirements. For example, the California Integrated Waste Management Board specifies depth of monitoring for the windrow, static pile, and other systems to ensure that the time-temperature criteria are met.

Although there are no specific USEPA or state regulations relating to worker health in composting facilities, a standard developed by U.S. Department of Labor, Occupational Safety and Health Administration (OSHA), pertains to workers who deal with solid waste. This is the General Industry Safety Order 29 CFR 1910.1030, "Bloodborne Pathogens" (Geyer, 1994). This standard especially pertains to two primary pathogens, human immunodeficiency virus (HIV) and hepatitis type B virus (HBV). This regulation may be applicable to recycling and composting facilities handling MSW.

The composting process is capable of disinfecting wastes. However, either due to poor design or poor operations some pathogens may survive the composting process (Yanko, 1988). This chapter will provide an understanding of composting as a waste management tool and show how composting allows beneficial reuse of biosolids or other wastes without compromising the health of facility workers, compost users, or residents adjacent to facilities.

Although composting is a pathogen-reduction process, poor facility design or operations can result in pathogen survival in final products (Yanko, 1988). The fate of these pathogens once compost is applied to soil, their survival, and the potential for transmission of diseases to humans are also covered in this chapter.

PRIMARY PATHOGENS IN WASTES AND COMPOST

Primary pathogens are found in various waste streams. The primary pathogens found in wastes fall into four major categories:

* bacteria
* enteric viruses
* protozoa
* helminths: nematodes (roundworms) and cestodes (tapeworms)

Most of the data are from biosolids, since little information is available on solid waste, yard waste, food wastes, and other waste streams. Tables 8.1a, 8.1b, and 8.1c show the various primary pathogens found in waste streams that are composted and the associated diseases. For composting to be effective for disinfection, it must be able to reduce the pathogens to a level of very low potential risk or eliminate them.

TABLE 8.1a. Some Bacteria Found in Biosolids and
the Diseases They Transmit.

Bacteria	Disease
Salmonella (approximately 1700 types)	Salmonellosis Gastroenteritis
Salmonella typhi	Typhoid fever
Mycobacterium tuberculosis	Tuberculosis
Shigella (4 species)	Shigellosis Bacterial dysentery Gastroenteritis
Escherichia coli	Gastroenteritis
Yersinia sp.	Yersinosis
Vibrio cholerae	Cholera

Sources: Epstein and Donovan, 1992; Akin et al., 1978; Ward et al., 1984.

TABLE 8.1b. Some Viruses Found in Biosolids and
the Diseases They Transmit.

Virus	Disease
Adenovirus (31 types)	Conjectivitis, respiratory infections, gastroenteritis
Poliovirus	Poliomyelitis
Coxsackie virus	Aseptic meningitis, gastroenteritis
Echovirus	Aseptic meningitis
Reovirus	Respiratory infections, gastroenteritis
Hepatitis virus	Infectious hepatitis
Rotavirus	Gastroenteritis, infant diarrhea

Sources: Epstein and Donovan, 1992; Ward et al., 1984.

TABLE 8.1c. Some Protozoa and Helminth Parasites Found in Biosolids and the Diseases They Transmit.

Organism	Disease
Protozoa	
Entamoeba histolytica	Amoebic dysentery, ameobiasis
Giardia lamblia	Giardiasis
Balantidium coli	Balantidiasis
Naegleria fowleri	Meningoencephalitis
Toxoplasma gondii	Toxoplasmosis
Helminths—Nematodes	
Ascaris lumbricoides	Ascariasis
Ancylostoma duodenale	Hookworm, Ancylostomiasis
Necator americanus	Hookworm
Ancylostoma braziliense (cat hookworm)	Cutaneous larva migrans
Ancylostoma caninum (dog hookworm)	Cutaneous larva migrans
Enterobius vermicularis (pinworm)	Enterobiasis

(continued)

TABLE 8.1c. (continued).

Organism	Disease
Helminths—Nematodes *(continued)*	
Strongyloides stercoralis (threadworm)	Strongyloidiasis
Toxocara cati (cat roundworm)	Visceral larva migrans
Toxocara canis (dog roundworm)	Visceral larva migrans
Trichuris trichiura (whipworm)	Trichuriasis
Helminths—Cestodes	
Taenia saginata (Beef tapeworm)	Taeniasis
Taenia solium (pork tapeworm)	Taeniasis
Hymenolepis nana (dwarf tapeworm)	Taeniasis
Echinococcus granulosus (dog tapeworm)	Unilocular echinococcosis
Echinococcus multi-locularis	Alveolar hydatid disease

Sources: Akin et al., 1978; Epstein and Donovan, 1992.

There are two principal sources of pathogens in MSW, domestic animal wastes and discarded diapers containing feces and urine. Gaby (1975) found high levels of fecal coliforms and fecal streptococcus organisms in raw MSW as well as five types of salmonellae. Figure 8.1 shows the levels of total coliforms, fecal coliforms and fecal streptococcus organisms in MSW, sewage sludge and raw refuse-sewage sludge mixtures prior to composting. As illustrated, there was little seasonal variation in the levels of indicator organisms or fecal streptococcus. Gaby (1975) also found that species of *Proteus* and coagulase-positive staphylococci were isolated in large numbers from raw refuse but were never found in raw biosolids. These organisms disappeared rapidly with increased compost temperatures. Parasites were also identified in biosolids.

Cooper and Golueke (1979) reported finding large numbers of indicator microorganisms in MSW, equalling or surpassing those found in raw sewage. This was attributed to domestic animal waste. Similar data were reported by Donnelly and Scarpino (1984).

The main sources of primary pathogens in MSW at a facility are diapers, discarded contaminated medical material from the home (band-aids, gauze, etc.), personal care products, tissues, toilet paper, feminine hygiene products, syringes, diabetic utensils, domestic animal fecal matter, and vectors (insects, fleas, ticks, rodents, flies).

Pahren (1987) reported that excreta of household pets (e.g., cat litter or dog feces placed in refuse containers) can contaminate the solid waste stream with human pathogens. Lundgren et al. (1968) found 164 viruses, including

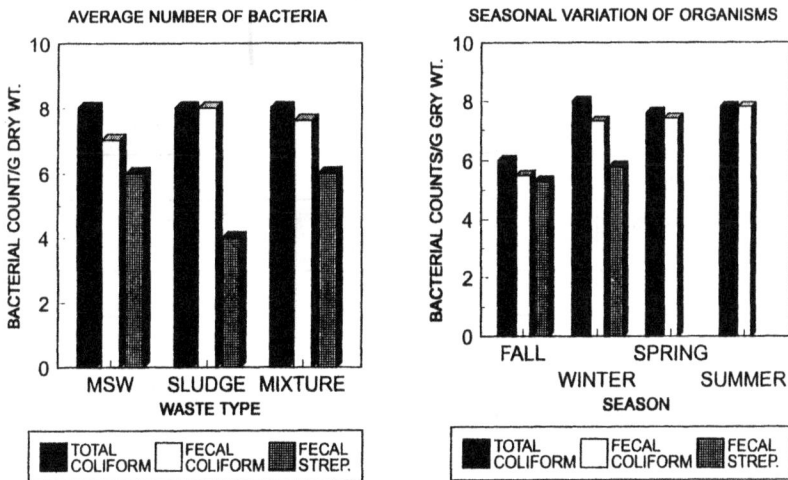

FIGURE 8.1. Bacteria in raw wastes prior to composting. (From Gaby, 1975.)

Coxsackie virus and echovirus collected from nasal, throat, and rectal swabs of beagle dogs. Most were identified as either Coxsackie viruses B1, B3, B4, or echoviruses type 6. Grew et al. (1970) noted poliovirus type 1, Coxsackie virus A-20, and adenovirus in human and animal stools. All of these are human pathogens.

Table 8.2 shows the levels of total coliforms, fecal coliforms and fecal streptococci bacteria in various components of the solid waste stream. The highest levels were found in paper and garden wastes. Pahren (1987) speculated that the high bacterial levels in paper could be from discarded facial tissue and wrappings containing domestic animal feces. Bacterial growth could also occur on food waste and paper while the refuse awaits collection.

The high levels of indicator organisms and pathogens in yard waste (Table 8.3) may be the result of fecal matter from domestic animals. Pathogens in food waste can come from eggs, raw chicken and other contaminated foods. Since the level of antagonistic organisms in yard and food wastes is much lower than the level found in biosolids, fecal coliform and fecal streptococ-

TABLE 8.2. Estimate of Bacteria in Various Components of Solid Waste.

Category	Total Coliforms (%)		Fecal Coliforms (%)		Fecal Streptococci (%)	
	1974	1975	1974	1975	1974	1975
Paper	66.4	33.2	29.5	13.5	14.6	51.4
Garden wastes	18.8	18.3	6.0	7.4	54.0	14.6
Metal	0.5	4.1	10.5	5.3	4.2	7.3
Glass	0.4	0.1	4.3	0.5	0.9	0.1
Food wastes	2.1	19.6	2.9	8.4	2.5	13.4
Plastic, rubber, leather	0.6	7.4	1.8	10.6	0.9	5.6
Textiles	6.0	3.1	1.6	4.0	1.4	0.5
Fines	4.8	8.2	42.6	4.1	16.4	2.6
Ash, rock, dirt	0.3	1.5	0.2	10.2	3.6	2.3
Diapers	0.06	4.5	0.6	35.9	1.3	2.2
Wood	0.01	0.0	0.0	0.1	0.2	0.0

Source: Pahren, 1987.

TABLE 8.3. Microbiological Concentration in Food Waste, Yard Waste, and Wood Wastes.

Organism	Food Waste	Yard Waste	Yard Waste and Waste Paper	Wood Waste
Indicator Organism				
Total coliform	5.00×10^6	8.0×10^5	5.00×10^5	1.30×10^6
Fecal coliform	2.00×10^4	8.00×10^5	5.00×10^5	1.30×10^6
Escherichia coli	3.50×10^3	8.00×10^5	3.00×10^5	1.30×10^6
Fecal streptococcus	8.00×10^6	1.60×10^6	1.60×10^6	1.60×10^6
Enterococcus	1.30×10^5	2.30×10^5	1.30×10^5	3.00×10^5
Pathogens				
Salmonella spp.	<0.002	<0.002	0.36	<0.002
Staphylococcus	32.2	0.8	4.4	3.8
Listeria spp.	<0.02	0.02	<0.02	<0.02
Parasites	Protozoa	negative	negative	negative

Source: E&A Environmental Consultants, Inc., 1994.

cus organisms can grow and reach high levels. For example, biosolids normally contain salmonellae organisms at a level of 10^2–10^3/g dry weight. After composting, these organisms are not detected. However, if a sterilized compost is reinoculated with salmonellae, the organism often reaches levels of 10^6/g dry solids or greater. Brandon et al. (1977) found that the repopulation of *Salmonella enteritidis* in sterile compost reached a level of 10^9 organisms per gram dry weight. However, salmonellae contamination of compost is generally not of concern since there is a rapid die-off within a short period of time (Burge et al., 1987).

Another source of pathogens in MSW is disposable diapers (Pahren, 1987; Huber et al., 1994). Disposable diapers may contain enteric viruses that are excreted in the feces of infants and young children (Huber et al., 1994). Peterson (1974) indicated that since infants are the most effective carriers of viruses, viruses found in diapers represent a potential health risk to persons handling solid waste and a potential pollution of groundwater.

In addition to identifying disease-causing organisms, it is important to evaluate the potential risk as related to dose response. The infective dose is not only specific to each pathogen, but individuals' susceptibility also varies (Cliver, 1980; Bryan, 1977; Kowal, 1982). Table 8.4 shows dose response relationships for several organisms. Workers in composting facilities handling raw sludge or MSW can be exposed to a variety of pathogens. As pointed out earlier, there is no evidence of worker infections or health problems, however.

Akin (1983) reviewed the literature on infective dose data for enteroviruses and other pathogens. The widest dose response range was found for enteric bacteria. *Salmonella* spp. required 10^5 to 10^8 cells to produce a 50% disease rate in healthy adults. Three species of *Shigella* produced illness in subjects administered 10 to 100 organisms. Administering small doses, 1 to 10 cysts, of *Entamoeba coli* and *Giardia lamblia* caused amoebic infections. Very low doses of enteric viruses were found to produce infection.

Administering various doses of *Salmonella typhi* to adult volunteers, Hornick et al. (1970) found that none of the 14 volunteers showed any symptoms when 1000 organisms were administered. Twenty-eight percent of the adults became ill when a dose of 100,000 organisms was administered, and 95% of the subjects were ill when 1,000,000,000 organisms were administered.

Kowal (1982) summarized the data on oral infective doses to man from enteric bacteria. No infection or illness was reported from several bacteria at the levels indicated below:

- *Escherichia coli* 10^4
- *Salmonella typhi* 10^3
- *Salmonella derby* 10^5-10^6
- *Streptococcus faecalis* var. *liquefaciens* 10^8

Pahren (1987) reviewed the literature on infective doses for bacteria and

TABLE 8.4. *Approximate Dose Response Relationship for Several Pathogens Found in Wastes.*

Pathogen	No. of Organisms to Produce Disease in 25–75% of Subjects Tested	No. of Organisms to Produce Disease in Any of the Subjects
Shigella spp.	10^2-10^5	10^1
Salmonella spp.	10^5-10^9	10^4
Escherichia coli	10^6-10^{10}	10^6
Vibrio cholerae	10^3-10^{11}	10^3

Source: Bryan, 1977.

viruses. In addition to the infective dose, other factors such as age and general health are important. Pahren (1987) stated "However, people do not live in a germ- nor risk-free society. Microorganisms are present almost everywhere—in the air, the soil, and on objects that people touch."

WORKER HEALTH RISKS OF SOLID WASTE COMPOSTING

No bacterial, viral, or parasitic diseases have been reported among workers engaged in composting, nor has there ever been reported or documented evidence of public health effects from wastewater treatment facilities. Since these facilities represent the most exposed individuals, evaluating those data can provide insight into the potential infection to compost workers handling raw wastes containing pathogens.

The potential health impact on workers in composting facilities is highest during the receiving and handling of the raw materials prior to composting. This may occur in the handling of raw or digested biosolids during transportation and mixing or from dust and aerosols around conveyor belts, mixers and other equipment. Workers in MSW and recycling facilities are exposed to pathogens during sorting, operating of equipment, and maintenance of equipment. The exposure is greatest on the tipping floor and the handling of waste during sorting.

In 1978, contractors working at a wastewater treatment plant were infected by *Giardia* cysts while handling raw sludge for a composting demonstration project. The mode of infection most likely was ingestion, which probably occurred due to poor hygienic practices.

Although Geyer (1994) listed a multitude of pathogenic organisms that may be found in solid waste, he did not indicate any data or evidence of worker infections. Table 8.5 provides a list of diseases and sources that may be associated with handling of solid waste.

Clark et al. (1976) reviewed the literature on disease risks to workers associated with sewage. The authors cited an evaluation of sewer workers in Berlin. No typhoid and paratyphoid fever was found. The study examined 430 sewer workers for shigellosis and found no incidence of the disease. Gastrointestinal disorders were reported for sewage farm workers in India (Krishnamoorthi et al., 1973; Clark et al., 1976).

Although bacteria and viruses were isolated from samples of sludge, effluent, and air, Sekla et al. (1980) noted few significant findings upon the examination of 77 sewage workers at two secondary sewage treatment plants in Canada. No pathogenic bacteria or viruses were isolated from throat swabs of personnel. As a group, the employees did not differ significantly from other Manitobans tested as controls.

TABLE 8.5. Diseases and Their Sources Associated with Municipal Solid Waste.

Disease	Source
Hepatitis B	Materials contaminated with human blood
HIV	Contamination with fresh human blood
Hepatitis A and E	Materials contaminated with fecal matter
Hemorrhagic fevers	Contaminated human blood; aerosols or rodent excreta; bites by ticks, fleas, and other vectors from infected animals
Meningococcal	Aerosols from respiratory discharges
Arboviruses	Arthropod-borne viral diseases transmitted by ticks or mosquito bites
Tetanus	Spores from contaminated soil or animal and human fecal material
Infectious pulmonary tuberculosis	Respiratory secretions from humans and cattle
Multiple drug-resistant tuberculosis	Respiratory secretions from humans and cattle
Staphylococcus	Respiratory, gastroenteritis, skin infections
Diplococcus	Pneumonia, meningitis
Escherichia coli	Diarrhea
Klebsiella pneumonia	Respiratory infections

Source: Geyer, 1994.

Several studies have been conducted to determine the health effects of populations surrounding wastewater treatment plants (Johnson et al., 1978; Fannin et al., 1978; Carnow et al., 1979). These studies involved the communities of Chicago, Illinois; Tecumseh, Michigan; and Skokie, Illinois. None of the studies found evidence of serious adverse health effects on the various populations. Some evidence was found in the Chicago study of minor skin diseases, nausea, weakness, diarrhea, and respiratory discomfort to workers (Johnson et al., 1978).

No incidence of diseases has been reported from workers dealing with solid waste at different types of composting facilities. Workers at biosolids composting facilities, especially those working with raw biosolids, need to exercise prudent hygienic practices to reduce potential infection. Individuals engaged in handling solid waste at various facilities [e.g., landfills, transfer stations, collection and transportation equipment, materials recycling facilities (MRF), solid waste composting facilities, and tipping floors at incinerator and refuse-derived fuel plants] should maintain certain hygienic and health prevention practices.

These include:

* tetanus vaccinations
* hepatitis B vaccination
* dust control
* personal protective equipment (dust masks, gloves, outer gear, etc.)
* cleaning facilities

Supervisory personnel should maintain good housekeeping and cleanliness of working areas. Personnel should be instructed on hygiene and be discouraged from eating or smoking except in areas where they can wash their hands.

THE EFFECT OF COMPOSTING ON PATHOGEN DESTRUCTION

The principal mode of disinfection of wastes through composting is based on temperature-time relationships that destroy pathogens. One of the best examples of this relationship is the pasteurization of milk. Heating of milk to 60–63°C (140–145°F) for 20 to 30 minutes results in pasteurization and destruction of pathogenic bacteria.

Table 8.6 shows the temperature-time required for the destruction of several pathogens in biosolids (Stern, 1974). The destruction of poliovirus in 40% compost solids as related to temperature-time is shown in Table 8.7. At 35°C poliovirus survived much longer than at 47°C. Data on heat inactivation of total coliforms, fecal coliforms, fecal streptococcus, and *Salmonella enteritidis* serotype Montevideo also showed great reduction of organisms when the temperature exceeded 55 to 65°C. Data on heat inactivation of coliform bacteria, fecal streptococcus and *Salmonella enteritidis* are shown in Figures 8.2, 8.3, and 8.4 (Ward and Brandon, 1977).

The temperature-time relationship is the basis of the USEPA biosolid "Processes to Further Reduce Pathogens" as stated in 40CFR Part 503. During composting, temperature variation can occur throughout the mass. Therefore, USEPA as well as many states specify different temperature-time criteria for different composting systems.

Based on a review of some of the early literature on pathogen destruction by composting, Wiley (1962) indicated that pathogen destruction during composting is the result of thermal kill and antibiotic action or the decomposing organisms or their products.

TABLE 8.6. Temperature-Time Required for
Pathogen Destruction in Biosolids.

Organism	Exposure Time (in minutes) for Destruction at Various Temperatures				
	50°C	55°C	60°C	65°C	70°C
Entamoeba histolytic cysts	5				
Ascaris lumbricoides eggs	60	7			
Brucella abortus		60		3	
Corynebacterium diphtheriae		45			4
Salmonella typhi			30		4
Escherichia coli			60		5
Micrococcus pyogenes var. *aureus*					20
Mycobacterium tuberculosis					20
Shigella spp.	60				
Mycobacterium diphtheria	45				
Necator americanus	50				
Taenia saginata					5
Viruses					25

Source: Stern, 1974.

TABLE 8.7. Inactivation of Poliovirus in Composted
Biosolids at 60% Moisture.

Treatment	Percentage Recovery of Plaque-Forming Units
35°C, 200 minutes	30
39°C, 20 minutes	7.2
43°C, 20 minutes	0.087
47°C, 5 minutes	0.003

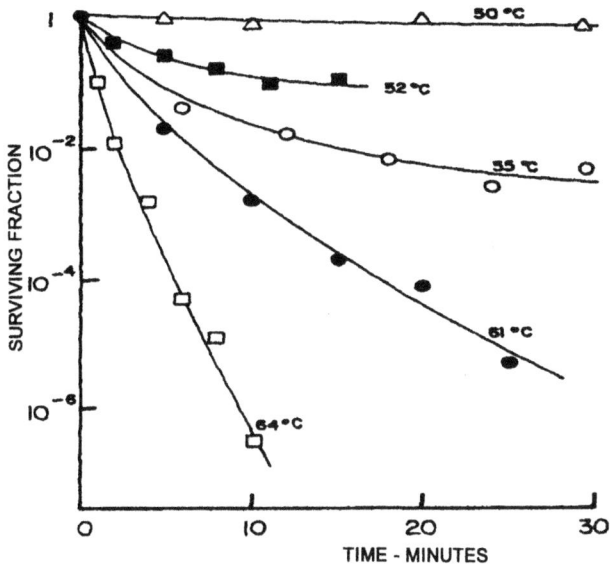

FIGURE 8.2. Heat inactivation of coliform bacteria in composted biosolids. (Data from Ward and Brandon, 1977.)

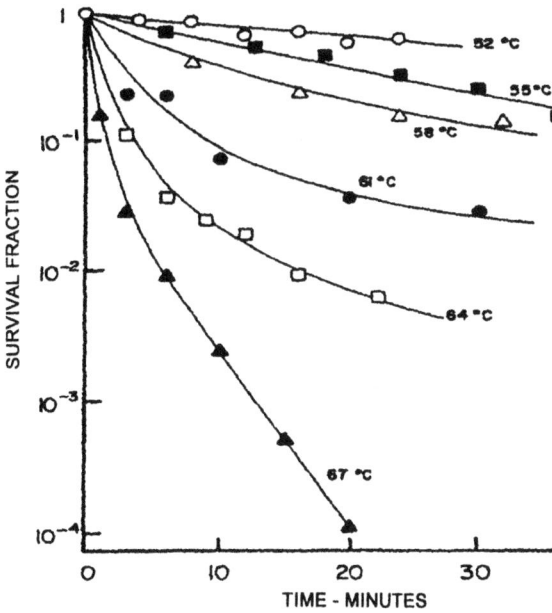

FIGURE 8.3. Heat inactivation of fecal streptococcus bacteria in composted biosolids. (Data from Ward and Brandon, 1977.)

227

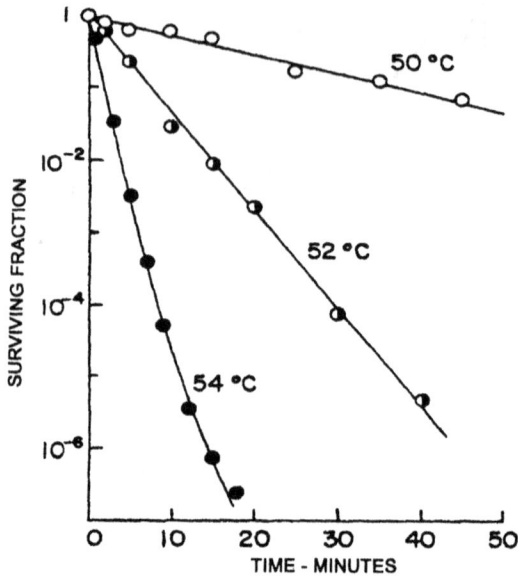

FIGURE 8.4. Heat inactivation of *Salmonella enteritidis* serotype Montivideo in composted biosolids. (Data from Ward and Brandon, 1977.)

The theory that antibiotic substances besides temperature result in pathogen destruction by composting was tested by Knoll (1961), who extracted a solution from composted material at different stages of the process. In an extract taken between the 7th and 16th day, no inoculated *Salmonella cairo* were able to grow. He indicated that the development of inhibitors originated in the presence of actinomycetes and molds and concluded that this phenomenon is due to an unknown antibiotic-producing organism. Although the optimum temperature that produces the antibiotic was not determined, Knoll (1961) indicated that 50–55°C appeared to be the temperature at which these substances are generated.

Gaby (1975) found no oppressive or antagonistic material in compost. Golueke (1983) pointed out that indigenous organisms are in a better position to compete for nutrients than pathogenic microorganisms. Furthermore, he indicated that since thermal destruction is not instantaneous, time acts as a factor, providing for the combination of several inhibitory factors to act on pathogenic organisms.

Pathogenic microbial antagonism has been studied by several researchers (Brandon and Neuhauser, 1978; Millner et al., 1987). For example, Millner et al. (1987) studied the growth suppression of salmonellae by compost microflora. They indicated that the types and amounts of different microorganisms affect the growth of salmonellae organisms. The presence of coliforms only or metabolically active bacteria and actinomycetes resulted in the death of salmonellae in compost.

Knoll (1961) described several experiments where he subjected different salmonellae strains to composting temperatures at the Baden-Baden biosolids-refuse composting plant. After 14 days of reactor time with temperatures of 55–60°C and a moisture content of 40–60%, the product did not contain pathogens.

Morgan and Macdonald (1969) investigated the fate of *Mycobacterium tuberculosis* during open windrow composting of biosolids and refuse. The research was conducted at the U.S. Public Health Service–Tennessee Valley Authority Research and Demonstration Compost Plant at Johnson City, Tennessee. These researchers found that the organism was destroyed after 10 days of composting when the temperature averaged 60°C. They also noted that if the temperatures remained low, the bacteria survived for long periods of time until the temperature exceeded 44°C for long periods of time. Equipment used to handle compost that was less than 17 days old should not be used with finished compost to prevent reintroduction of pathogens into the finished product.

Gaby (1975) reported on a series of studies on pathogen reduction during windrowing of refuse-biosolids composting. Salmonella and shigella either originally present or introduced into refuse-biosolids mixtures were not found within 7 to 21 days. Enteroviruses were not found in the raw refuse, the biosolids or a mixture of the two materials. Poliovirus Type 2 was introduced into the windrows but were inactivated after three to seven days. Human parasitic cysts and ova were introduced into the center of the windrow but they disintegrated after seven days. However, dog parasitic ova survived for 35 days. *Leptospira philadelphia*, a spirochaete, did not survive for more than two days after introduction into the windrow.

Krogstad and Gudding (1975) inoculated solid waste and biosolids with *Salmonella typhimurium, Serratia marcescens* and *Bacillus cereus* and made periodic measurements to determine the die-off rate. The data for *Salmonella typhimurium* are shown in Figure 8.5. The organism could not be detected after four days when the temperature in a horizontal drum composter was kept around 65°C. The researchers concluded that three to five days in a reactor vessel with temperatures of 60–65°C would destroy the pathogens.

Walke (1975) monitored *Escherichia coli, Salmonella eidleberg*, and *Candia albicans* during windrow composting of bark-biosolid mixtures. The initial compost contained these organisms at a level of 10^6 microbes per dry gram. After 24 hours the levels were 11, 130 and 620 microbes per dry gram of solids for *E. coli, Salmonella* sp., and *Candida albicans,* respectively. No organisms were detected after 36 hours.

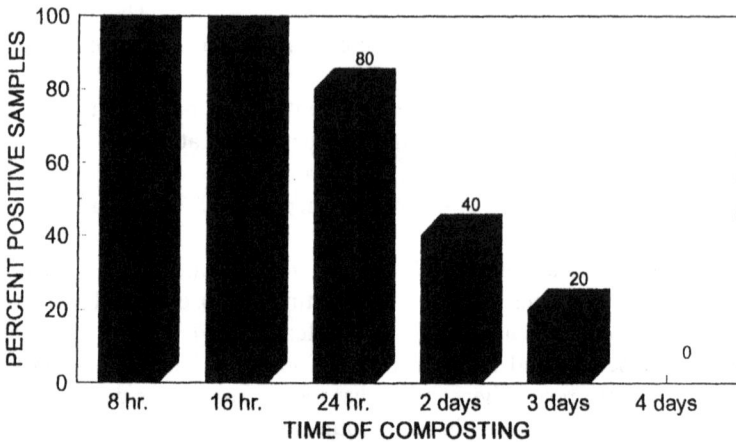

FIGURE 8.5. Effect of composting on survival of Salmonella typhimurium. (Data from Krogstad and Gudding, 1975.)

From 1973 to 1978, the U.S. Department of Agriculture conducted numerous studies on pathogens' survival during composting by both the windrow and the aerated static pile method (Burge et al., 1978; Burge and Cramer, 1974; Epstein et al., 1976). The results showed that salmonellae increased in growth initially but was destroyed within 10 days of composting in the static pile and 15 days in the windrow method (Figures 8.6 and 8.7). The destruction of an indicator virus, f2 bacteriophage, took 45 to 70 days in the windrow method and approximately 13 days using the aerated static pile method. This indicator virus was selected because it was more resistant to inactivation by heat than enteric pathogens including viruses, bacteria, protozoa cysts and helminth ova.

Cooper and Golueke (1979) found that during the first 43 to 46 days of windrow composting, 81 to 92% of the total coliforms were eliminated; 77 to 96% of fecal coliform organisms were eliminated; and 98 to 99.8% of the fecal streptococci were eliminated. The levels were 10^2, 10^2, and 10^4 organisms for total coliforms, fecal coliforms, and fecal streptococci, respectively, after 35 to 40 days of composting.

Pereira-Neto et al. (1986) evaluated the efficiency of the aerated static pile method in destroying the *Escherichia coli*, fecal streptococci, and salmonellae in a refuse-biosolids mixture. Salmonellae were destroyed in 7 to 15 days; *Escherichia coli* decreased from 10^7 to $<10^2$ in 15 days; and fecal streptococci decreased from 10^7 to $<10^2$ in 30 days.

In a second study, Pereira-Neto et al. (1986) found that the aerated static pile method was more effective than the windrow system for pathogen

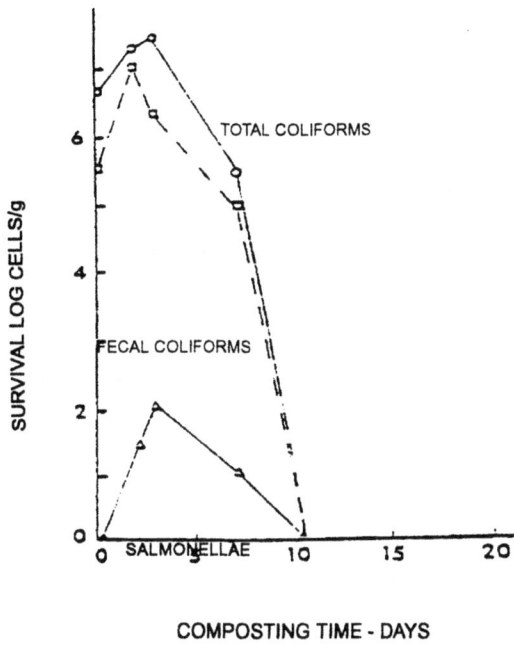

FIGURE 8.6. Destruction of salmonellae, fecal coliform, and total coliforms during aerated static pile composting of raw biosolids. (Data from Burge et al., 1978.)

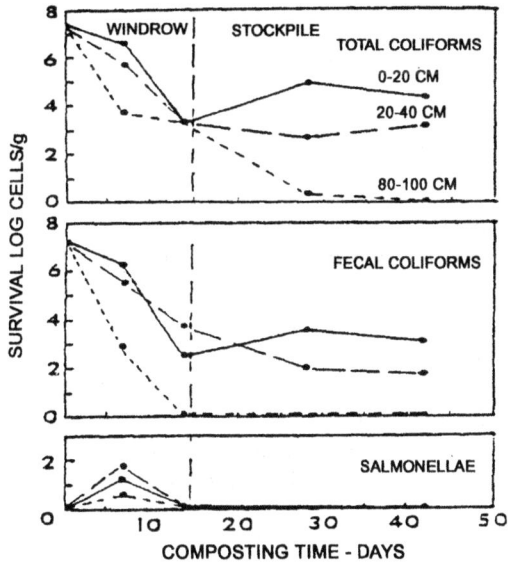

FIGURE 8.7. Destruction of salmonellae, fecal coliforms, and total coliforms at three depths during windrow composting of digested biosolids. (From Burge and Cramer, 1974.)

231

inactivation. *Escherichia coli*, fecal streptococci and salmonellae were evaluated in both windrows and static piles. In the static pile, salmonellae and *Escherichia coli* were not detected after 16 days of composting. After 32 days, fecal streptococci were below 10^2 colony forming units/gram wet weight (cfu/gww). Salmonellae and *Escherichia coli* were still present in windrows after 60 days and fecal streptococci were present at levels of 10^2–10^3 cfu/gww.

Over a number of years the city of Windsor, Ontario, Canada, has been composting biosolids using the aerated static pile method. During that time considerable testing has been performed on the pathogenic microorganisms in biosolids and the resultant compost. Generally, in the biosolids total coliform, fecal coliform and fecal streptococci were found at levels up to 5.5 $\times$ 10^8, 5.1 $\times$ 10^7, 3.4 $\times$ 10^6 organisms, respectively. The compost contained <1000/100 g of these organisms. No *Salmonella*, *Shigella*, or *Yersinia enterocolitica* were found in the biosolids or compost. *Salmonella derby* and *Salmonella oranienburg* were isolated in the biosolids. None were isolated in the compost after 28 days of composting. At later dates *Salmonella hadar*, *Salmonella infantis*, and *Salmonella oranienburg* were isolated in the biosolids but none was found in the resultant compost.

Yanko (1988) evaluated pathogen levels in compost from several static pile and windrow composting operations (Table 8.8). The study showed that indicator organisms and salmonella were present in many of the compost products. A review of the data in relation to the facilities clearly indicated that facilities that were operated properly (e.g., Washington Suburban Sanitary Commission Site II in Maryland) had excellent pathogen kill. By contrast, poorly run facilities where temperature control was inadequate or where recontamination occurred showed indicator or pathogenic organisms. Some facilities were able to modify their operations and produce a disinfected product.

One major concern over the years has been the potential for regrowth of salmonellae since that organism can survive and grow without a human or animal host. Regrowth can result from recontamination of the compost within the windrow or pile if organisms survived in cooler portions or if the compost was contaminated.

Figure 8.8 shows that fecal streptococcus bacteria grow rapidly in sterilized compost since many of their antagonisms are not present. High levels were also obtained in compost saturated with *Salmonella* spp. In nonsterilized compost, regrowth was restricted by competition for nutrients (Ward and Brandon, 1977).

Russ and Yanko (1981) studied the repopulation of *Salmonella* sp. in previously composted sewage sludge. They found that growth occurred in the mesophilic temperature range of 20 to 40°C when moisture content was

TABLE 8.8. Salmonellae Detected in Compost from
Various Composting Systems.

Compost Sampling Location	Composting System	Number of Organisms Detected	Number of Samples Measure	Percent Detected in Samples
LACSD bulk[1]	Windrow	3	54	6
LACSD bag[2]	Windrow	9	52	17
LACSD BA[3]	Windrow	58	102	57
LACSD BA[3]	Windrow	36	52	69
Philadelphia	Static pile	23	45	51
Philadelphia	Static pile	36	45	80
I-B-1	Static pile	0	6	0
III-B-3	Static pile	0	6	0
III-B-4	Static pile	1	6	17
IV-B-1	Static pile	3	6	50
V-B-1	Static pile	0	6	0
IX-B-1	Static pile	1	6	17
II-C-1	In-vessel	1	6	17
X-C-1	In-vessel	1	6	17
III-J-1	Aerated windrow	0	6	0
VIII-J-1	Aerated windrow	0	6	0
VII-A-2	Windrow	0	6	0
IX-A-2	Windrow	0	6	0
VIII-H-1	Proprietary	1	6	17

[1]Los Angeles County Sanitation Districts, bulk.
[2]Bagged compost.
[3]BA = bulking agent, woodchips, or rice hulls.
Source: Yanko, 1988.

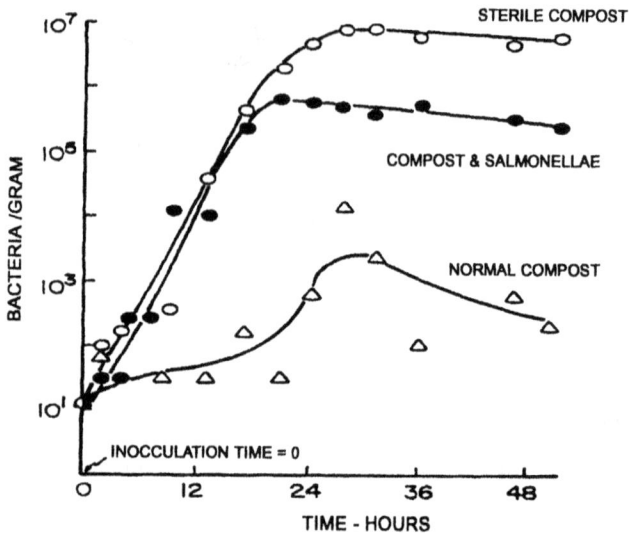

FIGURE 8.8. Growth of fecal streptococcus bacteria in normal composted biosolids, in sterilized composted biosolids, and in composted biosolids saturated with *Salmonella* spp. Following sterilization of the compost, *Salmonella* spp. were allowed to grow to approximately 100 million per gram. (Data from Ward and Brandon, 1977.)

greater than 20% and the carbon/nitrogen (C/N) ratio was greater than 15:1. This happened in spite of the presence of competing organisms. Repopulation of salmonellae was transient, with peak growth occurring after five days followed by subsequent die-off.

The authors stated that the moisture and nutrient variables were controlled within a fairly tight range. There was no statistical evaluation and the ranges for total solids, volatile solids and C/N ratio were extremely close. Total solids ranged from 67.2 to 70.9; volatile solids from 14.3 to 17.7; and C/N ratio from 13.1 to 15.1. Considering the methodologies and sample variability, one can only distinguish some trends from these data.

Burge et al. (1987) also studied the potential for regrowth of salmonellae in composted biosolids. Sampling biosolids compost from 30 municipalities, these researchers found that 12% of the samples contained salmonellae. The organism, however, did not survive unless the compost was sterilized. The authors indicated that the indigenous microflora suppressed the growth of salmonellae. When bacteria, actinomycetes, fungi, and protozoa, all typical inhabitants of a stabilized but not sterilized compost, were introduced into sterile compost they fully suppressed the growth of salmonellae. Based on these data, Burge et al. (1987) concluded that the potential for salmonellae regrowth was negligible.

Microbial suppression of salmonellae in compost is the result of: (1) growth suppression (i.e., inhibition resulting in reduced growth rate) and (2) death (Millner et al., 1987). Salmonellae suppression did not occur in compost subjected to 70°C. This finding was attributed to the lack of antagonistic microorganisms. Compost subjected to 55°C that had large numbers and kinds of microbes suppressed salmonellae. Millner et al. (1987) concluded that with proper curing, negligible regrowth of salmonellae occurs. Curing at mesophilic temperatures encourages the growth of numerous microbes that would be antagonistic to salmonellae.

These data clearly indicate that proper composting can be very effective in the disinfection or sanitization of compost. It is also evident that on occasion or through poor management practices survival of pathogens in the compost is possible. What happens if a contaminated product is applied to land? The data on survival and movement of organisms in soils as a result of contaminated compost are meager. However, numerous studies have examined pathogens in soils as a result of biosolids application. A brief discussion of this subject follows.

SURVIVAL OF PATHOGENS IN SOILS AND ON PLANTS

The survival of pathogens in soils depends on several physical parameters. Soil moisture, soil moisture retention characteristics, pH, organic matter, soil colloidal matter, soil temperature, and competitive or antagonistic organisms all affect pathogen survival in soils. Further, organisms on the surface of soils are destroyed by temperature, desiccation, and ultraviolet light. Alexander et al. (1991) reported that ionic strength of the inflowing water, bacterial density, and velocity of flow influenced bacterial movement.

Reviews by Akin (1983), Rudolfs et al. (1950), Golueke (1983), Lance (1977) and Sorber and Moore (1986) have shown that pathogenic organisms may survive in soils and on plants from several days to weeks or even years. Data on the survival of pathogens in or on soils and plants are shown in Tables 8.9, 8.10, and 8.11.

Table 8.10 shows data on the survival of viruses in soils and on some vegetables. Although Larkin et al. (1978) indicated that poliovirus 1 survived on lettuce and radishes for 14–36 days, most (99%) was lost the first five to six days. Wellings et al. (1975) found that virus survives in the soil for at least 28 days. It can be adsorbed by the soil, desorbed, and moved with subsurface water.

In a laboratory study, Duboise et al. (1976) investigated the die-off rate of poliovirus in a forest soil as related to temperature. At 4°C the virus was reduced less than 1 log unit in 84 days whereas at 20°C a 5-log reduction

TABLE 8.9. Survival of Bacteria in Soils or Plants.

Organism	Media	Survival (days)
Salmonella sp.	Soil	15–7280
Salmonella typhimurium	Soil	<28–70
Salmonella typhi	Soil Vegetables, and fruits	30–120 <1–68
Salmonella cerro	Tomatoes	4
Coliform	Soil surface Vegetables Grass and clover	38 35 6–34
Streptococci	Soil	35–63
Fecal streptococci	Soil	23–67
Shigella	Vegetables	2–10
Shigella alkalescens	Tomatoes	6
Tubercle bacilli	Soil	>180

Sources: Rudolfs et al., 1950, 1951; Sepp, 1971; Golueke, 1983.

TABLE 8.10. Survival of Some Viruses in Soils and Plants.

Virus	Media	Survival (days)
Poliovirus	Sand	<77–91
Poliovirus 1	Loamy fine sand Soil Lettuce and radishes	84 <123 36
Poliovirus 1,2,3	Soil	78
Coxsackie 83	Clay	<161
Enteroviruses	Soil Vegetables	8 4–6

Sources: Golueke, 1983; Tierney et al., 1977; Parsons et al., 1975; Duboise et al., 1976.

TABLE 8.11. Survival of Some Parasites in Soil and Plants.

Parasite	Media	Survival (days)
Entamoeba histolytica	Soil	8–10
Ascaris ova	Soil Vegetables and fruit	up to 7 years 27–35
Hookworm larvae	Soil	<180

Sources: Golueke, 1983; Parsons et al., 1975.

took place in the same period of time. Damgaard-Larsen et al. (1977) used lysimeters to study the survival and movement of enteroviruses in soil as a result of biosolid application. It took 23 weeks during a Danish winter to inactivate Coxsackie virus seeded in a sand and clay soil. The die-off rate was 0.5–1.0 log units per month when seeded at 4.5 log concentrations per 0.1 ml.

Most of the bacteria survived in soils or on plants for relatively short periods of time. *Salmonella* sp. survived for nearly 20 years. Virus survived in soils for three to five months. Ascaris ova were viable for up to seven years in soil. Much of the literature is from studies where irrigated effluent was applied to soils. It is difficult to extrapolate those data to compost or biosolids applied as a semi-solid since movement through soils is a function of water movement. Data on virus survival in field soils are more limited. Most of the literature focuses on movement of viruses through soils. The predominance of data is from column studies using liquids.

Data on the survival of pathogens from biosolids applied to land could be indicative of survival when compost containing pathogens is applied to soil.

Based on a review of the literature prior to 1986, Sorber and Moore (1986) summarized the data on survival of microorganisms as related to biosolids application to soil. Their data are shown in Table 8.12.

Moisture is a key factor in the survival of organisms in soil. Beard (1940) reported that the survival of *Salmonella typhosa* was greatest during the rainy season. In sandy soil, which dried out rapidly due to its low water retention, survival time was between 4–7 days. By comparison, in loam soils or peat, which retained water, survival time was over 42 days.

Beard (1940) found that in a peat soil with a low pH of 3–4 survival time was very short. Similarly, Cuthbert et al. (1950) reported that *E. coli* and *Streptococcus faecalis* when inoculated into peat (pH 2.9–4.5) died out in a few days compared to a limestone soil (pH 5.8–7.8) where the organisms survived for several weeks.

TABLE 8.12. Summary of Data on the Survival of Microorganisms from Biosolids Applied to Soil.

Organism	Soil Depth (cm)	Die-off Rate (days, 90%)			Die-off Rate (days, 99%)		
		Min.	Max.	Med.	Min.	Max.	Med.
Salmonella	0–5	6	61	12	11	45	22
	5–15	4	22	15	7	45	30
Fecal streptococci	0–5	7	28	17	14	63	24
	5–15	NA	NA	NA	NA	NA	NA
Fecal coliform	0–5	7	84	25	12	165	60
	5–15	4	49	16	9	56	32
Total coliform	0–5	16	72	40	28	350	155
	5–15	35	70	42	NA	NA	NA
Viruses	0–5	<1	30	3	2	52	6
	5–15	30	56	30	60	100	60
Parasites	0–5	17	270	77	68	500	81
	5–15	NA	NA	NA	NA	NA	NA

Notes: Results of numerous observations.
Min. = minimum; max. = maximum; med. = median.
Source: Sorber and Moore, 1986.

In addition to moisture, soil temperature and exposure to sunlight have been shown to affect pathogen survival (Beard, 1940). Van Dorsal et al. (1967) reported that fecal coliform survival in the summer was 3.3 days whereas in the winter the organism survived up to 13.4 days. Rudolfs et al. (1951) studied the effect of sewage contamination of vegetables on the survival of several bacteria. The vegetables were sprayed with suspensions containing the organisms. *Salmonella cerro* survived for less than four days. By the third day, the reduction was 97.5%. *Shigella alkalescens* survived on tomatoes from three to six days. On the fourth day the reduction in organisms was 99.5%.

Bryanskaya (1966) showed that actinomycetes in soil suppressed the growth of salmonella and dysentery bacilli. The presence of antagonists or suppressive material in soil is supported by the evidence that in sterilized soil pathogens grow to much higher numbers than in unsterilized soil.

No information is available on the presence of viruses in compost or their survival when compost is applied to soil. Available data are primarily from sewage effluent or biosolids applied to soil. Several reviews have been conducted on the fate of viruses in soil and on crops (Bitton, 1975; Gerba et al., 1975).

Viruses are electrically charged protein particles. As such they have the potential to adsorb on clays, organic matter and other colloidal surfaces. Adsorption of viruses increases their survival potential. Therefore, virus migration is dependent on the soil's capacity to adsorb viruses.

Bitton (1975) indicated that virus removal by soils is primarily the result of adsorption. Other factors such as salt concentration, pH, organic matter, soil compaction, and water flow rates can affect the extent of retention of viruses. Duboise et al. (1976) showed that ionic strength and pH of soil water greatly affected poliovirus adsorption. The studies were conducted using sewage effluent.

There are no data showing survival of parasites in compost. According to data from Windsor, Ontario, when parasites were found in biosolids they were not detected in the compost. Parasites can survive in soils for long periods of time (Table 8.10). Specifically, Parsons et al. (1975) noted that *Ascaris* ova survived in soil for up to seven years. However Sorber and Moore (1986) found that 90% of *Ascaris* ova were inactivated in 17–24 days. After applying biosolids in the fall, however, the survival rate extended to 65 days. Data on *Toxocara* ova showed that 90% were inactivated in 77 days.

A summary of Sorber and Moore's data from the application of biosolids to soil is shown in Table 8.11. The lower survival rate for many pathogens was lower at the upper surface than at lower depths. This was probably due to exposure to sunlight or the upper layer drying out.

Between 1987 and 1989, 30 patients in South Australia were diagnosed as being infected by *Legionella longbeachae*. Epidemiological investigations by the South Australian Communicable Disease Control Unit indicated that gardening may be the environmental source and therefore a major risk factor. Steele et al. (1990a) isolated organisms resembling *Legionella longbeachae* from a number of potting mixes and soil surrounding plants collected from the home of four patients. In a subsequent study, Steele et al. (1990b) found *Legionella longbeachae* serogroup I and other *Legionella* spp. in potting soils in Australia. The organisms were found in 73% of 45 potting soils produced by 13 manufacturers but were not detected in 19 potting soils from Greece, Switzerland, and the United Kingdom. *Legionella* spp. were found in composted pine bark, fresh pine sawdust, and composted pine and eucalyptus sawdust. The organism survived in potting mixes from 3 to 10 months at temperatures from −20°C to 35°C but not at 43°C.

Hughes and Steele (1994) evaluated the presence of *Legionella* spp. in compost produced from large-scale composting facilities and from home composts (Table 8.13). Populations of legionellae ranged from 1×10^3 to 5×10^5 CFU/g. Similar levels were found in home compost. In contrast to the previous study, the predominant species was *Legionella pneumophila*, not *Legionella longbeachae*. The latter organism was found in small numbers in home compost and in only two of six large-scale facilities. Most of the *Legionella* species found in composts have not been shown to cause infections in South Australia.

An article titled "Trouble in the potting shed" (*Lancet,* 336:151, 1990) indicated that *Legionella* has been found in many parts of the world in association with diverse natural waters, cooling towers, evaporative condensers, humidifiers and domestic water supplies. Rowbotham (1980) found

TABLE 8.13. Detection of Legionella *in Compost Produced from Large-Scale Facilities and from Backyard Composting.*

Source of Compost	Test	No. (%) Samples with Legionella Species		
		Present	Not Detected	Overgrown
Compost from large-scale facilities	Initial	21 (64)	5 (15)	7 (21)
	Final	28 (85)	3 (9)	2 (6)
Home compost	Initial	34 (42)	19 (24)	27 (34)
	Final	45 (56)	24 (30)	11 (14)

Source: Hughes and Steele, 1994.

that *Legionella* multiply inside soil amoeba. There are no data on whether this organism survives the composting process. However, it appears that composting temperatures exceeding 43°C would destroy it.

None of these studies indicated the conditions and methodology of composting. As shown earlier, improper composting results in the survival of pathogens. The finding that the compost of only two of six facilities had *Legionella* suggests that these two facilities did not reach proper temperatures or otherwise contaminated the compost. No data are presented on the water used in composting either. If the water was contaminated, it could have contaminated the compost.

The data presented here clearly show that pathogens survive both in soils and on plants. Although the survival rate for most organisms in soils or on plants is relatively short, products that are applied to soils where humans or animals can ingest them should be free of pathogens or the pathogen levels should be extremely low. One major advantage of properly composting wastes is that pathogens are destroyed and the potential for contamination of soils or plants, therefore, is very low.

CONCLUSIONS

Many of the wastes that are being composted contain human pathogenic organisms. As expected, biosolids and municipal solid waste contain pathogens, but yard waste and food waste can also contain significant levels. In yard waste the major source of pathogens is domestic animal feces. In food waste eggs, chicken parts and other contaminated sources can result in significant levels of pathogens.

Composting, if carried out properly, is very effective in destroying pathogens. This is primarily the result of temperature-time relationships. However, other factors contribute to the demise of pathogens such as antagonistic organisms and ammonia. Regrowth of salmonellae is not a serious problem as these organisms die very quickly during curing and storage.

Studies of workers at composting facilities have not shown an increase in health risk related to composting activities. In order to protect workers from potential health risks, however, workers should be instructed about proper hygienic practices. Further, facilities should be designed and operated to minimize dust and exposure to pathogens through inhalation, and safety standards and protective equipment should be provided to reduce the risk of dermal contact and ingestion of pathogens.

Only the data from South Australia have indicated that application of compost has resulted in disease. In addition to the effectiveness of the composting process in disinfecting wastes, most pathogenic organisms survive for relatively short periods of time in soils and on plants.

REFERENCES

Akin, E. W. 1983. Infective dose of waterborne pathogens. In *Proc. 2nd Nat'l. Symp. Municipal Wastewater Disinfection.* Orlando, FL. EPA-600/9-83-009. Health Effects Res. Lab., Cincinnati, OH.

Akin, E. W., W. Jakubowski, J. B. Lucas, and H. R. Pahren. 1978. Health hazards associated with wastewater effluent and sludge: Microbiological consideration, pp. 9–26. In B. P. Sagik and C. A. Sorber (eds.). *Risk Assessment and Health Effects of Land Application of Municipal Wastewater and Sludge.* Univ. of Texas, San Antonio, TX.

Alexander, M., R. J. Wagenet, P. C. Baveye, J. T. Gannon, U. Mingelgrin, and Y. Tan. 1991. *Movement of Bacteria through Soil and Aquifer Sand.* USEPA, Robert S. Kerr Environ. Res. Lab. Ada, OK. EPA/600/S2-91/010.

Beard, P. J. 1940. Longevity of *Eberthella typhosus* in various soils. *Am. J. Public Health.* 30:1077–1082.

Bitton, G. 1975. Adsorption of viruses onto surfaces in soil and water. *Water Res.* 9:473–484.

Brandon, J. R., W. D. Burge, and N. E. Enkiri. 1977. Inactivation by ionizing radiation of *Salmonella enteritidis* serotype montevideo growth in composted sewage sludge. *Applied. Environ. Microbiol.* 33:1011–1012.

Brandon, J. R. and K. S. Neuhauser. 1978. *Moisture Effects on Inactivation and Growth of Bacteria and Fungi in Sludges.* Pub. No. SAND 78-1304. Sandia Lab., Albuquerque, NM.

Bryan, F. L. 1977. Disease transmitted by foods contaminated by wastewater. *J. Food Protection.* 40:45–52.

Bryanskaya, A. M. 1966. Antagonistic effect of actinomyces on pathogenic bacteria in soil. *Hygiene and Sanitation.* 31:123–125.

Burge, W. D. and W. N. Cramer. 1974. *Destruction of Pathogens by Composting Sewage Sludge.* Progress Report. Aug. 1, 1973 to April 1, 1974. Joint Project USDA Agricultural Res. Ser., Maryland Environmental Ser. and Water Resources Management Administration, Washington, DC.

Burge, W. D., W. N. Cramer, and E. Epstein. 1978. Destruction of pathogens in sewage sludge by composting. *Transactions of the ASAE.* 21(3):510–514.

Burge, W. D., P. D. Millner, N. K. Enkiri, and D. Hussong. 1987. *Regrowth of Salmonellae in Composted Sewage Sludge.* Project Summary. U.S. Environmental Protection Agency, Water Engineering Research Lab., Cincinnati, OH. EPA/600/S2-86/106.

Carnow, B., R. Northrop, R. Wadden, S. Rosenberg, J. Holden, A. Neal, L. Sheaf, and S. Meyer. 1979. *Health Effects of Aerosols Emitted from an Activated Sludge Plant.* Health Effects Res. Lab., Office of Res. and Dev. USEPA. Report 600/1-79-019.

Clark, C. S., E. J. Cleary, G. M. Schiff, C. C. Linnemann, Jr., J. P. Phair, and T. M. Briggs. 1976. Disease risks of occupational exposure to sewage. *J. Env. Eng. Div. of Amer. Soc. of Civil Eng.* 102:375–388.

Clark, C. S., H. S. Bjornson, J. W. Holland, T. L. Huge, V. A. Majeti, and P. S. Gartside. 1980. Occupational hazards associated with sludge handling, pp. 215–244. In C. Bitton, B. L. Damron, G. T. Edds, and J. M. Davidson (eds.). *Sludge—Health Risks of Land Application.* Ann Arbor Science Publishers Inc., Ann Arbor, MI.

Cliver, D. O. 1980. Infection with minimal quantities of pathogens from wastewater aerosols, pp. 78–89. In H. Pharen and W. Jakubowski (eds.). *Wastewater, Aerosols and Disease. Proc. of a Symposium.* EPA/600/9-80-028.

Cooper, R. C. and C. G. Golueke. 1979. Survival of enteric bacteria and viruses in compost and its leachate. *Compost Sci./Land Util.* 20(2):29–35.

Cuthbert, W. A., J. J. Panes, and E. C. Hill. 1950. Survival of *Bacterium coli* type 1 and *Streptococcus faecalis* in soil. *J. Applied Bacteriology.* 18:408–414.

Damgaard-Larsen, S., K. O. Jensen, E. Lund, and B. Nissen. 1977. Survival and movement of enterovirus in connection with land disposal of sludges. *Water Res.* 11:503–508.

Donnelly, J. A. and P. V. Scarpino. 1984. *Isolation, Characterization, and Identification of Microorganisms from Laboratory and Full-Scale Landfills.* EPA-600/2-84-119. Municipal Environ. Res. Lab., Cincinnati, OH.

Duboise, S. M., B. E. Moore, and B. P. Sagik. 1976. Poliovirus survival and movement in a sandy forest soil. *Applied and Environ. Microb.* 31(4):536–543.

E&A Environmental Consultants, Inc. 1994. *Food Waste Collection and Composting Demonstration Project for City of Seattle Solid Waste Utility.* Final Report.

Epstein, E. and J. F. Donovan. 1992. Pathogens in composting and their fate. In *Pathogens in Sludge: What Does It Mean?* Water Environ. Federation. Preconf. Seminar, New Orleans, LA.

Epstein, E., G. B. Willson, W. D. Burge, D. C. Mullen and N. K. Enkiri. 1976. A forced aeration system for composting wastewater sludge. *J. Water Pollut. Control Fed.* 48:688–694.

Fannin, K. F., K. W. Cochran, H. Ross, and A. S. Monto. 1978. *Health Effects of a Wastewater Treatment System.* Health Effects Res. Lab., Office of Res. and Dev., USEPA Report 600/1-78-062.

Gaby, W. L. 1975. *Evaluation of Health Hazards Associated with Solid Waste/Sewage Sludge Mixtures.* EPA-670/2-75-023. Nat'l. Environ. Res. Center, Office of Res. and Devel., USEPA, Cincinnati, OH.

Gerba, C. P., C. Wallis, and J. L. Melnick. 1975. Fate of wastewater bacteria and viruses in soil. *ASCE. J. Irr. Drain. Div.* 101(3):157–174.

Geyer, M. D. 1994. Bloodborne pathogens and solid waste. *Solid Waste Association of North America (SWANA) Nat'l. Conf.,* San Antonio, TX.

Golueke, C. G. 1983. Epidemiological aspects of sludge handling and management. Part two. *BioCycle.* 24(3):52–58.

Grew, N. R. Gohd, J. Argueas, and J. Kato. 1970. Enteroviruses in rural families and their domestic animals. *Am. J. Epidemiol.* 91:518.

Hornick, R. B., S. E. Greisman, T. E. Woodward, H. L. Dupont, A. T. Dawkins, and M. J. Snyder. 1970. Typhoid fever: Pathogenesis and immunological control. *New England J. Med.* 283(13):686–691.

Huber, M. S., C. P. Gerba, M. Abbaszadegan, J. A. Robinson, and S. M. Bradford. 1994. Study of persistence of enteric viruses in landfilled disposable diapers. *Environ. Sci. Technol.* 28:176–1772.

Hughes, M. S. and T. W. Steele. 1994. Occurrence and distribution of *Legionella* species in composted plant material. *Applied Environ. Microbiol.* 60(6):2003–2005.

Johnson, D. E., D. E. Camann, J. W. Register, R. J. Prevost, J. B. Tillery, R. E. Thomas, J. M. Taylor, and J. M. Rosenfeld. 1978. *Health Implications of Sewage Facilities.* Health Effects Res. Lab., Office of Res. and Dev., USEPA Report 600/1-78-032.

Knoll, K. H. 1961. Public health and refuse disposal. *Compost Sci.* 2:35–40.

Kowal, N. E. 1982. *Health Effects of Land Treatment: Microbiological.* EPA/600/1-82-007. Health Effect Research Laboratory, Cincinnati, OH.

Krishnamoorthi, K. P., M. K. Abdulappa, and A. K. Anwikar. 1973. Intestinal parasitic infections associated with sewage farm workers with special reference to helminths and protozoa pp. 347–355. *Proc. Symp. on Environ. Pollution,* India.

Krogstad, O. and R. Gudding. 1975. The survival of some pathogenic micro-organisms during reactor composting. *Acta Agric. Scandinavica.* 25:281–284.

Lance, J. C. 1977. Fate of pathogens in saturated and unsaturated soils. In *1977 Nat'l Conf. on Composting of Municipal Residues and Sludges.* Info. Transfer Inc. and Hazardous Materials Control Res. Institute, Silver Spring, MD.

Larkin, E. P., J. T. Tierney, J. Lovett, D. Van Donsel, and D. W. Francis. 1978. Land application of sewage wastes: Potential for contamination of foodstuffs and agricultural soils by viruses, bacterial pathogens, pp. 102–115. In B. P. Sagic and C. A. Sorber (eds.). *Risk Assessment and Health Effects of Land Application of Municipal Wastewater and Sludges.* Univ. of Texas at San Antonio, San Antonio, TX.

Lundgren, D. L., W. E. Clapper, and A. Sanchez. 1968. Isolation of human enteroviruses from beagle dogs. *Proc. Soc. Exp. Bio. Med.* 128:463–466.

Millner, P. D., K. E. Powers, N. K. Enkiri, and W. D. Burge. 1987. Microbial mediated growth suppression and death of *Salmonella* in composted sewage sludge. *Microb. Ecol.* 14:255–265.

Morgan, M. T. and F. W. Macdonald. 1969. Tests show MB tuberculosis doesn't survive composting. *J. Environ. Health.* 32(1):101–108.

Pahren, H. R. 1987. Microorganisms in municipal solid waste and public health implications. *CRC Critical Reviews in Environmental Control.* 17(3):187–228.

Parsons, D. C., C. Brownlee, D. Wetter, A. Maurer, E. Haughton, L. Kordner, and M. Slezak. 1975. *Health Aspects of Sewage Effluent Irrigation.* Pollution Control Branch, British Columbia Water Resources Service, Victoria, BC.

Peterson, M. L. 1974. Soiled disposable diapers a potential source of viruses. *Am. J. Public Health.* 64(9):912–914.

Pereira-Neto, J. T., E. I. Stentiford, and D. V. Smith. 1986. Survival of faecal indicator micro-organisms in refuse/sludge composting using the aerated static pile system. *Waste Management & Res.* 4:397–406.

Rowbotham, T. J. 1980. Preliminary report on the pathogenicity of *Legionella pneumophila* for fresh water and soil amoebae. *J. Clin. Pathol.* 33:1179–1183.

Rudolfs, W., L. L. Falk, and R. A. Ragotzkie. 1950. Literature review of the occurrence and survival of enteric pathogenic, and relative organisms in soil, water, sewage, and sludge, and on vegetation. *Sewage and Industrial Wastes.* 22(10):1261–1281.

Rudolfs, W., L. L. Falk, and R. A. Ragotzkie. 1951. Contamination of vegetables grown in polluted soil. *Sewage and Industrial Wastes.* 23(3):253–268.

Russ, C. F. and W. A. Yanko. 1981. Factors affecting salmonellae repopulation in composted sludges. *Applied and Environ. Microb.* 41(3):597–602.

Sekla, L., D. Gemmill, J. Manfreda, M. Lysyk, W. Stackiw, C. Kay, C. Hopper, L. VanBuckenhout, and G. Eibisch. 1980. Sewage treatment plant workers and their environment: A health study. *Proc. of a Symposium. Wastewater, Aerosols and Disease.* EPA-600/9-80-028.

Sepp, E. 1971. *The Use of Sewage for Irrigation—A Literature Review.* State of California Dept. Pub. Health, Bureau of Sanitary Engineering, Sacramento, CA.

Sorber, C. A. and B. E. Moore. 1986. Survival and transport of pathogens in sludge-amended soil. In *Proceedings of the National Conference on Municipal Treatment Plant Sludge Management.* Orlando, FL.

Stedman's Medical Dictionary. 1976. The Williams & Wilkins Co., Baltimore, MD.

Steele, T. W., J. Lanser, and N. Sanger. 1990a. Isolation of *Legionella longbeachae* serogroup I from potting mixes. *Applied Environ. Microb.* 56(1):49–53.

Steele, T. W., J. A. Lanser, and N. Sanger. 1990b. Distribution of *Legionella longbeachae* serogroup I and other *Legionella* in potting soil. *Applied Environ. Microbiol.* 56(10):2984–2988.

Stern, G. 1974. Pasteurization of liquid digested sludge. In *Proc. of the Nat'l. Conf. on Composting Municipal Sludge Management.*

Tierney, J. T., R. Sullivan, and E. Larkin. 1977. Persistence of poliovirus 1 in soil and on vegetables grown in soils previously flooded with inoculated sewage sludge or effluent. *Applied and Environ. Microb.* 33(3):109–113.

United States Environmental Protection Agency. 1994. *A Plain English Guide to the EPA Part 503 Biosolids Rule.* EPA/832/R-93/003. Office of Wastewater Management (4204), Washington, DC.

Van Dorsal, D. J., E. E. Geldreich, and N. A. Clarke. 1967. Seasonal variations in survival of indicator bacteria in soil and their contribution to storm-water pollution. *Applied Microbiol.* 15:1362–1370.

Vogt, G. and J. Walsh. 1984. *Evaluation of Landfilled Municipal and Selected Industrial Wastes.* Final Report. Contract No. 68-03-2758. Municipal Environmental Lab., USEPA, Cincinnati, OH.

Walke, R. 1975. The preparation, characterization and agricultural use of bark-sewage compost. Ph.D. Diss. Univ. of New Hampshire, Durham, NH.

Ward R. L. and J. R. Brandon. 1977. Effect of heat on pathogenic organisms found in wastewater sludge, pp. 122–134. In *1977 National Conf. on Composting of Municipal Residues and Sludges.* Information Transfer, Inc., Rockville, MD.

Ward, R. L., G. A. McFeters, and J. G. Yeager. 1984. *Pathogens in Sludge: Occurrence, Inactivation, and Potential Regrowth.* Sandia Report. Sand83-0557.TTC-0428. Sandia Nat'l Lab., Albuquerque, NM.

Wellings, F. M., A. L. Lewis, C. W. Mountain, and L. M. Stark. 1975. Virus consideration in land disposal of sewage effluent and sludge. *Florida Scientist.* 38(4):202–207.

Wiley, J. S. 1962. Pathogen survival in composting municipal wastes. *J. Water Poll. Control Fed.* 34:80–90.

Yanko, W.A. 1988. *Occurrence of Pathogens in Distribution and Marketing.* Report No. EPA/1-87/014 (NTIS PB #88-154273/AS).

CHAPTER 9

Bioaerosols

INTRODUCTION

Compost bioaerosols are organisms or biological agents that can be dispersed through the air and affect human health. Bioaerosols can contain living organisms including bacteria, fungi, actinomycetes, arthropods, and protozoa as well as microbial products such as endotoxin, microbial enzymes, β-1,3-glucans, and mycotoxins (Millner et al., 1994).

Table 9.1 lists the sources of bioaerosols as related to several common feedstocks used in composting. Many bioaerosols are present in our daily environment including our homes and places of employment. During composting, bioaerosols are not only present in the waste materials but can also be generated during the process. Bioaerosols can affect the health of workers or residents adjacent to composting facilities. The level and type of bioaerosols are a function of the feedstock. Level of bioaerosols in the atmosphere of a facility or in the air surrounding a facility can also be a function of facility design and operations.

Another term that encompasses bioaerosols is "aerobiology." This term refers to the study of airborne bioaerosols, especially pollens and spores, and their impact on living organisms (Lacey, 1991). Inhalation of fungal spores and fungal metabolites can produce respiratory symptoms. Several factors illicit response, including the immunological state of the receptor, the fungal species present, and extent of exposure. Many fungi have been implicated in occupational asthma and other pulmonary infections.

Table 9.2 shows some of the fungi associated with occupational illnesses (Lacey, 1991). It is obvious that workers in different occupations are exposed to a variety of fungal spores that can inflict allergies, asthma and other respiratory illnesses.

TABLE 9.1. Compost Feedstocks as Sources of Bioaerosols.

Feedstock	Most Important Biohazards							Other Specific Biohazards
	Af	ET	TA	Enteric Bact. and Virus	Fungi	Bact.	Mites	
Yard waste (including grass clippings, brush, leaves, wood chips)	+	+	+	−	+	+	+	Pollen, terpines, resins, lectins, phenols, toxalbumin
Food/household waste (MSW)	+	+	−	+	+	+	+	
Food processing, winery waste (pomace), and cheese	+	+	+	+	+	+	+	
Fisheries (shellfish)	+	+	+	+	+	+	+	Allergens of crustaceans
Agricultural wastes (cotton gin and textiles, garden waste, sugarcane/pineapple)	+	+	+	+	+	+	+	Deep mycoses, pollen, terpenes, resins, lectins
Biosolids	+	+	−	+	+	+	+	
Animal wastes								
Carcasses	−	+	−	+	−	−	−	
Manures	−	+	−	+				
Barnyard Manure	+	+	+	+	+	+	+	

+ = present; − = absent.
Source: Millner et al., 1994.

TABLE 9.2. Some Fungi Implicated in Occupational Illnesses in Outdoor Environments.

Allergen Source	Fungus
Cereal grains and straw	*Puccinia* spp.
	Ustilago spp.
	Tilletia caries
	Verticillium lecanii
	Aphanocladium album
	Paecilomyces farinosu
Reeds	*Apiospora montagnei*
Tomato growing	*Verticillium albo-atrum*
Mushroom culture	*Aspergillus fumigatus*
	Cladosporium spp.
	Scytalidium spp.
	Penicillium sp.
Sawmills, papermill	*Conidiosporium corticale*
	Alternaria spp.
Soup manufacture	*Agaricus bisporus*
	Boletus edulis
Flour	*Alternaria* spp.
	Aspergillus spp.
Cheese dairy	*Penicillium camembertii*
Tobacco	*Scopulariopsis brevicaulis*
Enzyme production	*Aspergillus flavus*

(continued)

TABLE 9.2. (continued).

Allergen Source	Fungus
Food protein culture	*Candida tropicalis*
Chiropody	*Trichophyton rubrum*
Greenhouse and plant nurseries	*Aspergillus fumigatus*
Grain harvesting	*Verticillium lecanii*
	Aphanocladium album
Cattle farmers	*Aspergillus ruber*
	Dermatophagoides farinae

Sources: Lacy, 1991; Van den Bogart et al., 1993; Schlueter et al., 1972; Warren, 1981.

Inhalation of dusts from organic materials can illicit inflammation, allergy, and infection. Organic dusts can contain microbes and other biological agents both living and nonliving. Table 9.3 shows the general relationship between types of response and symptoms associated with dust from various organic materials (Millner et al. 1994). The three major responses are inflammation, allergy and infection. The most common route of exposure is through inhalation. The principal receptors affected are workers. However, residents adjacent to composting facilities may also be impacted.

The two bioaerosols of greatest interest relating to worker health and the environment surrounding composting facilities are *Aspergillus fumigatus* and endotoxin. *Aspergillus fumigatus* is of concern to both worker health and populations surrounding composting facilities whereas endotoxin is primarily of concern to workers in composting, recycling and other solid waste handling facilities.

Other biological agents could potentially present health problems, including the thermophilic actinomycetes, glucans, and mycotoxins. However, these bioaerosols have not been shown to be significant or sufficiently prevalent in composting operations to cause concern.

Two comprehensive reviews of bioaerosols, especially *A. fumigatus,* in relation to composting were published in 1994:

(1) Ault, S. K. and M. Schott. 1994. *Aspergillus.* Aspergillosis and composting operations in California. Technical Bulletin No. 1. California Integrated Waste Management Board, Sacramento, CA

(2) Miller, P. D., S. A. Olenchock, E. Epstein, R. Rylander, J. Haines, J. Walker, B. L. Ooi, and M. Maritato. 1994. Bioaerosols associated with composting facilities. *Compost Sci. & Util.* 2(4):6–57

TABLE 9.3. Relationships between Types of Response, Specific Response, and Symptoms Resulting from Inhalation of Organic Dusts.

Type of Response	Specific Response	Major Symptoms	Source of Organic Dust	References
Inflammation (non-immunologically mediated, specific IgE's normal)	Mucous membrane irritation (MMI)	Irritation (itchy, watery) of eyes, nose and throat, including dry cough. No elevated IgE	Cotton, compost, grain dust	Richerson, 1990, 1993
	Acute bronchoconstriction	Contraction of bronchial smooth muscles resulting in reversible narrowing of the airways; acute decrease in FFV after exposure	Cotton, grain, animal and compost dust	Castellan et al., 1987
	Chronic bronchitis	Increased secretion of mucus with productive cough persisting over time	Wood, grain, cotton, tea, dust	Kilburn, 1986
	Toxic pneumonitis (toxic dust syndrome)	Influenza-like with fever, chills, muscle and joint pains, fatigue, headache	Cotton, grain, animal dust, "sick" buildings, mycotoxin	Rylander and Peterson, 1990, COD Report, 1991a, 1991b
	Non-allergic asthma (irritant receptor type)	Chest tightness/pressure, pulmonary eosinophilia, specific IgE's normal	Grain, wood, green/dried plant and coffee bean dust	Salvaggio et al., 1986

(continued)

251

TABLE 9.3. (continued).

Type of Response	Specific Response	Major Symptoms	Source of Organic Dust	References
Allergy (immunologically mediated, specific IgE's normal)	Hypersensitivity pneumonitis (allergic alveolitis)	Granulomatous pneumonitis, often with radiographic evidence of lung fibrosis when exposures are repeated frequently	Moldy hay and straw, thermophilic actinomycetes/fungi, amoebae, animal proteins, humidifier reservoirs	Fink, 1986; Pepys, 1969; Edwards et al., 1974; Van den Bogart et al., 1993; Van Assendeleft et al., 1979
	Allergic asthma	Chest tightness/pressure, pulmonary eosinophilia	Grain, wood, green/dried plant and coffee bean dust	Salvaggio et al., 1986
	Allergic rhinitis	Itchy, watery eyes, sinuses and throat; sneezing; histamine release	Pollen, fungal spores, proteins, enzymes insect debris	Salvaggio et al., 1986; Burge, 1985; Burge and Solomon, 1990
Infection	Necrotizing pneumonia invasive, systemic	Abnormal X-rays/CT scans; isolation of microbes from normally sterile tissues or body fluids	Specific bacteria and fungi	Denning and Steven, 1990; Meeker et al., 1991; Zuk et al., 1989
	Allergic bronchopulmonitis, non-invasive, non-systemic	Microgranulomas on chest X-rays; dyspnea 4–6 hrs. after exposure to agent; decreased FEV_1; Microbe-specific IgE in serum	*Aspergillus* and possibly other microbes	Stevens, 1992

Source: Millner et al., 1994.

Considerably more information is available on *A. fumigatus* in relation to potential impacts to workers and the general public. Data on the prevalence of *A. fumigatus* in the environment and data from the medical literature will be provided. This chapter will also present specific case histories from several facilities in the United States.

ASPERGILLUS FUMIGATUS

Aspergillus fumigatus is a member of the genus *Aspergillus,* which includes approximately 300 species that are very common in the environment (Rippon, 1974; Slavin and Winzenburger, 1977; Emmons et al., 1977). It is a fungus found in decaying organic matter and soil throughout the world (Rippon, 1974). As Rippon (1974) indicated "Aspergillus spores are airborne and constantly inhaled." Furthermore, he stated, "The omnipresence and ubiquity of *Aspergillus* organisms are a major factor in designating this plant as 'our moldy earth'." *Aspergillus fumigatus* is heat tolerant and thus survives the high temperatures normally found in composting.

One of the most interesting findings was reported by Polacheck et al. (1989), who noted that damage to the Aleppo Codex (Halab), the earliest text of the Hebrew Old Testament, was found to be caused by a member of the genus *Aspergillus*. The Codex was written in the 10th century A.D. on a parchment in Tiberias, Israel. The parchment was transferred to a synagogue in Aleppo, Syria, which was set on fire in 1947. Originally it was thought that the damage to the lower corners of all the pages was caused by the fire. However, phase-contrast and fluorescence microscopy revealed the presence of septate, hypha-like filaments that through specific fluorescent antibody stain were found to be *Aspergillus hyphae.*

Morphology

Aspergillus fumigatus (Fresenius) belongs to the plant kingdom, Thallophyta phylum, Ascomycetes class, Aspergillales order, and Aspergillaceae family. *Aspergillus fumigatus* colonies are velvety to floccose, blue green to gray green with development of the conidial heads, reverse colorless or showing yellow, green, or red-brown shades. Conidial heads are columnar up to 400 × 50 microns, but often much shorter or smaller. Conidiophores are smooth, short and often greenish arising from submerged hyphae or as very short branches from aerial hyphae. They enlarge gradually upwards to form the apical flask-shaped vesicle (Commonwealth Mycological Institute, 1966). Figure 9.1 shows the flask-shaped conidiophore (Zinsser, 1968). Colonies are shown in Figures 9.2 and 9.3 (courtesy of Dr. Pat Millner, USDA, Beltsville, MD).

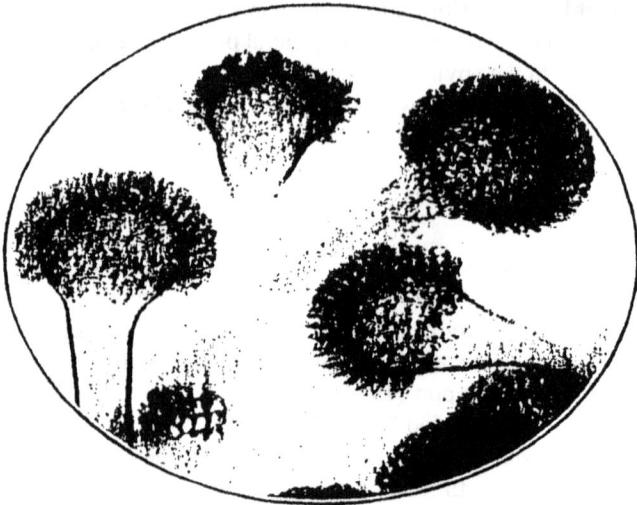

FIGURE 9.1. *Aspergillus fumigatus* (Zinsser, 1968).

Pathogenicity

Aspergillus fumigatus is a recognized pathogen of birds, animals, and humans. Aspergillosis caused by *Aspergillus fumigatus* is an acute or chronic inflammatory, granulomatous infection primarily of the respiratory tract (sinuses, bronchi, and lungs) and occasionally other tissues (e.g., the ear, otomycosis) (Burnett and Schuster, 1973; Rippon, 1974). Figure 9.4 shows the hypha in aspergilloma (courtesy of Dr. Jonathan Epstein, M.D., Johns Hopkins Medical Center, Baltimore, MD).

Aspergillosis (the disease caused by the aspergillus organism), was first recognized in animals. Rippon (1974) indicated that probably the first human case was reported by Sluyter in 1847. Subsequently, Virchow (1856) published a paper describing *Aspergillus fumigatus* as the causative agent in bronchial and pulmonary disease of humans. In 1984, Gaucher and Sergent (1984) noted that this organism was an important human pathogen in people employed to chew grain for the feeding of pigeons. Similarly, Renon in 1897 associated the disease with certain occupations such as pigeon handlers and handlers of fur and wigs (Rippon, 1974; Zinsser, 1968).

Although aspergillosis can be caused by several species, *Aspergillus fumigatus* accounts for almost all of the diseases both allergic and invasive (Rippon, 1974). *Aspergillus fumigatus* is often referred to as an opportunistic fungus or secondary pathogen since it invades and infects debilitated or immunocompromised individuals (Emmons et al., 1977).

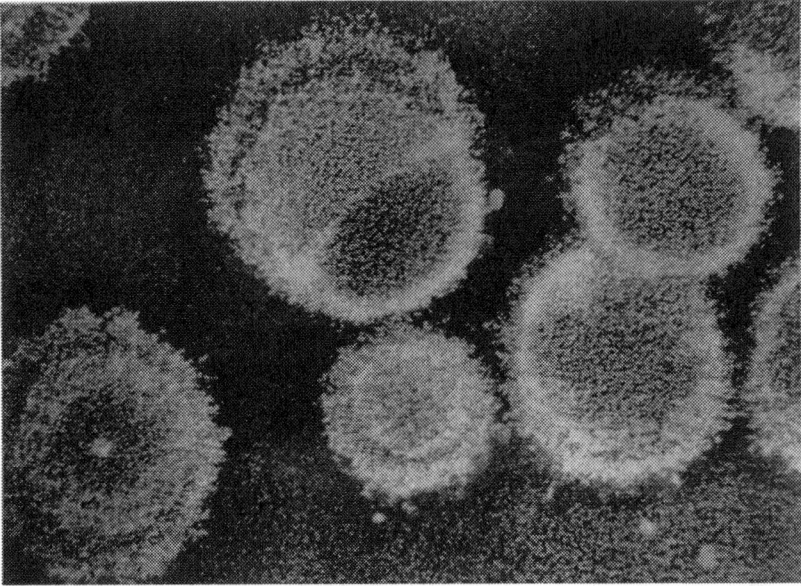

FIGURE 9.2. Colonies of *Aspergillus fumigatus* on oxgall agar, 45°C, 24–48 hr. (Courtesy of Dr. Pat Millner, USDA, Beltsville, MD.)

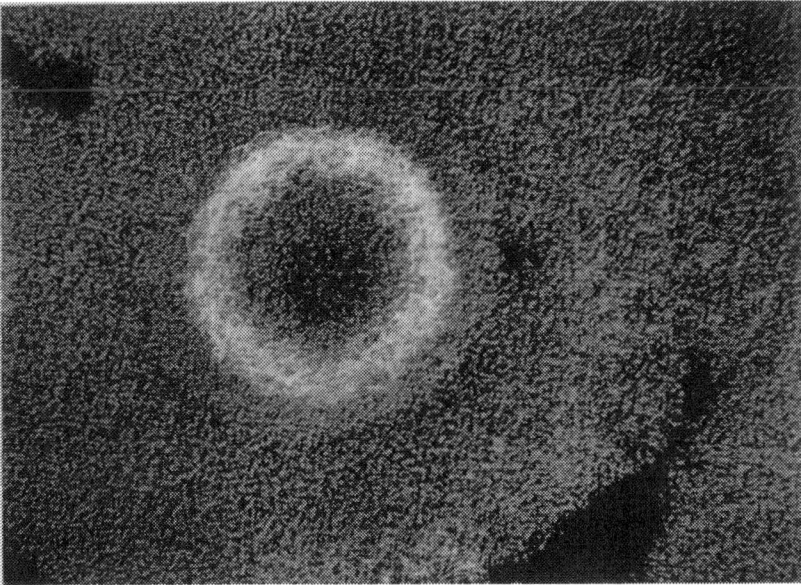

FIGURE 9.3. *Aspergillus fumigatus* colony. (Courtesy of Dr. Pat Millner, USDA, Beltsville, MD.)

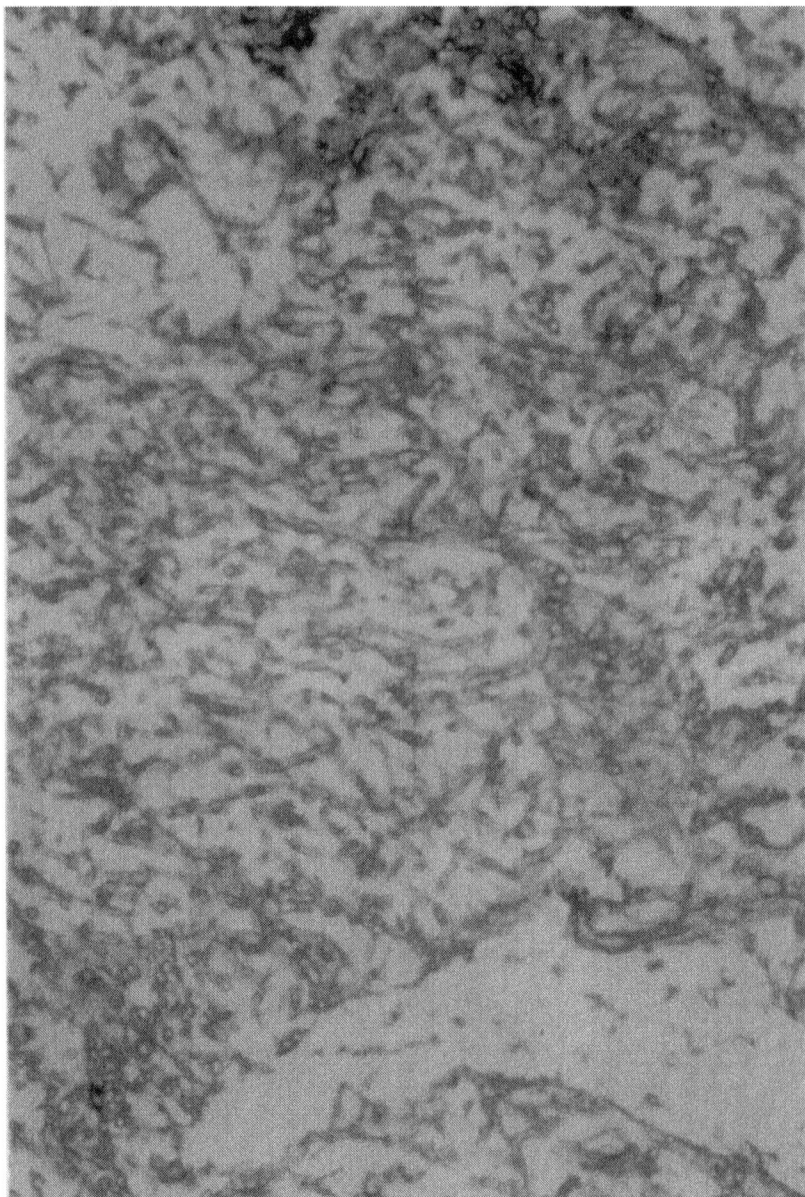

FIGURE 9.4.   (a) *A. fumigatus* hypha in aspergilloma. (Courtesy of Dr. Jonathan Epstein, M.D., Johns Hopkins Medical Center, Baltimore, MD.)

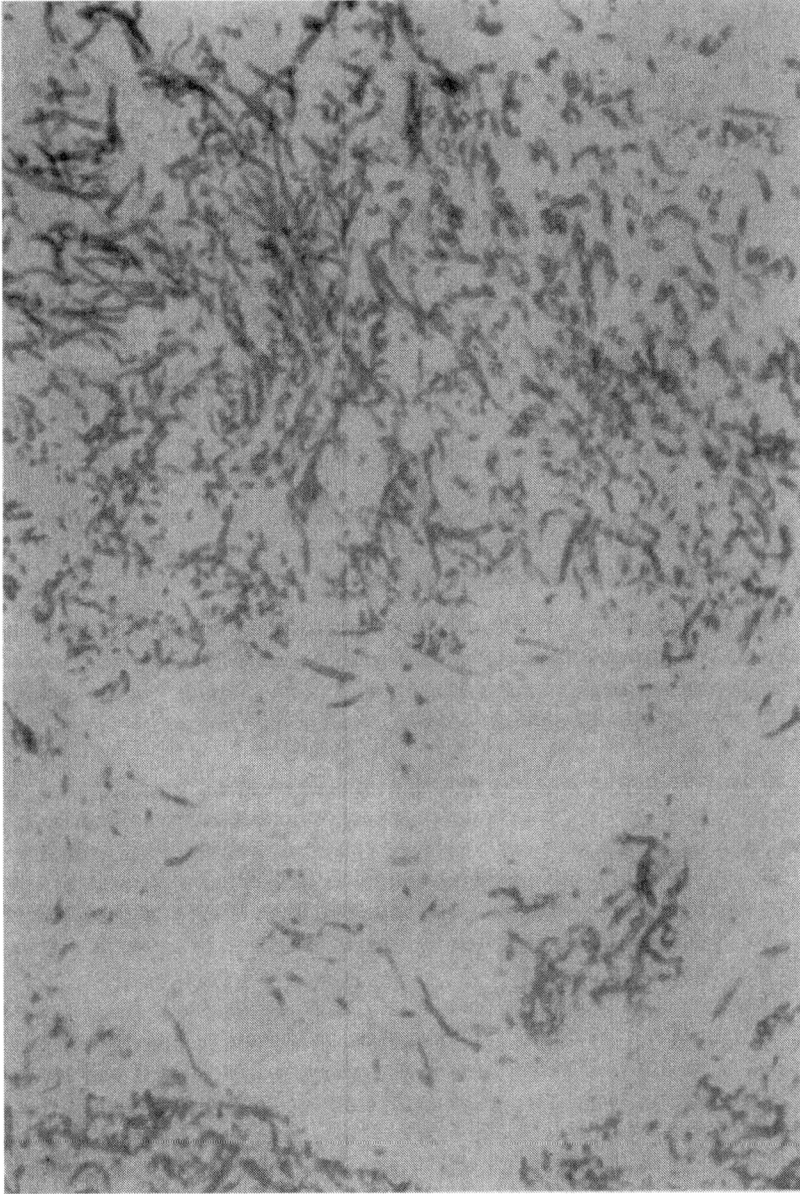

FIGURE 9.4 (continued). (b) Low power of MS stain of *A. fumigatus*. (Courtesy of Dr. Jonathan Epstein, M.D., Johns Hopkins Medical Center, Baltimore, MD.)

Rippon (1974) indicated that three factors are involved in disseminating aspergillosis: (1) lowering of resistance due to drugs or a debilitating disease, (2) an opportunity for invasion, and (3) disruption of the normal flora and inflammatory response by antibiotics or steroids.

Slavin and Winzenburger (1977) described five types of lung disease caused by *Aspergillus.*

1. Invasive or septicemic aspergillosis in which invasion of the bronchial wall produces a definite bronchitis often with systemic spread.
2. Saprophytic or aspergilloma in which the organism colonizes preexisting anatomic abnormalities such as congenital cysts or bronchiectatic cavities.
3. Extrinsic allergic alveolitis in which the organism produces a precipitating antibody and cell-mediated immunity response resulting in a hypersensitivity pneumonitis.
4. Bronchial asthma in which the mold produces an IgE-mediated bronchospasm.
5. Allergic bronchopulmonary aspergillosis (ABPA), a condition first described in 1952 and characterized by pulmonary infiltrates and peripheral blood and sputum eosinophilia.

According to Lacey (1991), allergic rhinitis and asthma occur in subjects who are predisposed to allergy and are readily sensitized by normal, everyday exposure to airborne allergens. The symptoms include rhinitis (inflammation of the nasal mucosa), wheeze, and asthma caused by ventilatory obstruction.

Extrinsic allergic alveolitis (pneumoconiosis resulting from exposure to organic dusts) is a T-lymphocyte-dependent granulomatous inflammatory reaction caused by a wide variety of organic dusts including fungi, actinomycetes, and avian and animal serum proteins. It is usually associated with exposure to a large concentration of spores (Lacey, 1991). Thus, Rylander (1986) indicated that a minimum of concentration of 10^8 spores is needed for sensitization.

There are eight diagnostic features of allergic bronchopulmonary aspergillosis (ABPA): (1) asthma, (2) immediate cutaneous reactivity to *Aspergillus*, (3) precipitating antibodies to *A. fumigatus*, (4) elevated total serum IgE, (5) history of roentgenographic infiltrates, (6) peripheral blood eosinophilia, (7) proximal bronchiectasis, and (8) elevated serum IgE-*A. fumigatus* and IgG-*A. fumigatus* compared with mold-sensitive asthmatic patients (Greenberger, 1984; Rosenberg et al., 1977; Wang et al., 1978).

Greenberger (1984) stated that when all of the diagnostic features are present, it is easy to establish the presence of *A. fumigatus*. However, Cunha (1995) noted that the mere recovery of *Aspergillus* from bronchoscopy or bronchial alveolar or sputum specimens indicates colonization rather than infection and, therefore, is not diagnostic.

A majority of the literature cited on aspergillosis is a result of infections of debilitated individuals or individuals taking immunosuppressant drugs in hospitals. The medical profession is concerned since in the past 15 years the fungal infection rate has nearly doubled in hospitalized patients (Wingard, 1995). The main reasons for this increase are largely attributable to advances in medical technology.

Although *Aspergillus* species account for only 1.3% of all invasive fungal infections, in oncology patients they are estimated to account for 5% to 35% (Wingard, 1995). Patterson (1995) pointed out that fungal infection remains a major limiting factor in the success of organ transplants. The likelihood of *Aspergillus* infection depends on host immune conditions and transplant site. According to Cunha (1995), diagnosis of aspergillosis infection is difficult. Although *Aspergillus* precipitants are highly elevated in aspergillomas and bronchial pulmonary aspergillosis, they are not helpful in the diagnosis. Also, blood cultures are almost always negative in invasive aspergillosis and are not helpful in making the diagnosis (Cunha, 1995).

Levels ranging from $15/m^3$ to $2400/m^3$ have been reported in hospitals (Noble and Clayton, 1963; Mullins et al., 1976). Only rarely have cases of non-debilitated individuals been cited. Rodenhuis et al. (1984) reported on a case involving a non-immunosuppressed patient. Bodey and Vartivarian (1989) discussed the potential for aspergillosis in hospitals from ventilation systems. Most of the infections were due to *Aspergillus fumigatus*. They stated that normal hosts rarely develop invasive disease.

Arnow et al. (1991) evaluated the environmental conditions of a new hospital that was opened in 1983. Monthly air sampling showed increasing concentrations of *Aspergillus fumigatus* and progressive increases in aspergillosis in patients. Heavy growth of the fungus was found on air filters. Removal of the contaminated air filters improved the situation.

Other potential sources of the fungus in a hospital environment include potted plants (Summerbell et al., 1989) and dust during renovation or construction (Aisner et al., 1976; Opal et al., 1986). Hopkins et al. (1989) reported that six patients in a campus hospital developed invasive aspergillosis. They also reported that over the past three years there were 19 cases.

Burton et al. (1972) reported on four renal transplant recipients who were infected by *Aspergillus fumigatus*. Other cases involving renal transplants were cited by Murray et al. (1975) and Rifkind et al. (1967). Open-heart surgery patients are also at risk. Infections caused by *A. fumigatus* occurred in six patients at Holston Valley Hospital and Medical Center in Kingsport, Tennessee (Richet et al., 1992).

Kilmowski et al. (1989) evaluated the incidence of aspergillosis in leukemia patients from 1964 to 1983. The incidence rate was 5.2 cases per 100 patients. Cases have also been reported in leukemia and cancer patients (Rose and Varkey, 1975; Aisner et al., 1976; Petheram and Seal, 1976).

More recently reported cases have involved acute immunodeficiency syndrome (AIDS) (Cox et al., 1990; Denning et al., 1991). Denning et al. (1991) described the predisposing factors, the clinical and radiologic features, and therapeutic outcomes of 13 patients with pulmonary aspergillosis. All of them had human immunodeficiency virus (HIV) and 12 had AIDS. The authors stated that pulmonary aspergillosis is unusual in patients with AIDS. Singh et al. (1991) reported on invasive pulmonary aspergillosis in two patients with the HIV infection. In reviewing the literature, they found 17 additional cases. Pursell et al. (1992) stated that invasive aspergillosis is an uncommon infectious complication in AIDS patients. They observed 972 patients with AIDS over a 10-year period. Upon autopsy, invasive aspergillosis was documented in four patients. Two other cases were suspected of having invasive aspergillosis.

It is clear from the review of the medical literature that the reported cases of aspergillosis and other diseases and symptoms related to *A. fumigatus* have occurred essentially in hospital patients. Thus, of the hundreds of reported cases few implicated gardening, composting, and related activities.

Vincken and Roels (1984) reported on a case of a healthy 20-year-old male who had a history of dyspnea, dry cough, and fever. The diagnosis was hypersensitivity pneumonitis. The individual started working in a vegetable compost plant where he turned piles with a pitchfork. Upon admission to the hospital he was diagnosed as having hypersensitivity pneumonitis. According to the authors, this may be the first report of acute hypersensitivity pneumonitis (allergic alveolitis).

Invasive pulmonary aspergillosis was found in a 34-year-old man who had worked as a gardener for 14 years, was heterosexual, and denied intravenous drug abuse (Zuk et al., 1989). Kramer et al. (1989) reported on a case of an asthmatic young man whose residence was located within 250 feet of a leaf composting site. The person developed allergic bronchopulmonary aspergillosis from inhalation of *A. fumigatus* spores released from the composting facility. Air samples at the patient's residence ranged from 1.2 to 2.4 CFU/m^3. At five other sites within 1.5 miles of the composting site, the concentration of *A. fumigatus* was 1.6 to 11.8 CFU/m^3. Samples collected from the composting site showed concentrations ranging from 60 to 385 CFU/m^3. No samples were taken between the patient's residence and the composting site.

Conrad et al. (1992) reported on the fatal outcome of an individual with chronic granulomatous disease who developed aspergillus pneumonia after shoveling wood chips in a landscaping operation. Further, Yoshida et al. (1993) described a case of hypersensitivity pneumonitis in a 37-year-old healthy, non-smoking female who had been cultivating vegetables in a poorly ventilated plastic greenhouse. The person developed cough, dyspnea, and high fever after tractor tilling and mixing straw and manure into the greenhouse soil.

Occurrence in the Environment

Numerous studies have been conducted to determine *A. fumigatus* occurrence in the environment. It is found universally and has been recovered in the upper atmosphere, from snow in the Antarctic and winds over the Sahara desert (Rippon, 1974). *A. fumigatus* has been found to grow on numerous substrates and media. The following are some examples:

- wood—Kaarik (1974) reported that *A. fumigatus* degraded cellulose, which resulted in "soft rot" of wood. Millner et al. (1977) found that wood chips used in the composting of sewage sludge had concentrations of *A. fumigatus.* Land et al. (1987), while assessing worker health in sawmills, found that *A. fumigatus* was present on the surface of kiln-dried wood.
- green leaves—Emmons et al. (1977) reported that when green leaves and branches from city streets are chipped, the resulting wood chips and vegetation provide an excellent medium for *A. fumigatus.*
- grass—Lacey (1975) found that *A. fumigatus* was the most abundant of the thermophilic fungi on pasture grass and that it was found on all grass species. Slavin and Winzenburger (1977) isolated *A. fumigatus* from lawnmower clippings.
- paper—Hughes (1968) found that *A. fumigatus* was present in paper-processing waters and in pulp piles.
- fabric—Turner (1967) reported that *A. fumigatus* grew on raw cotton fibers.
- leather—Turner (1967) and Orlita (1968) found that *A. fumigatus* grew on leather products.
- rubber and plastics—Cundell and Mulcock (1972) found *A. fumigatus* on rubber. The organism was not able to degrade rubber. Mills and Eggins (1974) reported that *A. fumigatus* could attack plasticizers.
- aviation fuel—Thomas (1977) reported that *A. fumigatus* was found in aviation fuel tanks.

- biosolids—Millner et al. (1977) noted that digested biosolids had levels of *A. fumigatus* ranging from 100 to 1000 colony-forming units (CFU) per gram dry weight.
- fireproofing material—Aisner et al. (1976) found that *A. fumigatus* was present in dust from a cellulose-base fireproofing material installed in a new hospital, resulting in the infection of cancer patients.
- compost—Kothary et al. (1984) detected 10^6 CFU/g in screened compost during the active phase of composting. In municipal solid waste compost Kane and Mullins (1973) found pure cultures of 304 isolates, which included *A. fumigatus*. Millner et al. (1977) reported that *A. fumigatus* was found in compost in every stage during biosolid composting. After four months of storage, *A. fumigatus* was detected in compost piles at the 10 cm depth but not at the 25 and 50 cm depths. Compost bagged and stored for one week to four months exhibited negligible or undetectable levels. Iacoboni et al. (1980) reported very low levels of *A. fumigatus* (<13 CFU/g) in windrow biosolids compost.
- commercial potting soil, manures, and mulches—Millner et al. (1977) reported that many of these products contained levels of *A. fumigatus* ranging from 100 to 100,000 CFU per gram dry weight.

Several studies have assessed the prevalence of *A. fumigatus* in homes, outdoor air, and the environment (Solomon, 1975; Baxter and Cookson, 1983; Hirsh and Sosman, 1976; Lumpkins et al., 1973; Sikora et al., 1985; Kodama and McGee, 1986).

For example, Solomon (1975) measured fungi in and immediately outside several Midwestern homes during two seasons, frost-free and subfreezing. *A. fumigatus* was found in 26 homes at a level of 40 isolates/m³ during the frost-free period and similar levels in 80 homes during subfreezing weather. A subsequent study (Solomon, 1976) reported that *A. fumigatus* was recovered in 31% of 47 homes. The range of isolates/m³ was from 1 to 946, with a mean of 24.4.

Hirsh and Sosman (1976) conducted a one-year survey of mold growth in 12 homes. *A. fumigatus* was one of the most common molds isolated. Specifically, it was the most common mold found in basements; the second most common in bathrooms; and the fourth and fifth most common in front rooms and bedrooms, respectively. Compared to other molds, the species was significantly more frequent in homes with pets. *A. fumigatus* was detected in 85% of all basements, 56% of bathrooms, 46% of living rooms and 42% of the bedrooms.

Lumpkins et al. (1973) conducted a survey of bioaerosols in homes and outdoors in 30 states from March 1971 to March 1973. They used a culture-plate technique, which underestimates the level of bioaerosols, since it depends on passive settling of spores. The total number of plates exposed was 2075 indoors and 448 outdoors. Of the bioaerosols found in homes, *Aspergillus* represented 9.4% in bedrooms, 27.6% in basements, 14.5% in family rooms, 10.8% in kitchens, 12.2% in living rooms, 11.8% in other rooms and 6.0% outdoors.

Sikora et al. (1985) evaluated the occurrence of *A. fumigatus* in various areas in Maryland during the fall, winter, spring, and summer of 1979 to 1980. Locations included lawns, wooded areas, agricultural areas, refuse containers, greenhouse, library stacks, attic, zoological area, boiler room, school playground, university parking lot, and a shopping center. Levels of *A. fumigatus* ranged from 0 to 740,000 CFU/m^3. The highest levels were found in a mushroom house. In an attic in Silver Spring, Maryland, levels in the spring were 1160 CFU/m^3. This was the result of shaking a feather duster located in the attic. The school playground, university parking lot, and shopping center had levels ranging from 1 to 12 CFU/m^3.

Larsen and Gravesen (1991) reported on an evaluation of outdoor air in Copenhagen, Denmark, since 1978. *Cladosporium, Alternaria, Penicillium* and *A. fumigatus* represented 84.1% of the total viable microflora. *Aspergillus* represented 0.8% of the total colonies obtained. From 0 to 204 CFU/m^3 were recorded. The highest average values were found in November and December.

Slavin and Winzenburger (1977) noted that the incidence of allergic bronchopulmonary aspergillosis is much lower in the United States than in Britain. It is the most common (approximately 80%) form of asthma and pulmonary eosinophilia in the United Kingdom. The also evaluated organic farmers in the St. Louis, Missouri, area and found that none of the organic farmers who were in close contact with compost piles were clinically sensitive to *A. fumigatus*.

These data clearly illustrate that *A. fumigatus* is universally present wherever decaying organic matter is found. Thus, individuals are constantly exposed to and inhale spores of *A. fumigatus*.

Occurrence in and around Composting Facilities

Most of the data on *A. fumigatus* in and around composting facilities in the United States have been collected at biosolids composting facilities. Recently, several studies have been conducted at yard waste facilities. The most comprehensive of these investigations have been conducted at the following facilities:

(1) Biosolid composting facilities
- Dickerson, Maryland
- Site II, Maryland
- Westbrook, Maine
- Windsor, Ontario, Canada
- Hampton Roads, Virginia
- Beltsville, Maryland
- Camden, New Jersey

(2) Yard waste composting facility

(3) MSW composting facility

A brief discussion of site conditions is presented, followed by data on the occurrence and dispersion of *Aspergillus fumigatus* at the various locations.

DICKERSON, MARYLAND

At the Dickerson site in Montgomery County, Maryland, the composting operations were carried on outside. All movement of materials was done using front-end loaders. The extended aerated static pile was employed. Wood chips were used as the bulking agent and were recovered by screening following composting. The site was paved to control runoff.

Prior to establishing the composting site, a study was conducted to determine the natural atmospheric concentrations of *Aspergillus fumigatus*. Jones and Cookson (1983) measured concentrations of *Aspergillus fumigatus* in and around the proposed biosolid composting site. In 17 locations they found that the concentration ranged from 0–71 CFU/m³ with a mean of 2.4 and 3.5 CFU/m³, respectively. The data on levels of *A. fumigatus* are shown in Figure 9.5 (Cookson et al., 1983).

Measurements were done with a two-stage Anderson sample before any activities took place and at different times of the day to evaluate the effect of site activity. Before daily activities maximum levels were 228 and 60 at the 1 mile and on-site locations, respectively. The lowest level of 27 was observed downwind at the one-half mile location. Activities increased *A. fumigatus* aerospores. The highest level of 553 was recorded on-site. There was little difference between the upwind and downwind measurements (JTC Environmental Consultants, Inc., 1981).

SITE II, MARYLAND

Several studies were conducted at Site II. The facility began operations in April of 1983. Initially, Site II consisted of enclosed composting and screening. Mixing was done with a windrow machine under a covered area

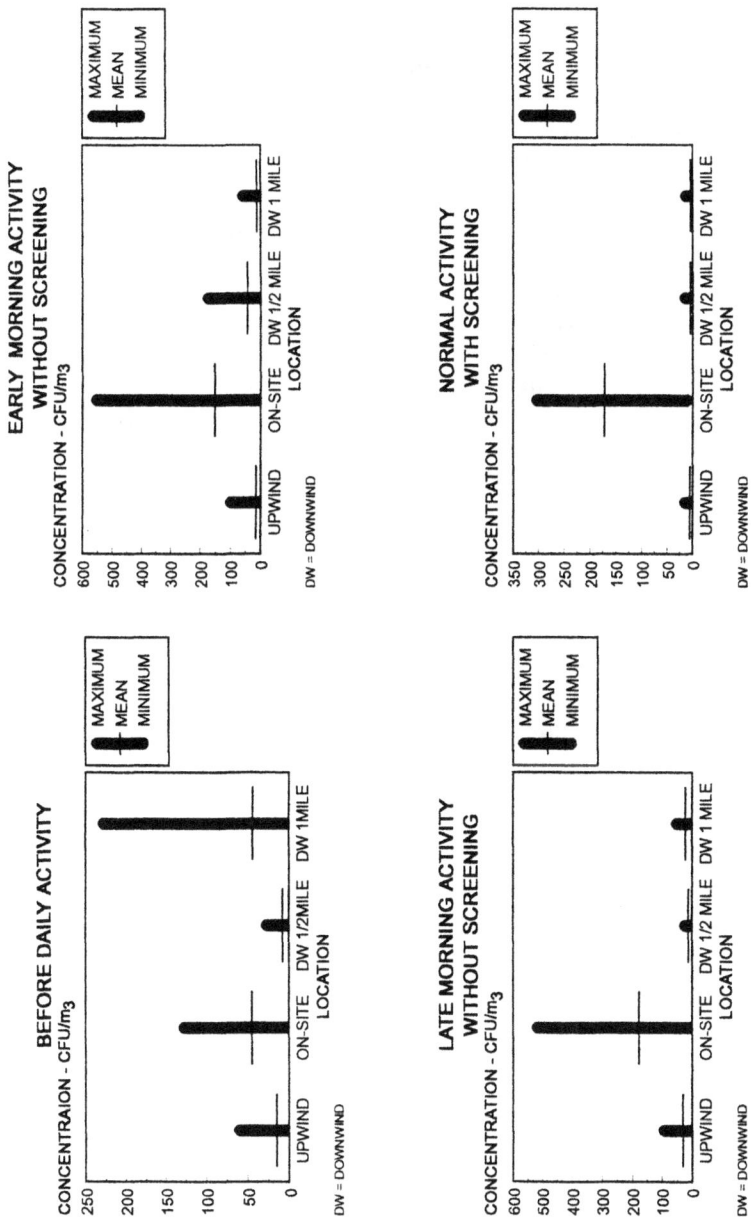

FIGURE 9.5. Concentration of *A. fumigatus* at Dickerson, Maryland, biosolids composting site, June to September 1981. (Data from Cookson et al., 1983.)

that was open on two sides. Curing and storage of materials took place outdoors. Later the mixing and curing areas were totally enclosed.

Monitoring and analysis was conducted by JTC Environmental Consultants, Inc., from 1983 to 1986 (Lees and Tockman, 1987). At the time of measurements, the site consisted of mixing sludge and wood chips in a covered area using front-end loaders and windrow turning machines. The mix was conveyed by front-end loaders to the totally enclosed composting building. Composting used the extended aerated static pile method. The piles were aerated by negative pressure (vacuum). At the completion of composting, the product was screened in a totally enclosed building. Wood chips were stored both outdoors and under a covered area. The compost was cured outdoors. Loading of trucks for distribution was done outdoors.

The study showed a slight increase in the frequency of detection of *A. fumigatus* at the sampling locations over time. The increase was not statistically significant, however.

During 1990 and 1991 additional studies were conducted at Site II (General Physics Corporation, 1991). At that time, the mixing area was enclosed. Figure 9.6 shows the data for *Aspergillus fumigatus*. The downwind sites were located 300 m to 2600 m from the site.

WESTBROOK, MAINE

The Westbrook facility was a totally open biosolid-composting facility. Mixing was done with front-end loaders on asphalt pavement. Static pile

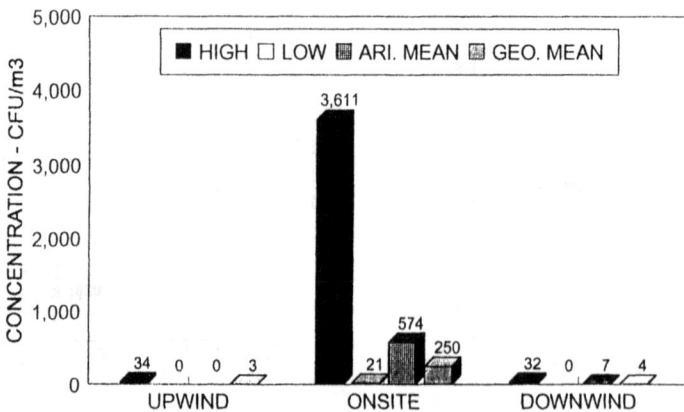

FIGURE 9.6. Concentration of *A. fumigatus* at or near WSSC Site II, 1990–1991. (Data from 12 locations. General Physics Corporation, 1991.)

building and teardown was done with front-end loaders. Wood chips were screened out for reuse and were stored outdoors as was the compost. The study was conducted for one year and published in a report by Passman (1980).

In the study, the criterion for judging aerospora levels to be significant was *A. fumigatus* concentrations equal to or exceeding 50 CFU/m^3. This number represents one logarithm increase above average background concentrations. Background measurements showed levels of 0 to 15 CFU/m^3 with a mean of 5 CFU/m^3. Figure 9.7 shows the data collected. Within 30 meters (90 feet) of the site center concentrations of *A. fumigatus* exceeded background levels, especially during August, September, and October. At 150 meters (450 feet), the concentrations were at ambient levels except for one measurement on July 18, 1979, where the levels reached 330 CFU/m^3. At 1500 meters (4500 feet) all the measurements were at ambient levels.

WINDSOR, ONTARIO

Windsor, Ontario, composted biosolids using wood chips in an open area adjacent to the wastewater treatment plant located in the city on the west side. Mixing of wood chips and biosolids was done under a three-sided covered building using a mix box. Front-end loaders transported the mix to an asphalt area where piles were built. The piles were torn down with front-end loaders and the material was screened in an open screen. Wood chips and finished compost were stored outdoors.

Clayton Environmental Consultants Ltd. (1983) conducted extensive studies throughout the wastewater treatment plant, composting site, and surrounding area. In contrast to the Maine study, these researchers deemed that a concentration exceeding 25 CFU/m^3 was significant. The data are summarized in Figure 9.8. Concentrations were highest near the mix area (120 CFU/m^3) and on the front-end loader (79 CFU/m^3). In all the other areas, concentrations of the fungus were less than 24 CFU/m^3. At the site fence, 575 feet from the compost piles, the concentration of *A. fumigatus* was less than 2 CFU/m^3.

HAMPTON ROADS, VIRGINIA

Hampton Roads Sanitation District (1981) undertook a study to determine *A. fumigatus* aerospores in and around a biosolids composting facility and several other industrial areas that were identified as potential sources because each had large quantities of raw wood products. The composting facility uses

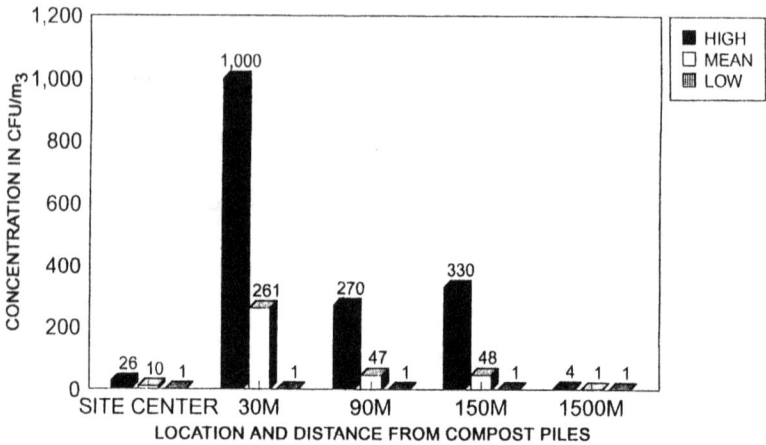

FIGURE 9.7. Concentration of A. fumigatus at the Westbrook, Maine, composting facility. (Data from Passman, 1980.)

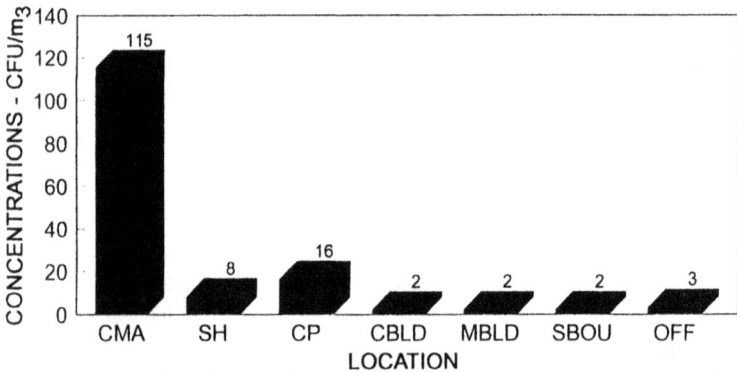

CMA = mix area: SH = shaker; CP = piles;
CBLD = compost building; MBLD = mix building;
SBOU = site boundary; OFF = office

FIGURE 9.8. Concentration of A. fumigatus at West Windsor, Canada. (Data from Clayton Environmental Consultants, Inc., 1983.)

268

an open static pile composting system. The only covered area contains the screen. Some mixing is also done under the roofed area. Composting increased the levels of *A. fumigatus* as shown in Figure 9.9 and Table 9.4. At 400 m downwind of the site, 13 colonies/m³ of air were collected. The industrial sources had maximum concentrations of 478 colonies/m³.

BELTSVILLE, MARYLAND

The U.S. Department of Agriculture maintained a biosolids composting research site at Beltsville, Maryland. All composting activities were done in the open. The studies conducted on the occurrence and dispersal of *A. fumigatus* were the first to be reported. Millner et al. (1980) reported on the dispersal of *A. fumigatus* as a function of mechanical agitation. These data are summarized in Table 9.5.

Pile agitation clearly resulted in the release of *A. fumigatus* spores. Additional information obtained in areas within and around the site showed that at 100 m (300 feet), most of the samples collected were at background levels.

CAMDEN, NEW JERSEY

The Camden composting facility was located at the wastewater treatment plant, which was adjacent to a licorice factory. Both of these are sources of

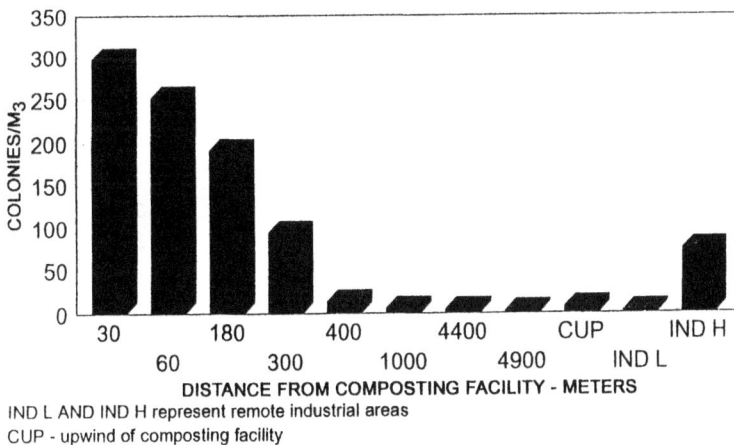

IND L AND IND H represent remote industrial areas
CUP - upwind of composting facility

FIGURE 9.9. *A. fumigatus* at several locations in relation to the Hampton Roads composting facility. (Data from Hampton Roads Sanitation District, 1983.)

TABLE 9.4. Aspergillus fumigatus *Levels in the Vicinity of the Hampton Roads, Virginia Biosolids Composting Facility.*

Distance from Facility (m)	No. of Samples	Geometric Mean (colonies/m³)	Range (colonies/m³)
Prior to operations	10–18	2	<2–336
30	36	297	<14–15, 960
60	27	251	<7–3775
180	18	191	<24–2332
300	10	94	<14–2597
400	25	13	<2–426
1000	26	5	<2–409
1200	26	5	<2–286
4400	26	4	<2–28
4900	26	2	<2–34
Upwind	26	7	<2–102
Industrial sites	1–18	<2–74	<2–478

Source: Hampton Roads Sanitation District, 1981.

TABLE 9.5. *Concentration of* Aspergillus fumigatus *as a Result of Mechanical Agitation.*

Sample Location	No. of Particles/m³
Non-compost background	0–24
Upwind of compost site	9–155
During compost pile agitation	
Upwind of FEL	2
3 m downwind of FEL	1390
30 m downwind of FEL	1820
60 m downwind of FEL	5020
15 min after pile agitation	
3 m downwind from pile	39
30 m downwind from pile	0

Source: Millner et al., 1980.

A. fumigatus. Operations took place outdoors using the aerated static pile method. There were no enclosures.

Sampling of *A. fumigatus* was done with a slit-air sampler. Levels of *A fumigatus* are shown in Table 9.6. Highest levels were found near the screen at the time of screening and southeast of the composting piles. Levels were much higher downwind than upwind. At 3 m (9 ft) downwind, levels as high as 3000 CFU/m³ were detected. At 40 m (120 ft), levels of *A. fumigatus* dropped considerably to less than 1000 CFU/m³. Agitation of the composting material by front-end loaders greatly increased dispersion of spores. During agitation, spore counts reached 1390 CFU/m³, which were reduced to 39 CFU/m³ 15 minutes after agitation stopped (Kothary et al., 1984).

Measurements of *A. fumigatus* taken at a distance of 250 m from actively composting biosolids were much lower on a day when screening did not take place than when screening occurred (Figure 9.10). Also, the levels of *A. fumigatus* were greatly reduced after rainfall (Figure 9.11) (Kothary et al., 1984). This finding indicates that *A. fumigatus* in the air can be reduced by misting or spraying during active composting activities (e.g., windrow turning). Enclosing a screen would also greatly reduce *A. fumigatus* spores in the air.

Yard Waste Composting Facilities

Zwerling and Strom (1991) reported on a study in four communities in New Jersey. They found high levels of *A. fumigatus* on-site during activity and background levels during periods of no activity. During high activity, the levels of *A. fumigatus* at the composting sites ranged from >5120 to >73,554 CFU/m³. However, during periods of average activity, the levels dropped significantly at 100 m (300 feet) and 500 m (1500 feet) downwind. At 100 m the level was at 354 CFU/m³ and at 500 m the level was 86 CFU/m³. These numbers were within the range of background levels.

A recent study at a Connecticut yard-waste composting site showed levels of *A. fumigatus* ranging from 0 to 2648 CFU/m³ on-site and 0 to 11 CFU/m³ downwind at distances of 500 feet to 1 mile (Figure 9.12). The downwind measurements were similar to levels found at background sites located remotely from the facility (E&A Environmental Consultants, Inc., 1993).

The most comprehensive yard-waste composting study evaluating bioaerosol levels and health symptoms was conducted at the Islip composting facility by the state of New York, Department of Health (NYDOH, 1994). The facility is located on an 18-hectare site with the nearest homes as close as 280 m. Air sampling was conducted from August to October and samples were analyzed for *A. fumigatus* and other fungal and bacterial spores. Periodic samples were collected for total suspended particulates

TABLE 9.6. *Total Thermotolerant Fungi and* Aspergillus fumigatus *in Air Samples at the Camden, New Jersey, Composting Facility.*

Sampling Location	Fungal Count (CFU/m³)		A. fumigatus as Percent of Total
	Total Fungi	A. fumigatus	
Northeast corner, near wood chips	3143	1850	59
Northeast corner near screen	1733	1616	93
Adjacent to screen when not operating	1213	1063	88
Adjacent to screen when operating	2575	2533	98
East of compost piles	1274	977	77
East of compost piles	733	713	93
East of compost piles	TNTC[1]	TNTC	—
Northwest of compost piles	79	62	78

[1]TNTC = too numerous to count.
Source: Kothery et al., 1984.

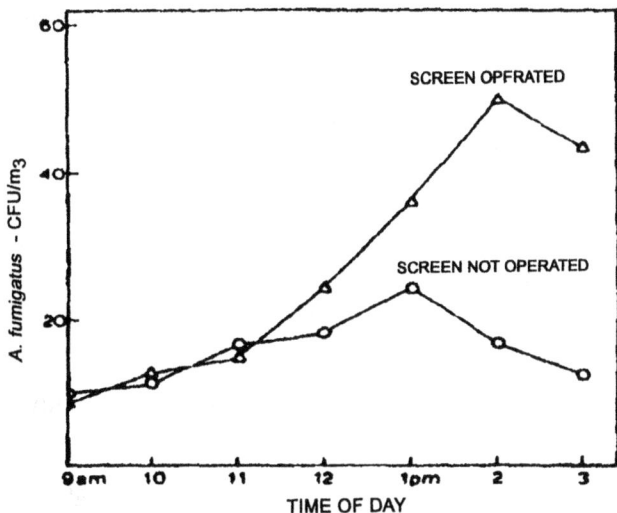

FIGURE 9.10. Levels of A. fumigatus in the air at a distance of 250 m on a day when the screen was operated and on a day when it was not operated during screening of biosolids compost. (Data from Kothary et al., 1984.)

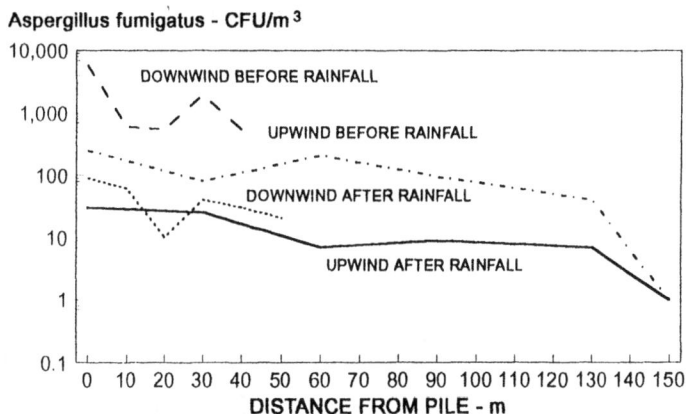

FIGURE 9.11. Levels of A. fumigatus before and after rainfall. (Data from Kothary et al., 1984.)

(dust). Four locations were evaluated: (1) study neighborhood nearest the facility, (2) the composting facility, (3) reference facility approximately 8 km southwest, and (4) MacArthur airport 460 m³ upwind from the composting facility.

Counts were highly variable, ranging from not detected to almost 1000 spores/m³ at the reference neighborhood and the airport. Average background spore counts at the reference neighborhood and the airport were about 50 spores/m³. *A. fumigatus* averaged about 100 spores/m³ at the study neighborhood and 500 spores/m³ at the compost facility. The increases in the study neighborhood were not statistically significant, indicating considerable variation at both the composting facility and the study neighborhood. The average count of *A. fumigatus* spores in the study neighborhood was four times the average background levels, and this difference was significant. A summary of the data for total spore count is presented in Table 9.7. The median and mean total airborne spore levels did not differ greatly among the four sites.

Table 9.8 shows the statistical data for *A. fumigatus*. Individual 4-minute spore counts of *A. fumigatus* ranged from non-detected at all sites to about 1000 spores/m³ at the airport and at the reference facility, to 14,000 spores/m³ at the study area and 22,000 spores/m³ at the composting facility. At the composting facility, the media spore count was 56 spores/m³ whereas the median at all the other locations was zero.

Table 9.9 shows the 4-minute spore counts for *A. fumigatus* made every two hours at the study neighbor. It is interesting to note that some of the highest counts occurred when the facility was not operated. For example, on

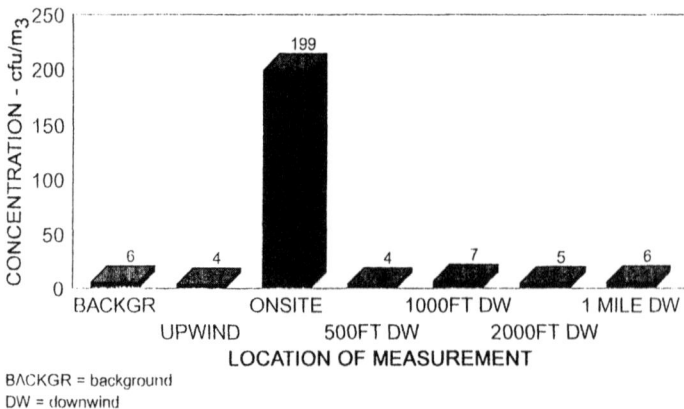

BACKGR = background
DW = downwind

FIGURE 9.12. Average concentration of *A. fumigatus* at or near a Connecticut yard waste composting facility. (Data from E&A Environmental Consultants, Inc., 1993.)

TABLE 9.7. Summary Statistics for Total Spore Counts
at the Four Locations Studied.

| Location | Spore Concentration (spores/m³) | | | | N |
	Mean	Minimum	Maximum	SD	
Airport	13,000	556	125,000	24,000	28
Reference neighborhood	8860	119	194,000	18,000	158
Composting fa-cility	7700	567	128,000	12,000	166
Study neighbor-hood	7340	352	54,700	8240	162

Source: NYDOH, 1994.

Saturday, September 5, the highest count was at 10 P.M. On Sundays, September 6 and October 18, the mean counts were higher than three other weekdays. In general, there were more non-detected hourly events during midweek than during Saturdays and Sundays. The highest number recorded was on a Wednesday at 6 P.M.

Normally one would have expected that the highest levels would have been recorded during weekdays from 8 A.M. to 4 P.M. However, this was not the case. Many of the high spore counts over the weekends and in the early evenings could have been attributed to mowing lawns or raking leaves.

MSW Composting Facility

A comprehensive study at a totally enclosed MSW-biosolids composting facility in the United States showed that the highest levels of *A. fumigatus* were found in the tipping floor, screen area, and composting area during pile turning. Very low levels were recorded outside the building, at the property line, or at various distances from the facility (Table 9.10) (E&A Environmental Consultants, Inc., 1994).

Other Studies

Clark et al. (1980) measured airborne levels of *A. fumigatus* at composting plants in the metropolitan Washington, D.C., and Camden, New Jersey areas (Table 9.11). At the composting sites, levels ranged from 0–475 CFU/m³ in the Washington, D.C. area, and 0–2940 CFU/m³ in Camden, New Jersey. At distances of 50 m to 150 m, the levels dropped to background.

TABLE 9.8. Summary Statistics for Aspergillus fumigatus Levels at the Four Locations Studied.

Location	Spore Concentration (spore/m^3)				N
	Mean[1]	Minimum	Maximum	SD	
Time-points common to the facility, reference neighborhood, and study neighborhood.					
Airport	38a	0	974	121	85
Reference neighborhood	55b	0	685	114	157
Composting facility	438a,b,c	0	22,100	1980	157
Study neighborhood	188c	0	14,200	1140	157
Study neighborhood (censored)[2]	116c	0	1600	245	156
All time-points.					
Airport	38	0	074	121	85
Reference neighborhood	55	0	685	114	158
Composting facility	457	0	22,100	1760	216
Study neighborhood	194	0	14,200	1010	208

[1]Mean values with matching letters are statistically significant ($p < 0.05$) by post-hoc multiple comparison.
[2]The highest value of 14,200 spores/m^3 was omitted from the calculations.
Source: NYDOH, 1994.

TABLE 9.9. Aspergillus fumigatus *Four-Minute Spore Counts Made Every Two Hours at the Study Neighborhood.*

Time of Day	Aspergillus fumigatus (spores/m³)							
	9/5 Sat.	9/6 Sun.	9/30 Wed.	10/1 Thurs.	10/7 Wed.	10/8 Thurs.	10/17 Sat.	10/18 Sun.
Midnight		93		0		0		0
2 A.M.		0		204		0		37
4		93		0		37		93
6	204	56	0	0	0	0	74	0
8	204	74	0	0	293	0	241	74
10	130	0	0	0	93	0	19	0
Noon	0	0	0	0	0	519	0	0

TABLE 9.9. (continued).

				Aspergillus fumigatus (spores/m³)				
Time of Day	9/5 Sat.	9/6 Sun.	9/30 Wed.	10/1 Thurs.	10/7 Wed.	10/8 Thurs.	10/17 Sat.	10/18 Sun.
2 P.M.	0	37	19	0	0	0	0	0
4	19	19	0	185	19	111	0	222
6	93	37	0	19	3222	0	0	0
8	93	0	37	19	19	0	0	93
10	833	93	56	56	0	0	0	0
Mean	175	42	12	40	405 (53)[1]	56	37	43
Times not detected	2	4	6	7	4	9	6	7

[1]Mean if the highest count of 3222 is deleted.
Source: NYDOH, 1994.

TABLE 9.10. Aspergillus fumigatus *at Various Locations in or around an MSW/Biosolids Composting Facility.*

Sample Location	Concentration Range (CFU/m³)
Tip floor	857–11,714
Primary screen area	2324–26,571
Compost floor during pile turning	>29,000
Post process screen	>29,000
Outside building—before working hours	2.4–5.9
Outside building—during working hours	0–4.7
Property boundary—upwind	5.9–27.6
Property boundary—downwind	1.2–11.8
150 m downwind from property boundary	0–10.6
3044 m downwind from property boundary	0–10.6
3.7 km downwind from property boundary	2.4–7.1

Source: E&A Environmental Consultants, Inc.

TABLE 9.11. Summary of Levels of Airborne A. fumigatus at Composting Sites in the Washington, D.C., and Camden, New Jersey, Areas.

	Concentrations of A. fumigatus (CFU/m³)		
Location	No. of Samples	Range	Average
Metro. Washington, D.C.			
Beltsville, compost	4	48–31	81
Blue Plains, compost	25	0–475	52
Blue Plains, other areas[1]	14	0–21	3
Piscataway plant	15	0–18	3
Camden, New Jersey			
Main plant, compost	13	0–2940	918
Main plant, other areas[2]	6	0–126	24
Baldwin Run plant	11	0–52	14

[1]At least 150 m from compost area.
[2]At least 50 m from compost area.
Source: Clark et al., 1980.

In another study, Clark et al. (1983) measured *A. fumigatus* at four Swedish composting plants. Three of the plants processed municipal refuse; the fourth composted biosolids. Outside sites where material was handled adjacent to compost piles and screening areas levels as high as 10^6 were found (Table 9.12). In general, lower levels of *A. fumigatus* were found in the biosolids-wood chip composting facility than in the MSW-biosolids composting plants.

Clark et al. (1984) evaluated the health risks of employees at several biosolids composting facilities. They indicated that there seems to be some evidence that workers at composting facilities had higher levels of IgG antibody against compost-derived endotoxin (lipopolysaccharide, LPS) than did workers with lower or no exposure to compost. They indicated that there was a higher rate of abnormal skin, ear and nose conditions among compost and workers associated with compost than those who were not exposed to composting operations. The authors could not explain the lower FVC (forced vital capacity) (i.e., an indication of pulmonary function) on Monday morning compared to Friday afternoon.

Epstein and Epstein (1985) reviewed the article and the data. In general, more symptoms occurred in intermediately exposed individuals than with compost workers. Often, more control individuals had symptoms than

TABLE 9.12. Highest Airborne Aspergillus fumigatus *Levels Indoors and Outside Specific Processing Units in Four Swedish Composting Plants.*

Location	Number of Samples	Concentration (10^3 CFU/m^3)	
		Median	Range
Stromstad[1]			
Inside refuse hopper	6	>640	380->1900
Outside long-term compost	4	>1500	670->6000
Landskrona[1]			
Inside bioreactor, operating	2		20->2600
Outside long-term compost	4	126	19->3700
Borlange[1]			
Inside fine compost screening area	3	85	51–94
Outside fine screening area	2	54	36–73
Gothenburg[2]			
Inside lunchroom and office	3	0.3	0.1–4
Outside screener hopper	3	>67	19->3500

[1]MSW and biosolids being composted.
[2]Biosolids and wood chips composted.
Source: Clark et al., 1983.

compost workers. Clark et al. (1984) found no consistent increases in antibody to *A. fumigatus* in compost-exposed workers by either the ELISA or CIE method. In their Table 9, antibody comparisons are listed as follows:

(1) *Aspergillus*—no consistent differences
(2) Histoplasma—no antibodies
(3) *Legionella*—no compost effect
(4) Endotoxin—suggested higher levels in compost workers

The lack of increased antibodies to *Aspergillus* supports the conclusion that, although *Aspergillus* colonization is more common in compost workers, infection with the organism is not. Blood tests were totally non-specific. The only significant positive finding was that fungal spores in the environment at the Site II compost site might have provoked the upper respiratory tract.

A comprehensive worker health evaluation was conducted at the Washington Suburban Sanitary Commission Site II biosolids composting facility from 1987 to 1991 (Chesapeake Occupational Health Services, 1991). During the five years, 242 individuals were screened, with an average of 46 workers participating annually. All 242 blood samples were reported as "None Detected." The data are shown in Figures 9.13 and 9.14.

Reports of asthma, bronchitis, earaches and shortness of breath were infrequent (Figure 9.13). Spirometric findings were also very low (Figure 9.14). Some of the spirometric findings could be related to smoking history. However, there was no excess frequency of abnormal spirometric findings. All biological laboratory analysis for *Aspergillus fumigatus* remained "None

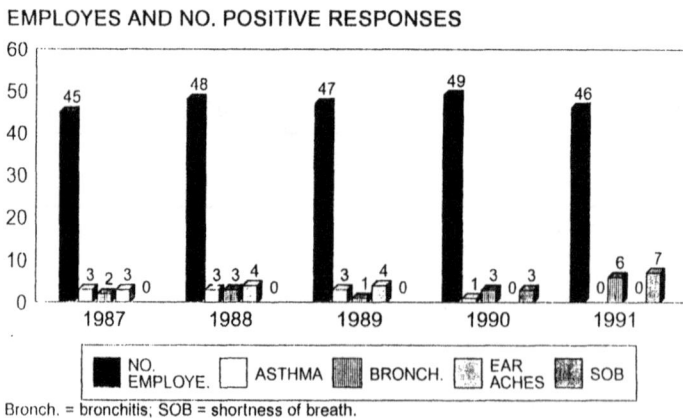

EMPLOYES AND NO. POSITIVE RESPONSES

NO. EMPLOYE.   ASTHMA   BRONCH.   EAR ACHES   SOB

Bronch. = bronchitis; SOB = shortness of breath.

FIGURE 9.13. Number of employees in health surveillance and positive responses to symptoms per year. (Data from Chesapeake Occupational Health Services, 1991.)

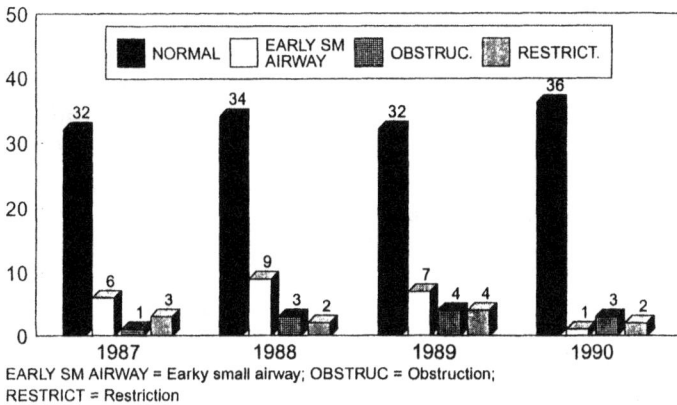

FIGURE 9.14. Spirometric findings of employees at Site II, biosolids composting facility, 1987–1990. (Data from Chesapeake Occupational Health Services, 1991.)

Detected." The study concluded that there was no evidence of adverse effect related to *Aspergillus fumigatus*.

ENDOTOXIN AND ORGANIC DUSTS

Endotoxins are relatively heat stable phospholipid-polysaccharide-protein complex macromolecules that form an integral part of the cell wall of gram-negative bacteria. Endotoxins are released into the environment during cell growth and after the cell dies when the integrity of the cell wall is ruptured (*Stedman's Medical Dictionary*, 1977; Bradley, 1979). Endotoxin can be toxic to humans and animals. The data on toxicity are based on direct injection of the substance into the animal bloodstream. There is little evidence that exposure to airborne endotoxin causes toxic conditions. However, inhaled endotoxin increases the activity of macrophages, which leads to a series of inflammatory conditions (Millner et al., 1994).

Endotoxins are found in organic dusts resulting from the processing of cotton, poultry facilities, MSW, biosolids, bagasse, hemp, hay, grain, biosolids drying and other vegetable dusts (Rylander and Vesterlund, 1982; Sheargren, 1986; Jones et al., 1984; Lenhart and Olenchock, 1984; Cavagna et al., 1969; Olenchock et al., 1983; DeLucca et al., 1984; Dutkiewicz et al., 1988). Endotoxins have been implicated in contaminated water and aerosolized during operation of a humidifier (Rylander and Haglind, 1984; Burge et al., 1985). It is often difficult to isolate specific response or symptoms to either organic dusts or endotoxin. (Table 9.3 provided some of the relationships between the inhalation of organic dusts and associated responses and symptoms.)

No regulatory levels are established for dose response or threshold limit values (TLVs) for endotoxin or organic dusts. However, international criteria have been developed based on the literature (Rylander et al., 1989). Thus, the International Commission on Occupational Health (ICOH) published suggested occupational health levels for acute effects of endotoxin (Rylander et al., 1989). The permissible endotoxin concentrations for two agricultural industries are as follows:

- cotton mills: $1.0–20$ ng/m^3
- animal feed: $0.2–470$ ng/m^3

Malmros et al. (1992) indicated that the TLV for gram-negative bacteria is $1000/m^3$ of air; for endotoxin the accepted level is 0.1 $\mu g/m^3$ of air.

Composting of MSW involves a preprocessing step to prepare the feedstock for effective and accelerated composting. This step is similar to activities at recycling facilities, refuse-derived fuel (RDF), incinerators or mass burn, and other solid waste processing facilities. The potential for release of endotoxin, gram-negative bacteria, and dust occurs on the tipping floor and during shredding, screening or trommeling and other activities involving materials handling.

Several studies have evaluated dust, endotoxin and biological agents at MSW facilities (Malmros et al., 1992; Sigsgaard et al., 1992; Malmros, 1990). Further, data from composting facilities have been published by Clark et al. (1983), Lundholm and Rylander (1980) and Rylander et al. (1983).

Lundholm and Rylander (1980) reported that workers at an experimental refuse compost facility had a higher incidence of subjective symptoms, such as nausea, headache, and diarrhea. Eleven employees were evaluated. Two reported nausea, one reported fever, five had headaches, and four had diarrhea. These symptoms were attributed to possible exposure to endotoxin.

Rylander et al. (1983) reported on studies of exposure to aerosols of microorganisms and toxins in sewage treatment and composting plants in the United States and Sweden. Data from composting plants showed average airborne dust levels ranging from 0.1 to 12.0 mg/m^3. The highest levels (median 10.6 mg/m^3) were found in the screening areas, the lowest near compost piles. The respirable proportion of gram-negative bacteria was in the range of 50% to 60%.

Rylander et al. (1983) concluded that too little information is available to establish dose-response relationships that could suggest standards. They suggested that a level of up to 1000 gram-negative bacteria/m^3 and 0.1 $\mu g/m^3$ of endotoxin should be considered as safe until additional information is available.

Clark et al. (1983) studied the airborne gram-negative bacteria endotoxin (lipopolysaccharide dust) at four Swedish composting plants. Three plants composted a mixture of solid waste and biosolids. The fourth composted biosolids and wood chips. Both indoor and outdoor sites were sampled at various operational locations. A considerable range of microbial concentrations were found in all plants. Gram-negative bacteria concentrations ranged from 0 to 370×10^3 CFU/m^3. Refuse hoppers, waste processing areas, and screening areas evidenced the highest concentration, ranging from 15×10^3 to 370×10^3 CFU/m^3 with median ranges of 43×10^3, 94×10^3, and 96×10^3, respectively for the processes cited. In most cases, the respirable size exceeded 50%. Endotoxin values ranged from 0.001 to 0.042 μg/m^3. These levels were below the suggested safe levels of 0.1 μg/m^3.

A study was conducted on 20 paper sorters, 8 garden compost workers, 44 garbage-handling workers and a control of 119 workers from the drinking water supply of Copenhagen (Sigsgaard et al., 1992). Conditions in the plant are described in Table 9.13. Health data revealed that there were no significant basic lung function parameters and bronchial reactivity across groups of workers. Garbage-handling employees had the greatest number of symptoms of chest tightness, influenza feeling, and itching of the eyes, nose, and throat. These symptoms are common to organic dusts and endotoxin. Endotoxin levels were within the criteria recommended by Rylander et al. (1989).

Malmros et al. (1992) evaluated total airborne viable microorganisms, total dust, airborne gram-negative bacteria, airborne fungi and endotoxin at an MSW sorting facility. The plant produced RDF and feedstock for composting. No information was provided on levels of the above parameters at the composting facility. The study provided data on the parameters studied before and after retrofitting the plant. Considerable reduction of the biological agents occurred as a result of plant modifications. Table 9.14a shows the conditions prior to plant improvements and Table 9.14b shows the reduction in biological agents after retrofitting.

Malmros et al. (1992) also obtained information on worker health. The plant employed 20 persons, 15 of whom were potentially exposed to occupational health risks. Nine employees had occupational disease. Upper-airway infections and acute bronchitis were the first symptoms. In three employees, organic toxic dust syndrome (ODTS) was suspected due to symptoms of light dyspnea, chills, feverishness, and general malaise. All nine workers had symptoms of cough, dyspnea. Other symptoms were not uniform across all workers. One of the nine was a non-smoker. He reported an influenza feeling and chest tightness. Eight of the nine exhibited bronchial asthma and one chronic bronchitis.

Nersting et al. (1991) evaluated the health risks associated with resource recovery recycling and composting as a result of endotoxin, dust, gram-nega-

TABLE 9.13. *Airborne Constituents in Several Danish Recycling Facilities.*

	Facility Type							
	Water Supply		Paper Sorting		Garden Compost		Garbage Handling	
Parameter	Mean	SD	Mean	SD	Mean	SD	Mean	SD
Total dust (mg/m³)	0.42	0.25	0.83	0.576	0.62	0.57	0.74	0.77[1]
Total airborne counts (CFU/m³)	76	38	4733	5891[1]	54,369	77,070[1,2]	46,133	125,503[1,2]
Gram-negative bacteria (CFU/m³)	10	1	4092	10,092	4521	5200[1]	4828	11,514[1]
Endotoxin (ng/m³)	2.5	3.9	1.3	1.5	0.8	1.1	2.5	4.4[2]
Molds (CFU/m³)	10	1	5229	5449[1]	26,884	67,556	14,372	30,990[1]
N	6		18		13		50	

[1]t-Test groups vs. water supply (control), p < 0.05.
[2]t-Test groups vs. paper sorting, p < 0.05.
Source: Sigsgaard et al., 1992.

TABLE 9.14a. Levels of Biological Agents at Various Locations in a Danish MSW Sorting Facility prior to Retrofitting the Plant.

Location/Process	Total Airborne Microorganisms (CFU/m³)	Gram-Negative Bacteria (CFU/m³)	Fungi (CFU/m³)	Endotoxin (µg/m³)
Tipping floor	>20,000	>6000	10,000	0.48
Manual sorting	>20,000	>6000	10,000	0.48
Office	3000	1300	2000	0.026
Secondary magnet	>10,000	8000	—[1]	0.55
Unpolluted exterior	1500	1200	4000	0.0003

[1]Numerous fungi on the plate providing uncertainty to the results.
Source: Malmros et al., 1992.

TABLE 9.14b. Levels of Biological Agents at Various Locations in a Danish MSW Sorting Facility after Retrofitting the Plant.

Location/Process	Total Airborne Microorganisms (CFU/m³)	Gram-Negative Bacteria (CFU/m³)	Fungi (CFU/m³)	Endotoxin (µg/m³)
Tipping floor	4800–7600	70–1600	1700–11,000	BDL[1]
Manual sorting	1000–51,000	0–2000	1400–16,300	0.110
Primary magnet	8400–77,000	1480–15,500	1900–18,000	BDL
Office	910–1800	250–410	60–1200	BDL
Outside	450–1000	35,250	30–260	—

[1]BDL = Below detection level.
Source: Malmros et al., 1992.

tive bacteria and fungi. The data for the composting plants are shown in Table 9.15. One plant composted garden and park waste; a second plant composted garden, park, animal and industrial waste; a third plant composted unsorted household wastes mixed with biosolids; and the fourth composted the organic fraction of household waste mixed with straw. Endotoxin levels were generally below levels shown to illicit response. Dust concentrations were below the TLV of 5 mg organic dust/m^3.

Marchand et al. (1995) evaluated bioaerosols in a MSW recycling and composting facility. The composting facility preprocessed MSW consisting of opening bags, manual sorting, mechanical grinding and composting in tunnels. The lowest counts of total bacteria, gram-negative bacteria, and fungi were found in the compost. Gram-negative bacteria ranged from 100 to 7200 in the morning and from 100 to 7900 in the afternoon. The highest levels were found in the sorting area, the magnetic conveyor, and the top and bottom of the tunnel composter. Unfortunately, there was no reported evaluation of *Aspergillus fumigatus* and endotoxin.

No reports are available on endotoxin impact on residences surrounding composting facilities. Dust control is very important to reduction of endotoxin levels.

Recently the National Institute for Occupational Health (NIOSH, 1995) issued a request for assistance in preventing organic dust toxic syndrome (ODTS), an acute respiratory illness. ODTS can occur when a large amount of organic dust is present in the air and appears to occur as a result of inhaling particles and toxins produced by microorganisms such as gram-negative bacteria (*Pseudomonas* species, *Enterobacter agglomerans*, and *Klebsiella* species), thermophilic organisms (*Aspergillus fumigatus* and *Micropolyspora faeni*), and other fungi (Schenker et al., 1991). NIOSH suggested that endotoxin, a component of organic dust, may be involved in the development of ODTS.

According to NIOSH, agricultural workers are at risk. They listed the following conditions:

- precipitin-negative farmer's lung disease (Edwards et al., 1974)
- grain fever in grain elevator workers (doPico et al., 1982)
- silo unloader's syndrome (Pratt and May, 1984)
- mill fever in cotton textile workers (Rylander et al., 1987)
- inhalation fever (Rask-Andersen and Pratt, 1992)

Cotton dust is the only specific agricultural dust for which OSHA has a standard: for byssinosis, a chronic lung disease (29 CFR 1910.1043). OSHS also has a standard for non-specific dusts: 15 mg/m^3 for total dust and 5 mg/m^3 for respirable dust (29CFR 1910.1000).

TABLE 9.15. *Range of Levels of Airborne Fungi, Endotoxin, and Gram-Negative Bacteria Found in Four Composting Plants in Denmark.*

Location	N	Total Microorganisms (CFU/m³)	Fungi (CFU/m³)	Gram-Negative Bacteria (CFU/m³)	Thermophilic Microorganisms
Reactor composting	2	$5 \times 10^2 - 2.7 \times 10^3$	$2 \times 10^2 - 2.6 \times 10^3$	$2 \times 10^2 - 5 \times 10^2$	$4 \times 10^2 - 2.1 \times 10^3$
Pile building/aeration	2	$9.22 \times 10^4 - 1.855 \times 10^5$	$1.09 \times 10^4 - 1.85 \times 10^4$	$1.6 \times 10^3 - 1.82 \times 10^4$	$3.9 \times 10^3 - 5.55 \times 10^4$
Indoor aeration of piles	1	8.357×10^5	5.7×10^3	5.86×10^4	2.28×10^4
Mattress building/aeration	4	$9.7 \times 10^3 - 2.505 \times 10^5$	$1.12 \times 10^4 - 2.505 \times 10^5$	$2.1 \times 10^3 - 1.14 \times 10^4$	$1.9 \times 10^3 - 3.11 \times 10^4$
Endotoxins: 0.11–3.49 ng/m³					
Dust <0.01–1.54 mg/m³					

Source: Nersting et al., 1991.

According to NIOSH, a total of 29 agricultural workers developed ODTS. They cited the following cases.

Case 1—Nine workers shoveling oats (Parker et al., 1988): Nine male workers, aged 15 to 60 years, developed chills and fever as a result of shoveling oats from a poorly ventilated storage bin. Airborne dust generated in a laboratory from the oats contained 39.5 mg/m³ of respirable dust. Endotoxin content in the respirable dust was elevated to 325.7 endotoxin units/mg dust.

Case 2—One worker shoveling wood chips and leaves (Weber et al., 1990): In the laboratory, investigators recreated the conditions and found that peak exposure to respirable dust was greater than 80 mg/m³; endotoxin concentrations ranged from 244 to 16,300 endotoxin units/m³ (Olenchock et al., 1991).

Case 3—Fourteen workers (silo unloader's disease): NIOSH cited Pratt and May (1984) and May et al. (1986) concerning workers unloading silos over an 11-year period. These workers encountered total dust levels ranging from 0.2 to 138 mg/m³; respirable dust ranging from 0.2 to 24 mg/m³; and microorganisms counts ranging from 10^5 to 10^{10}/m³.

Case 4—Five workers affected (CDC, 1986): Five workers at a municipal golf course became ill with influenza-like syndrome within hours of manually unloading a trailer filled with wood chips. The workers did not wear respiratory protection devices. Within 72 hours following handling of the wood chips, the workers had completely recovered.

Conrad et al. (1992) reported a case of microgranulomatous aspergillosis in a patient who shoveled wood chips. The researchers described a patient who developed chronic granulomatous disease; despite aggressive therapy, the disease was fatal. The patient had chronic granulomatous disease when admitted for fever and cough. He had been shoveling cedar chips in a landscaping project. He had been ill during childhood and had been hospitalized prior to this episode. The condition was attributed to inhalation of *Aspergillus fumigatus*. Since the described syndrome did not conform to the commonly described aspergillosis, the authors proposed a term, "microgranulomatous aspergillosis." They concluded that susceptible immunosuppressed individuals should be advised to avoid occupations where they might encounter high spore counts.

CONCLUSION

Bioaerosols are airborne organisms or biological agents that can affect the health of individuals. In composting, two bioaerosols can affect worker and public health, *Aspergillus fumigatus* and endotoxin. *A. fumigatus* is a very common fungus, associated with decaying organic matter. It is found throughout the environment including homes, places of work, hospitals, gardens, agricultural activities, parks, and other outdoor sites.

Aspergillus fumigatus is a secondary or opportunistic pathogen in that it primarily affects debilitated or immunocompromised individuals. Nearly all of the cases reported have occurred in hospitals. There is no known or established dose-response relationship or TLV levels for *A. fumigatus*.

Aspergillus fumigatus and other bioaerosols are found in composting feedstocks and are also generated during composting. The levels of *A. fumigatus* in the air of composting facilities are related to the type of process used and the specific activity. For example, windrow and agitating processes release greater numbers of bioaerosols. With aerated static pile systems, more bioaerosols are generated during pile teardown and material movement. Screening also generates considerable bioaerosols. High levels are found at a composting site but levels generally drop to background levels between 100 m and 200 m from the center of the site.

Endotoxin and organic dust from composting operations have not been shown to be of concern to the general public. These bioaerosols can impact worker health, however, presently, most of the available data are from MSW processing facilities in Denmark. Exposure to endotoxin in recycling, pre-processing of solid waste in composting facilities, and handling of MSW on tipping floors has affected worker health in Europe. To date, there are no reported cases in the United States. Proper ventilation and the use of dust masks reduce worker exposure to bioaerosols.

Workers are the most exposed individuals after nearly 20 years of intensive composting of biosolids in the United States and over 40 years of MSW composting in Europe. Not only are workers exposed to higher concentrations of bioaerosols, but during working hours they are more frequently exposed to bioaerosols than the residents or the general public adjacent to composting facilities. No reports or published literature indicate infections or disease from bioaerosols. Composting facilities must be designed to minimize worker exposure and release of bioaerosols to the atmosphere. Proper management of operations can greatly diminish release of bioaerosols to the environment.

REFERENCES

Aisner, J., S. C. Schimpff, J. E. Bennett, V. M. Young and P. H. Wiernik. 1976. *Aspergillus* infections in cancer patients, association with fireproofing materials in a new hospital. *J. Am. Med. Assoc.* 235(4):411–412.

Arnow, P. M., M. Sadigh, C. Costas, D. Weil, and R. Chudy. 1991. Endemic and epidemic aspergillosis associated with in-hospital replication of *Aspergillus* organisms. *J. of Infectious Diseases.* 164(5):998–1002.

Ault, S. K. and M. Schott. 1994. *Aspergillus,* aspergillosis and composting operations in California. *Tech. Bull. No. 1.* California Integrated Waste Management Board, Sacramento, CA.

Baxter, L. J. and J. T. Cookson. 1983. Natural atmospheric microbial conditions in a typical suburban area. *Appl. and Environ. Microb.* 45(3):919–934.

Bodey, G. P. and S. Vartibvarian. 1989. Aspergillosis. *European J. of Clinical Microbiology & Infectious Diseases.* 8(5):413–437.

Bradley, S. G. 1979. Cellular and molecular mechanisms of action of bacterial endotoxin. *Ann. Rev. Microb.* 33:67–94.

Burge, H. A. 1985. Fungus allergens. *Clin. Rev. Allergy.* 3:319–329.

Burge. H. A. and W. R. Solomon. 1990. Outdoor allergens, pp. 51–68. In R. F. Lockey and S. C. Bukantz (ed.). *Allergen Immunotherapy.* Marcel Dekker, Inc., New York.

Burge, P. S., M. Finnegan, N. Horsfield, D. Emery, P. Austwick, P. S. Davies, and C. A. C. Pickering. 1985. Occupational asthma in a factory with a contaminated humidifier. *Thorax.* 40:248–254.

Burnett, G. W. and G. S. Schuster. 1973. *Pathogenic Microbiology.* The C. V. Mosby Co., St. Louis, MO.

Burton, J. R., J. B. Zachery, R. Bessin, H. K. Rathbun, W. B. Greenough, S. Sterioff, J. R. Wright, R. E. Slavin, and G. M. Williams. 1972. Aspergillosis in four renal transplant recipients. *Ann. Intern. Med.* 77:383–388.

Castellan, R. M., S. A. Olenchock, K. B. Kinsey, and J. L. Hankinson. 1987. Inhaled endotoxin and decreased spirometric values: An exposure-response relation for cotton dust. *New Engl. J. Med.* 317:605–610.

Cavagna, G., V. Foa, and E. C. Vigliani. 1969. Effects in man and rabbits of inhalation of cotton dust or extracts and purified endotoxin. *Brit. J. Indust. Med.* 26:314–321.

CDC (Centers for Disease Control). 1986. Acute respiratory illness following occupational exposure to wood chips—Ohio. *MMWR* 35(30):483–490.

Chesapeake Occupational Health Services. 1991. *Health Surveillance Program for Compost Workers: An Epidemiologic Review. Report Submitted to Washington Suburban Sanitary Commission.* Site II, Silver Spring, MD. Chesapeake Occupational Health Services, Baltimore, MD.

Clark, C. S., H. S. Bjornson, J. W. Holland, T. L. Huge, V. A. Majeti, and P. S. Gartside. 1980a. Occupational hazards associated with sludge handling, pp. 215–244. In C. Bitton et al. (eds.). *Sludge Health Risks of Land Application.* Ann Arbor Sci. Pub. Inc., Ann Arbor, MI.

Clark, C. S., H. S. Bjornson, J. W. Holland, V. J. Elia, and V. A. Majeti. 1980b. *Evaluation of the Health Risks Associated with the Treatment and Disposal of Municipal Wastewater and Sludge.* Health Effects Research Lab, Cincinnati Univ., Ohio. R-805445 USEPA 600/1-80-030. 252 pp.

Clark, C. S., R. Rylander, and L. Larsson. 1983. Levels of gram-negative bacteria, *Aspergillus fumigatus,* dust and endotoxin at compost plants. *Applied and Environ. Microb.* 45(5):1501–1505.

Clark, C. S., H. S. Bjornson, J. Schwartz-Foulton, J. W. Holland, and P. S. Gartside. 1984. Biological health risks associated with composting of wastewater treatment plant sludge. *J. Water Poll. Control Fed.* 56(12):1269–1276.

Clayton Environmental Consultants, Ltd. 1983. *Air Sampling Program for Total Coliforms, Particulate, and Fungal Spores at Selected Areas in the Windsor West Pollution Control Plant.* Windsor, Ontario.

COD Report. 1991a. The committee on organic dusts—A presentation. International Commission on Occupational Health.

COD Report. 1991b. Health effects of grain and other organic dusts. *Proc. of an International Workshop.* Perth, W. Australia.

Commonwealth Mycological Institute. 1966. *Descriptions of Pathogenic Fungi and Bacteria.* No. 92. Eastern Press Ltd., London.

Conrad, D. J., M. Warnock, P. Blanc, M. Cowan, and J. A. Golden. 1992. Microgranulomatous aspergillosis after shoveling wood chips: Report of a fatal outcome in a patient with chronic granulomatous disease. *Amer. J. Ind. Med.* 22:411–418.

Cookson, J. T. J., R. B. Smith et al. (1983). Microbiological health significance of aerosols at sewage sludge composting facilities, pp. 148–167. In E. I. Stentiford (ed.). *Proc. International Conference on Composting of Solid Wastes and Slurries.* Univ. of Leeds, England.

Conrad, D. J., M. Watoock, P. Blanc, M. Cowan, and J. A. Golden. 1992. Microgranulomatous aspergillosis after shoveling wood chips: Report of a fatal outcome in a patient with chronic granulomatous disease. *Am. J. of Ind. Med.* 22:411–418.

Cox, J. N., F. di Dio, G.-P. Pizzolato, R. Lerch, and N. Pochon. 1990. *Aspergillus* endocarditis and myocarditis in a patient with the acquired immunodeficiency syndrome (AIDS). *Virchows Arichiv A Pathol. Anat.* 417:255–259.

Cundell, A. M. and A. P. Mulcock. 1972. Microbiological deterioration of vulcanized rubber. *Inter't. Biodeter. Bull.* 8:119–125.

Cunha, B. A. 1995. *Invasive Aspergillosis. Aspergillosis Special Report.* Internal Medicine. Medical Economics, Montvale, NJ.

DeLucca, A. J., M. A. Godshall, and M. S. Palmgren. 1984. Gram-negative bacterial endotoxin in grain elevator dusts. *Am. Ind. Hyg. Assoc. J.* 45:336–339.

Denning, D. W., S. E. Follansbee, M. Scolaro, S. Norris, H. Edelstein, and D. A. Stevens. 1991. Pulmonary aspergillosis in the acquired immunodeficiency syndrome. *The New England J. of Med.* 324(10):654–662.

Denning, D. W. and D. A. Stevens. 1990. The treatment of invasive aspergillosis. *Rev. Infect. Dis.* 12:1147.

doPico, G. A., D. Flaherty, P. Bhansali, and N. Chavaje. 1982. Grain fever syndrome induced by inhalation of airborne grain dust. *J. Allergy Clin. Immunol.* 69(5):435–443.

Dutkiewicz, J., L. Jablonski, and S. A. Olenchock. 1988. Occupational biohazards: A review. *Am. J. Ind. Med.* 14:605–623.

E&A Environmental Consultants, Inc. 1993. *Internal Report.* Canton, MA.

E&A Environmental Consultants, Inc. 1994. *Internal Report.* Canton, MA.

Edwards, J. H., J. T. Baker, and B. H. Davies. 1974. Precipitin test negative farmer's lung—Activation of the alternative pathway of complement by mouldy hay dusts. *Clin. Allergy.* 4:379–388.

Emmons, C. W., C. H. Binford, and J. P. Utz. 1977. *Medical Mycology.* Lea & Febiger, Philadelphia, PA.

Epstein, E. and J. I. Epstein. 1985. Health risks of composting. *BioCycle.* 26(4):38 40.

Fink, J. 1986. Hypersensitivity pneumonitis, pp. 481–499. In J. Merchant (ed.). Occupational

Respiratory Diseases. Dept. Health and Human Services (NIOSH), Publ. No. 86-102. U.S. Gov. Printing Office, Washington, DC. 801 pp.

Gaucher, E. and E. Sargent. 1984. Un cas de pseudotuberculose aspergillaire simple chez un gaveur de pigeons. *Bulletins et memoirs de la Societé Medicale de Hopitaux de Paris.* 11:512–521.

Greenberger, P. A. 1984. Allergic bronchopulmonary aspergillosis. *J. Allergy and Clinical Immunology.* 74(5):645–653.

General Physics Corporation. 1991. Environmental Services Divison. *Data Results: Revised Bioaerosol Monitoring Program for the Washington Suburban Sanitary Commission.* Montgomery County Regional Composting Facility. General Physics Corporation, Silver Spring, MD.

Hampton Roads Sanitation District. 1981. Aspergillus fumigatus: *A Background Report.* Hampton Roads, VA.

Hirsch, S. R. and J. A. Sosman. 1976. A one-year survey of mold growth inside twelve homes. *Annals of Allergy.* 36:30–38.

Hopkins, C. C., D. J. Webber, and R. H. Rubin. 1989. Invasive aspergillus infection: Possible non-ward source within the hospital environment. *J. of Hospital Infection.* 13(1):19–25.

Hughes, R. L. 1968. Microbiological degradation in the paper, printing and packaging industries, pp. 281–190. In A. H. Walters and J. J. Elphick (eds.). *Biodeterioration of Materials.* Elsevier Publ. Co., Ltd.

Iacoboni, M. D., T. J. LeBrun, and J. Livingston. 1980. Deep windrow composting of dewatered sewage sludge. In *Proc. National Conf. on Municipal & Industrial Sludge Composting.* Hazardous Materials Control Res. Institute, Rockville, MD.

Jones, B. L. and Cookson, J. T. 1983. Natural atmosphere microbial conditions in a typical suburban area. *Appl. and Environ. Microb.* 45(3):919.

Jones, W., K. Morring, S. A. Olenchock, T. Williams, and J. Hickey. 1984. Environmental study of poultry confinement buildings. *Am. Ind. Hyg. Assoc. J.* 45(11):760–766.

JTC Environmental Consultants, Inc. 1981. *Determinative Microbiological and Health Significance of Aerosols of a Sewage Sludge Composting Facility.* Report to Washington Suburban Sanitary Commission and Maryland Environ. Serv. JTC Environmental Consultants, Inc., Bethesda, MD.

Kaarik, A. A. 1974. Decomposition of wood. In C. H. Dickinson and C. J. F. Pugh (eds.). *Biology of Plant Litter Decomposition, Vol. 1.* Academic Press.

Kane, B. E. and J. T. Mullins. 1973. Thermophilic fungi in a municipal waste compost system. *Mycologia.* 65:1087–1100.

Kilburn, K. 1986. Chronic bronchitis and emphysema, pp. 503–529. In J. Merchant (ed.). *Occupational Respiratory Diseases.* Dept. Health and Human Services (NIOSH), Publ. No. 86-102. U.S. Gov. Printing Office, Washington, DC.

Kilmowski, L. L., C. Rothstein, and K. M. Cummings. 1989. Incidence of nosocomial aspergillosis in patients with leukemia over a twenty-year period. *Infection Control & Hospital Epidemiology.* 10(7):299–305.

Kodama, A. M. and R. I. McGee. 1986. Airborne microbial contaminants in indoor environments: Naturally ventilated and air conditioned homes. *Arch. Env. Health.* 41(5): 306–311.

Kothary, M. H., T.Chase, Jr., and J. D. Macmillan. 1984. Levels of *Aspergillus fumigatus* in air and in compost at a sewage sludge composting site. *Environ. Pollut.* (Series A) 34:1–14.

Kramer, M. N., V. P. Kurup, and J. N. Fink. 1989. Allergic bronchopulmonary aspergillosis from a contaminated dump site. *Am. Rev. Respir. Dis.* 140:1086–1088.

Lacey, J. 1975. Airborne spores in pastures. *Trans. Brit. Mycol. Soc.* 64:265–281.

Lacey, J. 1991. Aerobiology and health: the role of airborne fungal spores in respiratory disease, pp. 157–185. In D. L. Hawksworth (ed.). *Frontiers in Mycology.* Honorary and general lectures from the Fourth International Mycological Congress, Regensburg, Germany, C.A.B. International.

Land, C. J., K. Hult, R. Fuchs, S. Hagelberg, and H. Lundstrom. 1987. Tremorgenic mycotoxins from *Aspergillus fumigatus* as a possible occupational health problem in sawmills. *Appl. and Environ. Microb.* 53(4):787–790.

Larsen, L. and S. Gravesen. 1991. Seasonal variation of outdoor airborne viable microfungi in Copenhagen, Denmark. *Grana.* 30:467–471.

Lenhart, S. W. and S. A. Olenchock. 1984. Sources of respiratory insult in the poultry processing industry. *Am. J. Ind. Med.* 6:89–96.

Lees, P. S. J. and M. S. Tockman. 1987. *Evaluation of Possible Public Health Impact of WSSC Site II Sewage Sludge Composting Operations.* Report for Maryland Dept. of Health and Mental Hygiene. Johns Hopkins U., Baltimore, MD.

Lumpkins, E. D., Sr., S. L. Corbit, and G. T. Tiedeman. 1973. Airborne fungi survey. 1. Culture-plate survey of the home environment. *Annals of Allergy.* 31:361–370.

Lundholm, M. and R. Rylander. 1980. Occupational symptoms among compost workers. *J. Occup. Med.* 22(4):256–257.

Malmros, P. 1990. *Problems with the Working Environment in Solid Waste Treatment.* The National Labour Inspection of Denmark. Report Nr. 10/1990. Copenhagen, Denmark.

Malmros, P., T. Sigsgaard, and B. Bach. 1992. Occupational health problems due to garbage sorting. *Waste Management & Res.* 10:227–234.

Marchand, G. J. Lavole, and L. Lazure. 1995. Evaluation of bioaerosols in a municipal solid waste recycling and composting plant. *J. Air & Waste Manag. Assoc.* 45:778–781.

May, J. J., D. S. Pratt, L. Stallibes, P. R. Morey, S. A. Olenchock, and W. Deep. 1986. A study of silo unloading; the work environment and its physiologic effects. *Am. J. Ind. Med.* 10:318.

Meeker, D. P., G. N. Gephaardt, E. M. Cordasco Jr., and H. P. Wiedemann. 1991. Hypersensitivity pneumonitis versus invasive pulmonary aspergillosis: Two cases with unusual pathologic findings and review of the literature. *Amer. Rev. Respiratory Dis.* 143:431–436.

Millner, P. D., D. A. Bassett, and P. B. Marsh. 1980. Dispersal of *Aspergillus fumigatus* from sewage sludge compost piles subjected to mechanical agitation in open air. *Applied Environ. Microb.* 39:1000–1009.

Millner, P. D., P. B. Marsh, R. B. Snowden, and J. F. Parr. 1977. Occurrence of *Aspergillus fumigatus* during composting of sewage sludge. *Appl. and Environ. Microb.* 34(6):765–772.

Millner, P. D., S. A. Olenchock, E. Epstein, R. Rylander, J. Haines, J. Walker, B. L. Ooi, E. Horne, and M. Maritato. 1994. Bioaerosols associated with composting facilities. *Compost Sci. & Util.* 2(4):6–57.

Mills, J. and H. O. W. Eggins. 1974. The biodeterioration of certain plasticisers by thermophilic fungi. *Internat. Biodeterior. Bull.* 10:9–44.

Mullins, J., R. Harvey, and A. Seaton. 1976. Sources and incidence of airborne *Aspergillus fumigatus* (Fres). *Clinical Allergy.* 6:209–217.

Murray, H. W., J. O. Moore, and R. D. Luff. 1975. Disseminated aspergillosis in renal transplant patient: Diagnostic difficulties re-emphasized. *Johns Hopkins Med.* 137: 235–237.

Nersting, L., P. Malmros, T. Sigsgaard, and C. Petersen. 1991. Biological health risks associated with resource recovery, sorting of recycled waste and composting. *Grana.* 30:454–457.

NIOSH. 1995. *Request for Assistance in Preventing Organic Dust Toxic Syndrome.* 1995. Publ. No. 94-102. National Institute for Occupational Safety and Health, Cincinnati, OH.

Nobel, W. C. and Y. M. Clayton. 1963. Fungi in the air of hospital wards. *J. of Gen. Microb.* 32:397–402.

NYDOH. 1994. *A Prospective Study of Health Symptoms and Bioaerosol Levels Near a Yard Waste Composting Facility.* Islip Composting Facility, Town of Islip, Suffolk County, New York, State of New York, Department of Health, Albany, NY.

Olenchock, S. A., J. C. Mull, and W. J. Jones. 1983. Endotoxin in cotton: Washing effects and size distribution. *Am. J. Ind. Med.* 4:515–521.

Olenchock, S. A., W. G. Sorenson, G. J. Kullman, W. G. Jones, and J. J. Marx. 1991. Biohazards in composted wood chips, pp. 481–483. In H. W. Rossmoore (ed.). *Biodeterioration and Biodegradation 8.* Elsevier Applied Science, London, England.

Opal, S. M., A. A. Asp, P. B. Cannady, Jr., P. L. Morse, L. J. Burton, and P. G. Hammer II. 1986. Efficacy of infection control measures during a nosocomial outbreak of disseminated aspergillosis associated with hospital construction. *J. Infectious Dis.* 153:634–637.

Orlita, A. 1968. Biodeterioration in leather industry of materials. In Walters, A. H. and J. J. Elphick (eds.). *Biodeterioration of Materials.* Elsevier Publ. Co. Ltd., New York.

Parker, J. E., R. M. Castellan, S. A. Olenchock, W. G. Sorenson, and J. J. Marx. 1988. Organic dust toxic syndrome in farm workers following exposure to stored oats. *Amer. Rev. Respir. Dis.* 137(4)(suppl):297.

Passman, F. J. 1980. *Monitoring of* Aspergillus fumigatus *Associated with Municipal Sewage Sludge Composting Operations in the State of Maine. Final Report.* Portland Water District, Portland, ME.

Patterson, T. F. 1995. *Invasive Aspergillosis in Organ Transplantation. Aspergillosis Special Report.* Internal Medicine, Medical Economics, Montvale, NJ.

Pepys, J. 1969. Hypersensitivity diseases of the lungs due to fungi and organic dusts. *Monogr. Allerg. Vol. 4.* S. Karger, Basel.

Petheram, I. S. and R. M. E. Seal. 1976. *Aspergillus* prosthetic valve endocarditis. *Thorax.* 31:380–390.

Polacheck, I., I. F. Salkin, D. Schenhav, L. Ofer, M. Maggen, and J. H. Haines. 1989. Damage to an ancient parchment document by *Aspergillus. Mycopathologia.* 106:89–93.

Pratt, D. S. and J. J. May. 1984. Feed-associated respiratory illness in farmers. *Arch. Environ. Health.* 39(1):43–48.

Pursell, K. J., E. E. Telzak, and D. Armstrong. 1992. *Aspergillus* species colonization and invasive disease in patients with AIDS. *Clin. Infect. Dis.* 14(1):141–148.

Rask-Andersen, A. and D. S. Pratt. 1992. Inhalation fever: A proposed unifying term for febrile reactions to inhalation of noxious substances. *Br. J. Ind. Med.* 49:40.

Richerson, H. B. 1990. Unifying concepts underlying the effects of organic dust exposure. *Amer. J. Indus. Med.* 17:139–142.

Richerson, H. M. 1993. Hypersensitivity pneumonitis. In R. R. Rylander and R. R. Jacobs (eds.). *Organic Dust: Exposure, Effects and Prevention.* Lewis Publ., Chicago.

Richet, H. H., M. M. McNeil, B. J. Davis, E. Duncan, J. Strickler, D. Nunley, W. R. Jarvis, and O. C. Tabian. 1992. *Aspergillus fumigatus* sternal wound infections in patients undergoing open heart surgery. *Am. J. of Epidemiology.* 135(1):48–58.

Rifkind, D., T. L. Marchioro, S. A. Schneck, and R. B. Hill. 1967. Systemic fungal infections complicating renal transplantation and immunosuppressive therapy. *Amer. J. Med.* 43:28–38.

Rippon, J. W. 1974. *Medical Mycology. The Pathogenic Fungi and the Pathogenic Actinomycetes.* W. B Saunders, Philadelphia.

Rodenhuis, S., F. Beaumont, H. F. Kauffman, and H. J. Sluiter. 1984. Invasive pulmonary aspergillosis in a non-immunosuppressed patient: Successful management with systemic amphotericin and flucytosine and inhaled amphotericin. *Thorax.* 39:78–79.

Rose, H. D. and B. Varkey. 1975. Deep mycotic infection in the hospitalized adult: A study of 123 patients. *Medicine.* 54(6):499–507.

Rosenberg, M., R. Patterson, R. Mintzer, B. J. Cooper, M. Roberts, and K. E. Harris. 1977. Clinical and immunologic criteria for the diagnosis of allergic bronchopulmonary aspergillosis. *Ann. Intern. Med.* 86:405.

Rylander, R. 1986. Lung diseases caused by organic dusts in the farm environment. *Am. J. of Indust. Med.* 10:221–227.

Rylander, R., D. C. Christiani, and Y. Peterson. 1989. Committee on Organic Dusts of the International Commission on Occupational Health (A Report). Institutionen fuo hygien Goteborgs Universitet, Gothenburg, Sweden.

Rylander, R. and P. Haglind. 1984. Airborne endotoxin and humidifier disease. *Clin. Allerg.* 14:109–112.

Rylander, R., M. Lundholm, and C. S. Clark. 1983. Exposure to aerosol of micro-organisms and toxin during the handling of sewage sludge, pp. 69–78. In P. M. Wallis and D. L. Lehemann (eds.). *Biological Health Risk of Sludge Disposal to Land in Cold Climates.* Calgary University, Alberta, BC, Canada.

Rylander, R. and Y. Peterson. 1990. Organic dust and lung diseases. *Amer. J. Indust. Med.* 17:1–148.

Rylander, R., R. S. F. Schilling, C. A. C. Pickering, G. B. Rooke, A. N. Dempsey, and R. R. Jacobs. 1987. Effects after acute and chronic exposure to cotton dust: The Manchester criteria (editorial). *Br. J. Ind. Med.* 44:577–579.

Rylander, R. and J. Vesterlund. 1982. Airborne endotoxin in various occupational environments, pp. 399–409. In Watson et al. (eds.). *Endotoxins and Their Detection with Limulus Amebocyte Lysate Test.* Alan R. Liss, New York.

Salvaggio, J. E. 1970. Hypersensitivity pneumonitis: "Pandora's box." *New England J. of Medicine.* 283(6):314–315.

Salvaggio, J. E., G. Taylor, and H. Weil. 1986. Occupational asthma and rhinitis, pp. 461–477. In J. Merchant (ed.). *Occupational Respiratory Diseases.* Dept. Health and Human Services (NIOSH) Publ. No. 86-102. U.S. Govt. Printing Office, Washington, DC.

Schenker, M. T., T. Ferguson and T. Gamsky. 1991. Respiratory risks associated with agriculture. *Occup. Med: State of the Art Reviews.* 6(3):414–428.

Schlueter, D. P., N. F. Jordan, and G. T. Hensley. 1972. Wood-pulp workers' disease: A hypersensitivity pneumonitis caused by *Alternaria. Annals of Inter. Med.* 77:907–914.

Sheargren, J. N. 1986. Role of inhaled endotoxin symptoms suffered by workers in the West Windsor Pollution Control Plant. Paper submitted to the City of Windsor.

Sikora, L. J., P. D. Millner, and W. D. Burge. 1985. *Chemical and Microbial Aspects of Sludge Composting and Land Application.* Final Report submitted to USEPA, Municipal Environmental Research Laboratory, Cincinnati, OH.

Singh, N., V. L. Yu, and J. D. Rihs. 1991. Invasive aspergillosis. *So. Med. J.* 84(7):822–827.

Sigsgaard, T., P. Malmros, L. Nersting, and C. Petersen. 1992. *Respiratory Disease and Atopy in Danish Resource Recovery Workers.* Unpublished Report. Institute of Environ. and Occupational Med., U. of Aarhus, Denmark.

Slavin, R. G. and P. Winzenburger. 1977. Epidemiological aspects of allergic aspergillosis. *Annals of Allergy.* 38(3):215–218.

Solomon, W. R. 1975. Assessing fungus prevalence in domestic interiors. *J. Allergy Clin. Immunol.* 56(3):235–242.

Solomon, W. R. 1976. A volumetric study of winter fungus prevalence in the air of Midwestern homes. *J. Allergy Clin. Immunol.* 57(1):46–55.

Stedman's Medical Dictionary. 1977. The Williams & Wilkins Company, Baltimore, MD.

Summerbell, R. C., S. Krajden, and J. Kane. 1989. Potted plants in hospitals as reservoirs of pathogenic fungi. *Mycopathologia.* 106:13–22.

Stevens, D. A. 1992. Aspergillosis, pp. 1901–1903. In J. B. Wyngaarden, L. H. Smith, Jr., and J. C. Bennett (eds.). *Cecil Textbook of Medicine.* 19th Edition. W.B. Saunders Co., Philadelphia, PA.

Thomas, A. R. 1977. The genus *Aspergillus* and biodeterioration, pp. 453–479. In J. E. Smith and J. A. Pateman (eds.). *Genetics and Physiology of* Aspergillus. Brit. Mycol. Soc. Symp. Ser. No. 1, Academic Press. London.

Turner, J. N. 1967. *The Microbiology of Fabricated Materials.* J. and A. Churchill Ltd. London.

Van Assendeleft, A., K. O. Forsen, H. Keskinen, and K. Alanko. 1979. Humidifier-associated extrinsic allergic alveolitis. *Scan. J. Work Environ. & Health.* 5:35–41.

Van den Bogart, H. G., G. Van den Ende, P. C. Van Loon, and L. J. Van Griensven. 1993. Mushroom worker's lung: Serologic reactions to thermophilic actinomycetes present in the air of compost tunnels. *Mycopathologia.* 122(1):21–28.

Vincken, W. and P. Roels. 1984. Hypersensitivity pneumonitis due to *Aspergillus fumigatus* in compost. *Thorax.* 39:74–75.

Virchow, R. 1856. Beitrage zur lehre von den beim menschen vorkommenden pflanzlichen parasiten. *Archiv Pathologische Anatomie und Physiologie Virchow's.* 9:557–593.

Wang, J. L. F., R. Patterson, M. Rosenberg, M. Roberts, and B. J. Cooper. 1978. Serum IgE and IgG antibody against *Aspergillus fumigatus* as a diagnostic aid in allergic bronchopulmonary aspergillosis. *Am. Rev. Respiratory Dis.* 117:917.

Warren, C. P. W. 1981. Respiratory disorders in Manitoba cattle farmers. *CMA J.* 125:41–46.

Weber, S., E. Petsonk, G. Kullman, W. Jones, S. A. Olencock, and W. Sorenson. 1990. Hypersensitivity pneumonitis (HP) or organic dust toxic syndrome (ODTS): The clinical dilemma in organic dust exposure. *Am. Rev. Respir. Dis.* 141(4):A588.

Wingard, J. R. 1995. Aspergillus *Infections in Immunocompromised Cancer Patients.* Aspergillus *Special Report.* Internal Medicine, Medical Economics, Montvale, NJ.

Yoshida, K., A. Ueda, H. Yamasaki, K. Sato, K. Uchida, and M. Ando. 1993. Hypersensitivity pneumonitis resulting from *Aspergillus fumigatus* in a greenhouse. *Archives Environ. Health.* 48(4):260–262.

Zinsser, H. 1968. *Microbiology.* 14th Ed. Appleton-Century-Crofts, New York.

Zuk, J. A., D. King, H. D. Zakhour, and J. C. Delaney. 1989. Locally invasive pulmonary aspergillosis occurring in a gardener: An occupational hazard? *Thorax.* 44:678–679.

Zwerling E. M. and P. F. Strom. 1991. Levels of *Aspergillus fumigatus* at yard waste composting facilities in New Jersey. *Theobald Smith Soc. Pub. No. K-07526-1-91.* Rutgers U., New Brunswick, NJ.

CHAPTER 10

Odors and Volatile Organic Compounds

INTRODUCTION

One of the major problems associated with composting is odors. Several large MSW composting facilities, including those in Portland, Oregon, Dade County, Maryland, and Pembroke Pines, Florida, have been shut down because of odors. These facilities were poorly designed with little or no odor control. Similarly, several large yard waste and food waste facilities have had either to curtail operations or invest considerable funds in retrofitting or modifying their operations due to odors. Odors are a nuisance to neighboring receptors. Further, the presence of odors may focus public attention on health issues, as people often associate malodors with negative health impacts.

In many cases biosolids or MSW composting occurs in partially or entirely enclosed buildings. Under these conditions any odors emitted by the process can be collected and treated. Furthermore, the methodologies used, such as the aerated static pile and several mechanical systems, can be managed with a fairly high degree of odor management. Therefore, in many cases when odors have occurred, physical retrofitting of the facility has often reduced odors to acceptable levels.

Composting of yard wastes invariably occurs outdoors in turned windrows. The frequency of windrow turning, especially in the early phases of composting, can result in odor generation.

Today our knowledge of odor generation, odor management, and odor control allows us to design composting facilities that result in minimal odor impacts on adjacent communities.

A major factor affecting odor generation is the feedstock involved. Materials that decompose rapidly are apt to produce odors at high concentrations.

For example, raw biosolids produce odors at a higher intensity than digested biosolids. Grass and green waste generally produce more odors than brush or leaves. Further, compacted wet leaves produce a stronger unpleasant fermentation odor than fresh dry leaves.

The compounds that produce odors also differ, depending on the condition of the feedstock and the stage of composting. For example, odors emitted when materials are anaerobic are different from odors emitted during aerobic conditions. The generation and level of odors during composting are a function of the type of feedstock, composting system, system design, and facility operations.

Odor problems have occurred in biosolids, municipal solid waste, yard waste, and food waste composting facilities. During decomposition, odorous compounds are generated that, when emitted into the atmosphere, are a nuisance to populations living near composting facilities. As a result of the potential for odors, several states in the United States have restricted the operations of composting facilities in proximity to residences and businesses.

New York, for example, requires a buffer of 500 feet for municipal solid waste and biosolid composting facilities and 200 feet for yard waste composting facilities. Maine requires a buffer of 500 feet from residences. New Jersey requires a distance of 150 feet between leaf composting sites and residences. Finally, the state of California is proposing in the green material composting regulations (January 27, 1993 draft) a setback of at least 300 feet from any residence or hospital.

Odors are characterized in several ways (Walker, 1993):

(1) Odor quantity—dilutions to threshold: Odor levels are usually expressed as dilution-to-threshold (D/T) ratio rather than concentration. D/T values are determined by an odor panel of eight to ten people; they express the number of dilutions required by 50% of the panel to detect the odor (ED_{50}).

(2) Odor intensity: This is a measure of the perceived strength of the odor and is related to concentration. A standard compound such as n-butanol is used and is expressed as milligrams per liter. The relationship between perceived intensity and concentration is expressed as follows (Hooper and Cha, 1988):

$$S = KI^n$$

where

S = perceived intensity of sensation
I = physical intensity of odorant concentration

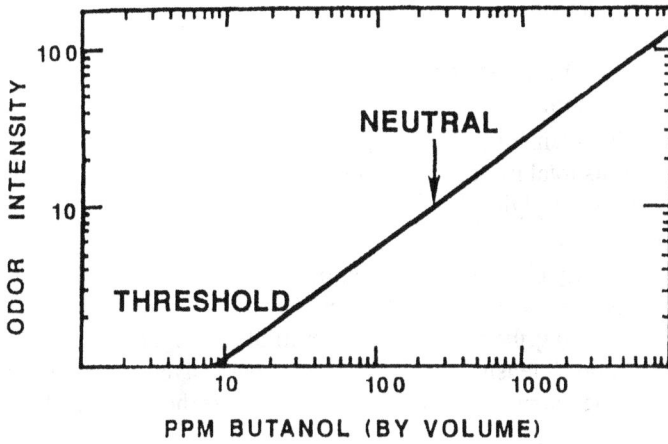

FIGURE 10.1. Odor intensity as related to concentration of butanol used as a standard. (Data from Hooper and Cha, 1988.)

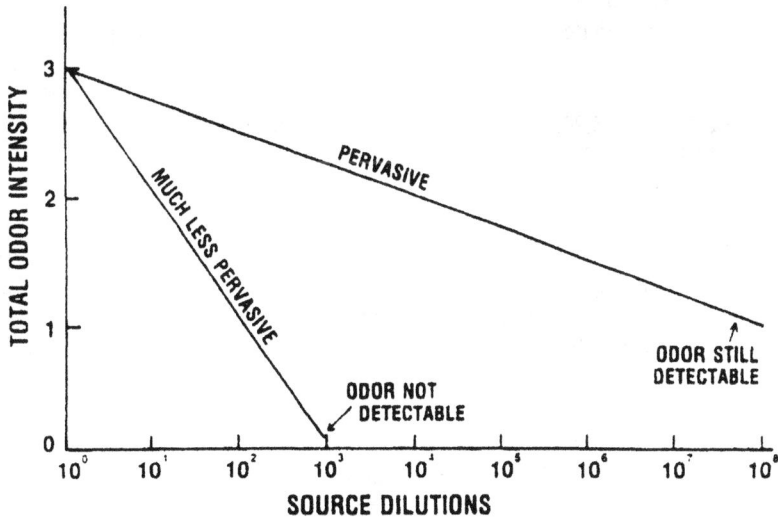

FIGURE 10.2. Dilution characteristics as related to total odor intensity for pervasive and less pervasive odors. (Data from Walker, 1993.)

n = slope of psychophysical function
K = y-intercept

Figure 10.1 illustrates the relationship between odor intensity and concentration.

(3) Once the intensity (butanol equivalent concentration) of an odorant is known, its total mass per unit time can be used to estimate the mass of odor by multiplying it times the volume of odorous air per unit time (Walker, 1993).

(4) The intensity characteristic of an odor is used to calculate odor pervasiveness. Pervasiveness is determined by progressively diluting the odor and measuring the intensity at each dilution. Figure 10.2 illustrates a plot of odor intensity versus dilution of pervasive and lesser pervasive odors. The flatter the slope, the more pervasive the odor (Walker, 1993).

This chapter will provide information on the type of compounds that may cause odors; levels measured as a result of different feedstocks and processing methods; and current work with modeling of odor emissions from composting facilities. The formation and emission of volatile organic compounds (VOCs) during composting will also be discussed. Odor treatment and control will be dealt with in *The Technology of Composting.*

ODOROUS COMPOUNDS AND ODORS EMITTED AT COMPOSTING FACILITIES

Considerable investigation and evaluation have been conducted at biosolids composting facilities in the United States (Van Durme et al., 1990; Wilber and Murray, 1990; Williams and Miller, 1993; Hentz et al., 1992; Walker, 1993). As a result, composting at these facilities is often highly controlled, and the design of the facilities incorporates extensive mitigation measures (USEPA, 1992). By comparison, little data are available on odorous compounds generated from food waste or yard waste facilities.

Table 10.1 shows some of the odorous compounds identified during composting (Miller, 1993; Williams and Miller, 1993). The major odor groups include: fatty acids; ammonia and other nitrogen-containing compounds; ketones; aromatics; and inorganic and organic sulfur compounds (Wilber and Murray, 1990; Williams and Miller, 1993).

Sulfur compounds and ammonia have been found in many biosolids and animal waste composting facilities. Sulfur compounds are also high in food wastes as they are a component of several amino acids. The odorous compounds most commonly found are hydrogen sulfide, dimethyl sulfide,

TABLE 10.1. *Compounds Identified in Composting Odors.*

Compound	Odor Characteristic	Odor Threshold µg/m³ Low[1]	Odor Threshold µg/m³ High	Odor Threshold µg/m³ ADL[2]
Sulfur Compounds				
Hydrogen sulfide	rotten egg	0.7	14	6.7
Carbon oxysulfide	pungent			
Carbon disulfide	disagreeable, sweet	24.3	23,000	665
Dimethyl sulfide	rotten cabbage	2.5	50.8	2.5
Dimethyl disulfide	sulfide	0.1	346	—
Dimethyl trisulfide	sulfide	6.2	6.2	—
Methamethiol	sulfide, pungent	0.04	82	4.2
Ethanethiol	sulfide, earthy	0.032	92	2.6
Ammonia and Nitrogen Containing Compounds				
Ammonia	pungent, sharp	26.6	39,600	33,100
Aminomethane	fishy, pungent	25.2	12,000	—
Dimethylamine	fishy, amine	84.6	84.6	88.1
Trimethylamine	fishy, pungent	0.8	0.8	0.52
3-Methylindole (skatole)	feces, chocolate	4.0×10^{-5}	268	—

(continued)

TABLE 10.1. (continued).

Compound	Odor Characteristic	Odor Threshold		
		$\mu g/m^3$ Low[1]	$\mu g/m^3$ High	$\mu g/m^3$ ADL[2]
Volatile Fatty Acids				
Methanoic (formic)	biting	45.0	37,800	—
Ethanoic (acetic)	vinegar	2500	250,000	2500
Propanoic (progionic)	rancid, pungent	84.0	60,000	—
Butanoic (butyric)	rancid	1.0	9000	3.7
Pentanoic (valeric)	unpleasant	2.6	2.6	—
3-Methylbutanoic (isovaleric)	rancid cheese	52.8	52.8	—
Ketones				
Propanone (acetone)	sweet, minty	47,500	1,610,000	241,000
Butanone (MEK)	sweet, acetone	737	147,000	30,000
2-Pentanone (MPK)	sweet	28,000	45,000	—
Other Compounds				
Benzothiozole	penetrating	442	2210	—
Ethanol (acetaldehyde)	green sweet	0.2	4140	385
Phenol	medicinal	178	2240	184

[1]Low threshold indicates the lower limit of detection to most sensitive persons. High threshold value means that it is odorous to most persons.
[2]Values recalculated from volume/volume data assuming 20°C and 1 atmosphere.
Source: Williams and Miller, 1993.

dimethyl disulfide, ammonia, limonene, and pinene (Hentz et al., 1992; Van Durme et al., 1992). The latter two are aromatic compounds released from wood chips used as a bulking agent in biosolids operating facilities.

Ammonia is also often released during composting operations involving animal wastes, food wastes, and sewage sludge. Feedstocks with low carbon-to-nitrogen ratios (lower than 20:1) release ammonia during composting. This has been a major problem with the composting of grass clippings, for example. As the C/N ratio increases, the ammonia levels decrease. Volatile fatty acids (VFAs) can also be a source of odors during the decomposition of organic matter (Kissel and Henry, 1992).

Iacoboni et al. (1980) evaluated odor emissions during the composting of biosolids in windrows. Figure 10.3 shows data for a single windrow; however, similar data were obtained for six windrows. Peak emissions generally occurred from 7 to 23 days into the composting cycle. The highest odor emissions were measured immediately after turning. After 15 minutes, the odors reached the levels prior to turning (Figure 10.4).

The majority of odors during a composting cycle were the result of surface emissions. Windrow turning accounted for only 15% of the odor units measured (Table 10.2). Windrows that were turned more frequently had lower odor emissions, but the increased number of turnings made the overall odor emissions the same as windrows that were turned less frequently. Overall, odor emissions are not as important to monitor as peak emissions, since peak emissions are most likely to be detected by receptors.

Van Durme et al. (1990) evaluated the odor and ammonia levels during active biosolids composting and curing at the Hampton Roads, Virginia,

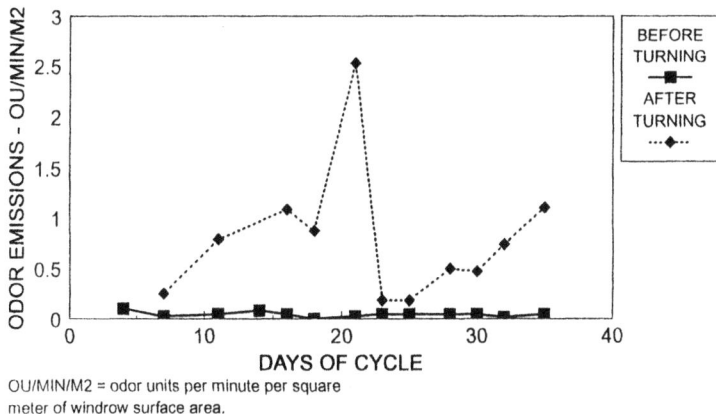

OU/MIN/M2 = odor units per minute per square
meter of windrow surface area.

FIGURE 10.3. Odor emissions before and after windrow turning during composting of biosolids. (Data from Iacoboni et al., 1980.)

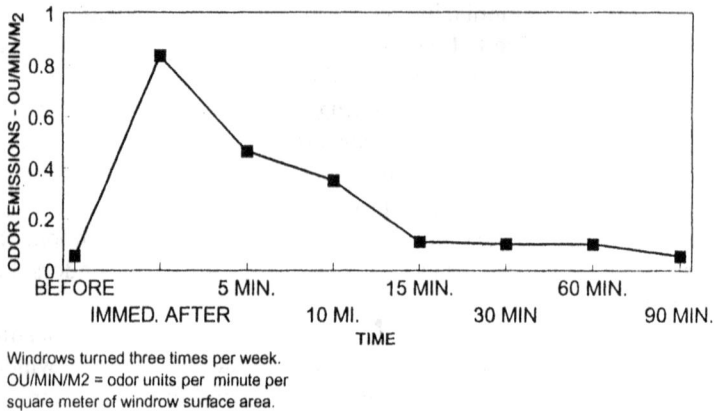

Windrows turned three times per week.
OU/MIN/M2 = odor units per minute per
square meter of windrow surface area.

FIGURE 10.4. Average odor emissions during a turning cycle. (Data from Iacoboni et al., 1980.)

composting facility. Odors were determined by collecting the gases in Tedlar bags and using the ASTM E-679 odor panel technique. Results are reported as ED_{50}. This method uses a panel to determine the number of dilutions of fresh air needed to reduce odor detection to the threshold level. The results are reported when 50% of the panel can detect an odor.

Van Durme et al. (1990) found different levels of odors at various times of the year. For example, odor levels measured in June were higher than in October, a difference that could not be explained. The study compared odors resulting from different aeration systems. Under negative aeration, surface odors were substantially less than during positive aeration (Figure 10.5). Further, ammonia levels decreased substantially with time for both surface and exhaust under negative aeration. Blower exhaust had the highest concentrations of odorous emissions, but exhaust can be captured as a point source and treated. Surface emissions, on the other hand, are a non-point source. Under negative aeration, surface odors were reduced during the third week and ammonia levels dropped during the second week. At the negative blower exhaust, there was a slight increase or no change in odors over time. Under positive aeration, ammonia surface emissions increased during the first two weeks of composting.

Studies have shown that odor generation is greatest during the first 9 to 10 days and is greatly reduced thereafter (Iacoboni et al., 1980). Experience by this author and other E&A personnel has shown that for the aerated static pile, negative aeration is best during the first 10 to 15 days. The exhaust air can be scrubbed in a biofilter or by means of other odor-scrubbing techniques. After that time, positive aeration can be used to reduce moisture for better screening and materials handling.

TABLE 10.2. *Odor Emissions from Windrow Surfaces and as Affected by Turning.*

Windrow No.	Ambient Odor Emission (ou/m²)	Peak Odor Emission after Turning (ou/min/m²)	Odor Emission per Turn (ou/m²)	No. of Turns	Odor Emissions Given off as a Result of Turning (ou/m²)	Total Odor Emissions for a Six-Week Compost Cycle (ou/m²)
1	2809	0.798	23.97	30	719	3528
2	2809	0.985	29.54	18	532	3341
3	2809	2.053	61.59	12	739	3548
4	2809	0.502	15.05	30	451	3261
5	2809	1.133	34.00	18	612	3421
6	2809	0.771	23.13	12	278	3087

Source: Iacoboni et al., 1980.

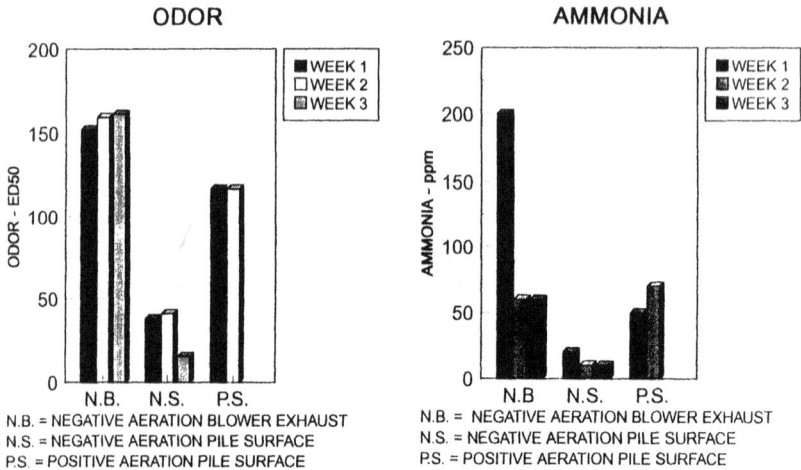

FIGURE 10.5. Odor and ammonia levels from an aerated static pile system. (Data from Van Durme et al., 1990.)

Table 10.3 shows the compounds that were found to be the major contributors to odors. The data represent the average of three sampling dates, showing considerable variation between dates. Dimethyl disulfide required the greatest dilutions to threshold in two of the three samples. The authors also compared their data to those obtained from other facilities (Table 10.4). Dimethyl disulfide was also highest for Site II and Denver and one of the highest for Sarasota. Similarly, most of the significant odor compounds identified at Hampton Roads were also found at other biosolids composting facilities.

Wilber and Murray (1990) found a direct correlation between oxygen demand and odor production at a biosolids composting facility. They noted that moisture was an important parameter; a higher moisture content of material going into the composting process resulted in higher generation of odors.

The type of biosolids also determined odor potential. For example, composting of secondary biosolids resulted in the highest level of odors. The order of odor production from highest to lowest was as follows: secondary biosolids > secondary + primary > primary > anaerobically digested and nitrified biosolids.

Finally, composting temperatures were found to affect odors (Figure 10.6). Odors were least with the high temperatures generated during composting.

Toffey et al. (1995) reported that effective and uniform aeration resulted

TABLE 10.3. *Significant Odorous Compounds Detected in Blower Exhaust from an Aerated Static Pile Composting Facility.*

Significant Odorous Compounds	Mean (μg/m³)	Range (μg/m³)	Threshold Odor Conc. (μg/m³)	Dilution to Threshold Mean	Dilution to Threshold Range
Ammonia	121K	45–170K	4K	30	11–43
Hydrogen sulfide	936	405–1062	12	61.5	34–89
Dimethyl disulfide	1042	860–1311	5	208	172–262
Dimethyl sulfide	1342	0–2667	3	493	0–1026
Limonene	1064	45–2667	6	178	8–445
α-Pinene	221	78–333	64	3	1–5
ED$_{50}$				784	152–1700

Source: Van Durme et al., 1990.

TABLE 10.4. Odorous Compounds Emitted in Blower Exhaust for Several Biosolids Composting Facilities.

Compost Facility	Hampton Roads, VA	Site II, MD	Metro Denver, CO	Sarasota, FL	Cape May, NJ
Process	Aerated static pile	Aerated static pile	Aerated windrow	Vertical in-vessel	Vertical in-vessel
Type of biosolids	Digested	Lime	Digested	Raw	Raw
Sample location	Blower exhaust	Blower exhaust	Blower exhaust	Blower exhaust	Blower exhaust
	$\mu g/m^3$	$\mu g/m^3$	$\mu g/m^3$	$\mu g/m^3$	$\mu g/m^3$
Ammonia	45K	231K	315K	858	30K
Hydrogen sulfide	—	—	152	<660	400
Acetophe-none	29	621	—	945	—
Dimethyl disulfide	959	46,289	8584	12,600	2464
Dimethyl sulfide	—	1305	3480	1055	—
Dimethyl trisulfide	137	3621	522	2314	—
Limonene	45	4520	—	6209	—
α-Pinene	251	3760	754	12,660	—
β-Pinene	258	261	232	12,408	—
Measured odor	ED_{50}	ED_{50}	ED_{50}	ED_{50}	ED_{50}
	152	288–832		440	300

Source: Van Durme et al., 1990.

312

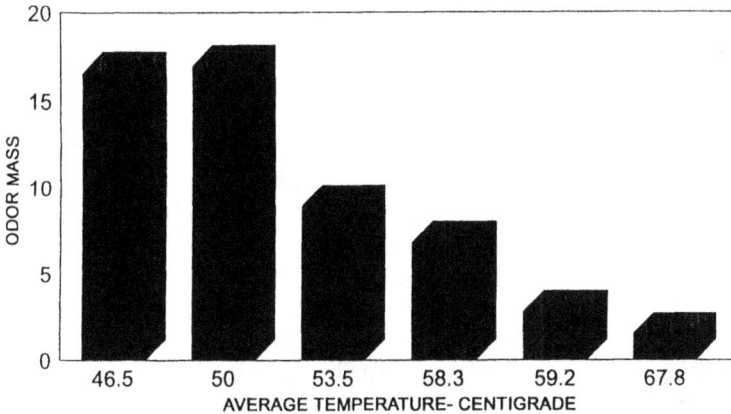

FIGURE 10.6. Effect of composting temperature on odor production. (Reprinted with permission from Wilber and Murray, 1990.)

in lower pile temperatures and reduced emissions of odors and VOCs. These researchers found that odor emissions increased with increased operating temperatures. Total odor emissions for various composting processes at the Biosolids Recycling Center (BRC) at the Southwest Water Pollution Control Plant in Philadelphia are shown in Figure 10.7. As illustrated, curing piles represented the greatest source of odors. Curing piles, sludge storage, and biofilters represented 35%, 26%, and 23% of the total mass emissions.

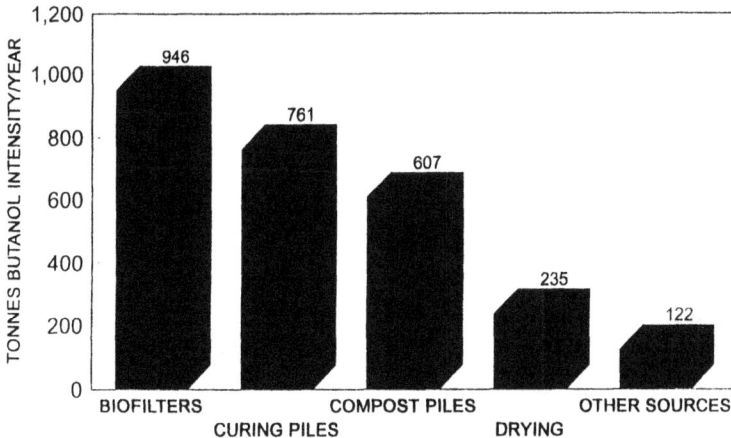

FIGURE 10.7. Total odor emissions at the biosolids composting facility in Philadelphia. (Data from Toffey et al., 1995.)

Modification and improvement of composting and curing resulted in the most effective reduction of odors (Figure 10.8).

Studying composting of food waste with varying proportions of yard waste, E&A Environmental Consultants, Inc., found that the odorous compounds were fatty acids, mercaptans, ketones, and ammonia (Table 10.5).

Many odorous compounds are formed during anaerobic periods. If a facility is designed properly, anaerobic conditions can be avoided through careful management of the composting process. Table 10.6 shows the effect of bulk density and porosity on oxygen levels and mercaptan production. Alone, yard waste, having the highest porosity, had high oxygen levels and very low mercaptan levels. However, increasing the percent of food waste or reducing the proportion of yard waste in the mix resulted in lower oxygen levels and higher mercaptan levels. Grinding, which reduced particle size and increased the bulk density, increased the mercaptan levels. The highest mercaptan concentration was observed with the lowest yard waste-to-food waste ratio and the smallest particle size.

Hovsenius (1987) found that high water content in the aerated piles of solid waste being composted resulted in free water at the bottom zone of the composting mass. This undoubtedly restricted aeration and resulted in anaerobic conditions and odor production.

The author visited Borlange, Sweden, where the research was conducted. The aeration system consisted of rectangular openings in the concrete floor. Composting was done outdoors. Air distribution was inadequate. A mix-box type machine was constructed, which was used to periodically mix the

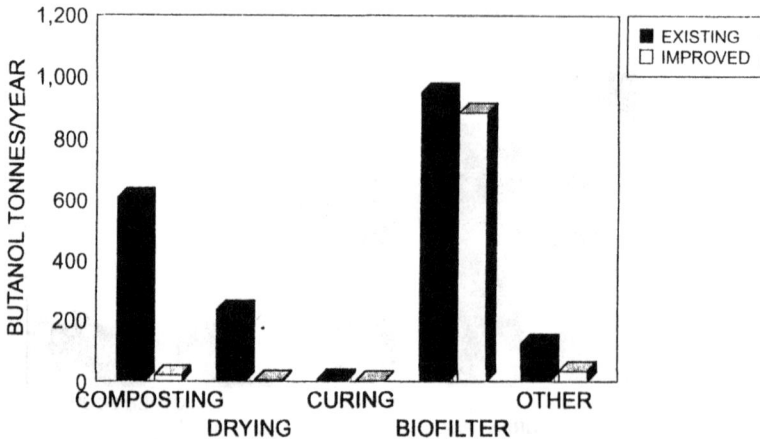

FIGURE 10.8. Annual odor emissions and potential reductions due to improvements of aerated static pile composting of biosolids. (Data from Toffey et al., 1995.)

TABLE 10.5. Odorous Compounds Detected at a
Composting Food Waste Facility.

Material or Compound and Decomposition Condition	Product	Laboratory Analysis (ppm)	Field Analysis (ppm)
Fatty acids, anaerobic decomposition	F formic acid Acetic acid	2.5 25.3	ND ND
Mercaptans and other organic sulfides, an-aerobic decomposition	Mercaptans Organic sulfides	0.32	ND 0.2–100 in pile
Ketones, anaerobic de-composition	Methyl ethyl ketone	600	—
Ammonia, aerobic, and anaerobic decom-position	Ammonia	ND	0.05–7 in exhaust air

ND = not detected.
Source: E&A Environmental Consultants, Inc., 1993.

TABLE 10.6. Effect of Initial Mix Density on Oxygen Levels
and Mercaptan Production.

Initial Mix Ratio: Yard Debris:Produce Waste	4:0	4:1	4:1	2:1
Preprocessing	None	None	Ground[1]	Ground[1]
Bulk density, kg/m³	151	252	755	704
Oxygen concentration percent[2]	19.9	18.8	0.3	0
Total mercaptans[3]	0.2	0.5	25	100

[1]Entire pile was ground with a hammermill.
[2]Air sample taken from a depth of 4 feet and a height of 3 feet.
[3]Air sample taken from the pile surface.
Source: E&A Environmental Consultants, Inc., 1993.

compost. Hovsenius (1987) recommended that moisture content not exceed 55% and that in this situation periodic turning is necessary.

VOLATILE ORGANIC COMPOUNDS

Organic compounds can be volatilized during composting, resulting in odors. Some compounds are the result of microbial degradation (Van Durme et al., 1990), others are compounds found in the feedstock.

Mercaptans and short-chain fatty acids emit highly objectionable odors at extremely low concentrations. Table 10.7 lists the odor-relevant VOCs found in the inlet and outlet of a chemical scrubber located at the Site II biosolids composting facility in Maryland (Lees and Tockman, 1987). Only those compounds that exceeded 5 ppb in the inlet gas are presented from the 151 compounds reported. The extensive list of compounds indicates the depth of the study in trying to identify odorous compounds.

The compounds in the exhaust gas were well below levels considered toxic. These concentrations represent maximum levels and, when released

TABLE 10.7. Odor-Relevant Volatile Organic Compounds in the Inlet and Outlet of a Chemical Scrubber at Site II, Maryland.

Compound	Inlet 2 (ppb)	Inlet 1 (ppb)	Outlet (ppb)
Vinyl ethyl ether	0.00	31.00	0.48
Trichlorofluoromethane	0.00	22.00	0.23
Dimethyl sulfide	0.00	22.00	0.00
Acetone	0.00	114.00	3.37
Methylethylketone	0.00	24.00	0.27
2-Methyl-butanol	0.00	5.00	0.00
3-Methyl-butanol	0.00	8.00	0.00
Benzene	0.00	5.00	0.14
Ethanone	0.00	7.00	0.00
Alpha-Pinene	0.00	74.00	0.00
Camphene	0.00	10.00	0.00
Dimethyl disulfide	0.78	106.00	0.07
Methypropylpentanol	0.00	10.00	0.00

TABLE 10.7. (continued).

Compound	Inlet 2 (ppb)	Inlet 1 (ppb)	Outlet (ppb)
Beta-pinene	0.00	5.00	0.00
Xylene	0.00	7.00	0.09
Decahydronaphthalene	0.00	14.00	0.00
Methylene octanenitrile	0.00	9.00	0.00
Limonene	0.21	29.00	0.03
Decahydromethylnaphthalene	0.00	7.00	0.00
Alkyl benzene	0.02	6.00	0.08
Methanol + cycloal	0.00	8.00	0.00
Alkylbenzene + alkane	0.00	7.00	0.00
Alkyl benzene	0.00	5.00	0.09
Alkylbenzene	0.00	5.00	0.00
Methyl hexanol	0.00	5.00	0.00
Ethyl hexanol	0.00	7.00	0.00
Acetophenone	0.05	5.00	0.21

Source: Lees and Tockman, 1987.

into the atmosphere, are highly unlikely to present a health hazard (Lees and Tockman, 1987).

Volatile organic compounds in the exhaust air during composting of biosolids in aerated static pile were characterized by Van Durme et al. (1992) (Table 10.8). Seventy-two different compounds were identified, and 29 of them with listed threshold limit values (TLV). The TLV is a time-weighted average concentration for a normal 8-hour work day and a 40-hour work week to which all workers are exposed daily without adverse effects.

In all cases, the TLVs were several magnitudes higher than the values found in the exhaust air. Comparison of the air emissions data to the state of Virginia allowable limits, which are 1/60 of the TLV for non-carcinogens and 1/100 of the TLV for carcinogens in ambient air, showed that the concentration of measured compounds was considerably lower than regulatory levels.

Toffey et al. (1995) evaluated VOC emissions for several odor sources at the Philadelphia Biosolids Recycling Center, which uses the aerated static

TABLE 10.8. *Volatile Organic Compounds in the Exhaust Air during Aerated Static Pile Composting of Biosolids.*

Compound	Concentration (μg/m³)	ACGIH-TLV[1] (μg/m³)
Acetaldehyde	60	180,000
Acetic acid	25	25,000
Acetone	2574	1,780,000
Benzene	104	30,000
Carbon disulfide	224	30,000
Chlorobenzene	9	350,000
Cyclohexane	327	1,050,000
Cyclohexanone	13	100,000
Cyclopentane	442	1,720,000
Dichlorobenzene	9	300,000
2-Ethoxyethanol	9	19,000
Ethylbenzene	16	435,000
Fluorotrichloromethane	1493	5,600,000

TABLE 10.8. (continued).

Compound	Concentration (μg/m³)	ACGIH-TLV[1] (μg/m³)
Heptane	39	1,600,000
Heptanone	46	230,000
Methanol	153	260,000
Methylacetate	144	610,000
Methyl chloride	16	175,000
Methyl ethyl ketone	974	590,000
Nonane	19	1,050,000
Octane	15	1,450,000
Pentane	884	1,800,000
Phenol	13	19,000
n-Propanol	64	500,000
Pyridine	47	15,000
Styrene	26	215,000
Toluene	488	375,000
1,1,2-Trichloroethane	27	45,000
Xylene	29	435,000

[1]American Conference of Governmental Industrial Hygienists—Threshold Limit Value: The time weighted average concentration for a normal 8-hour work day and a 40-hour work week, to which nearly all workers are exposed, day after day, without adverse effect.
Source: Van Durme et al., 1992.

pile composting method. The study showed that inadequate pile aeration resulted in high temperatures and inconsistency. Active composting took between 40 to 60 days; piles appeared to reheat, indicating instability. Biofilters were very dry, poorly constructed and ineffective.

The study evaluated a prototype aeration and biofilter and estimated the value of improvements. The results of the existing operations on VOC emissions and potential reductions due to improvements in aeration and biofilter design are shown in Figure 10.9. Biofilters, compost piles and curing piles were the highest sources of VOC emissions, representing 84% of the annual VOC emissions. Significant reductions in VOC from these sources could be achieved. The major VOC were toluene, 2-butanone (methyl ethyl ketone), *o*-xylene, ethyl benzene, styrene, and 1,1,2,2-tetrachloroethane.

Data on VOC concentrations at blower exhaust, biofilter exhaust, and curing and composting piles are shown in Figure 10.10. Improvements would greatly reduce VOC concentrations at these sources. With improvements in aeration and reduction in temperature during composting, it was predicted that annual VOC emissions could be reduced from 61.1 tonnes to 10.2 tonnes.

Total sulfur emissions for several sources are shown in Figure 10.11. Again, the major sources were the biofilters, compost piles, and curing piles. The predominant compounds were dimethyl disulfide, ethyl disulfide, dimethyl sulfide and carbon disulfide. Fifty-two percent of the sulfur emis-

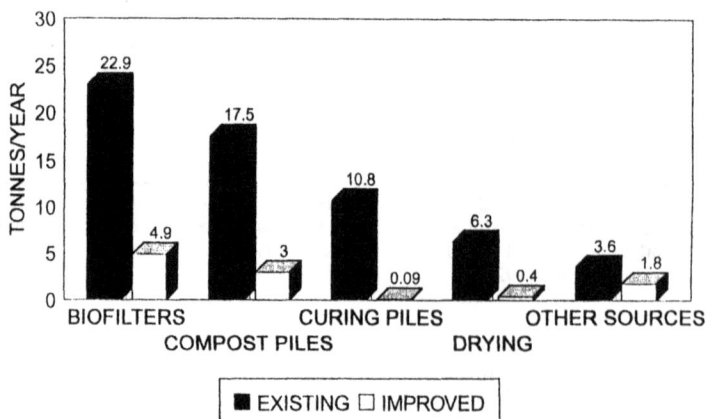

FIGURE 10.9. Existing and potential reductions in VOC as a result of improvements to a static pile composting system at the biosolids composting facility in Philadelphia. (Data from Toffey et al., 1995.)

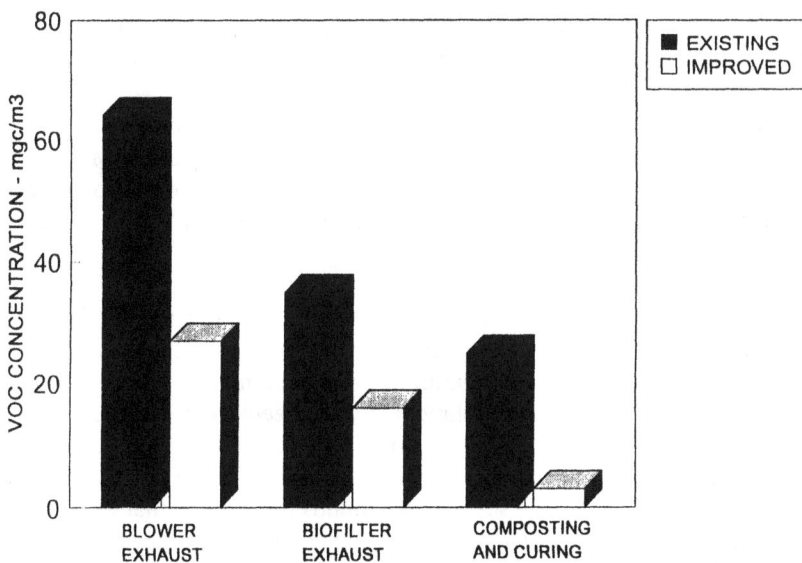

FIGURE 10.10. Effect of process modifications in an aerated static pile system on volatile organic compounds emissions. (Data from Toffey et al., 1995.)

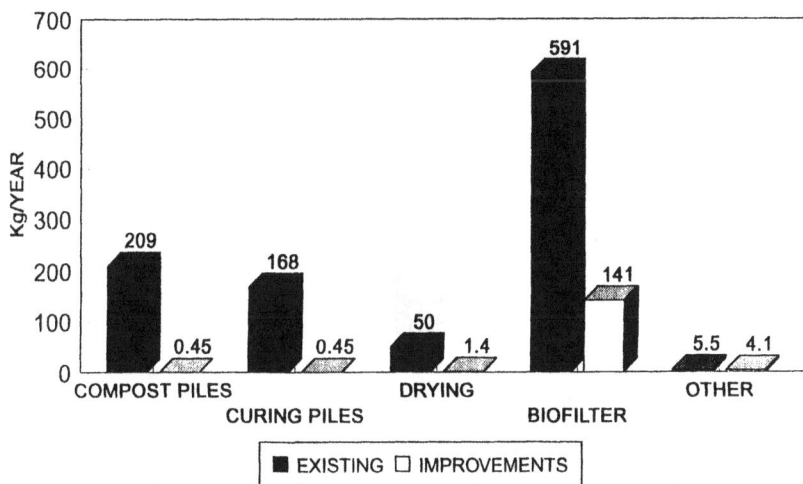

FIGURE 10.11. Total existing and potential emissions of sulfur compounds from an aerated static pile facility. (Data from Toffey et al., 1995.)

sions was caused by dimethyl disulfide (Figure 10.12). There was a good correlation between dimethyl disulfide emission concentrations and odors (ED_{50}). Dimethyl disulfide emission concentrations increased with composting temperatures.

Eitzer (1995) measured VOC emissions during the composting of MSW. Six facilities composted only MSW, one facility composted MSW and biosolids, while another composted MSW and yard waste. It would have been interesting to see the differences in VOC emission for the different waste streams.

The data provided in Table 10.9 show the concentration of targeted compounds at various processing functions. Generally, the tipping floor and shredding operations had the highest emissions of many of the compounds. Thus, many compounds are volatilized prior to composting. The air in the receiving (tipping floor) and shredding area needs to be exhausted and scrubbed to provide good working conditions as well as to treat for odor removal.

Biosolids and MSW had several compounds in common that were found at higher concentrations than many of the other compounds analyzed. These included trichlorofluoromethane, acetone, benzene, and ethyl benzene. Compounds not found in biosolids exhaust but found in MSW air were 2-butanone and trichloroethane. Dimethyl disulfide, dimethyl sulfide, limonene and pinene were found in biosolids composting operations but not

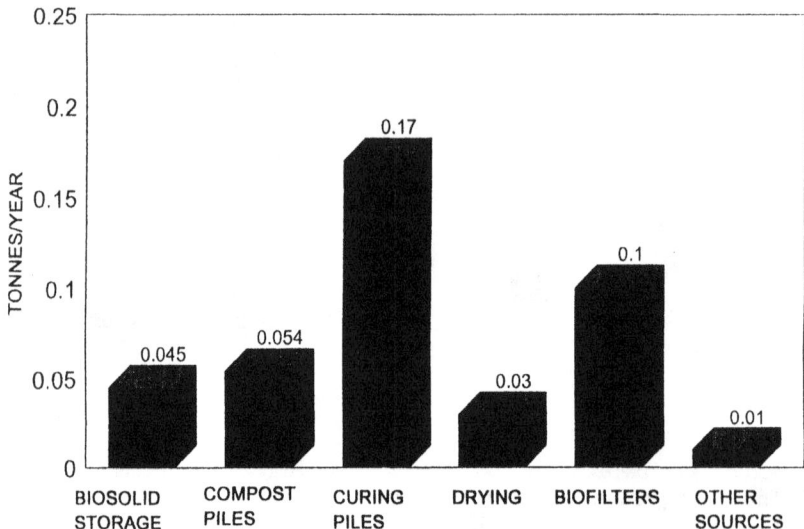

FIGURE 10.12. Total dimethyl disulfide emissions at the Philadelphia biosolids recycling center prior to improvements. (Data from Toffey et al., 1995.)

TABLE 10.9. Volatile Organic Compounds at Various MSW Composting Process Units.

Sample Type	Bkg. Air	Tip Floor	Shred	Indoor	Dig.	Fresh	Mid-Age	Old	Cure
Number of Samples	11	9	5	4	2	17	41	15	11
Compound									
Trichlorofluoromethane	3200	12,000	3500	2500	21,000	64,300	3500	4700	3700
Acetone	390	6100	0	7800	0	9200	9500	6100	2300
Carbon disulfide	0	0	0	0	0	5	9	8	6
Methyl chloride	1	2	2	0	0	1	0	0	25
1,1-Dichloroethane	0	0	0	0	0	0	0	0	0
2-Butanone	130	25,500	1400	9400	36,000	4300	8700	14,500	1400
Chloroform	5	6	7	1	0	1	2	0	0
1,1,1-Trichloroethane	10	1200	1100	410	2300	1500	49	99	30
Carbon tetrachloride	5	6	58	3	0	1	2	3	4
1,2-Dichloroethane	0	0	0	0	0	0	0	0	0
Benzene	34	12	8	150	130	35	22	50	120
Trichloroethane	0	4	16	5	21	98	3	3	2

(continued)

TABLE 10.9. (continued).

Sample Type	Bkg. Air	Tip Floor	Shred	Indoor	Dig.	Fresh	Mid-Age	Old	Cure
Number of Samples	11	9	5	4	2	17	41	15	11
Compound									
2-Hexanone	396	1700	1000	1000	1200	540	1000	700	130
Toluene	9	8800	11,500	1100	860	3900	700	400	88
Tetrachloroethane	2	170	130	23	360	210	37	78	4
4-Methyl-2-pentanone	22	28	1500	62	0	900	1100	830	60
Chlorobenzene	0	0	1	0	1	1	0	2	0
Ethylbenzene	11	38,100	23,400	2900	2900	3100	940	610	780
m,o-Xylene	2	2600	3700	250	150	260	190	120	49
Styrene	1	910	1600	110	250	250	76	31	4
Isopropyl benzene	0	2	44	51	21	45	9	10	17
n-Propyl benzene	0	65	130	42	23	69	47	7	2
4-Chlorotoluene	0	25	0	0	0	14	0	0	0
1,3,5-Trimethylbenzene	2	310	610	180	180	320	130	46	7
1,2,4-Trimethylbenzene	2	250	390	160	190	200	95	66	6

324

TABLE 10.9. (continued).

Sample Type	Bkg. Air	Tip Floor	Shred	Indoor	Dig.	Fresh	Mid-Age	Old	Cure
Number of Samples	11	9	5	4	2	17	41	15	11
Compound									
sec-Butylbenzene	0	0	0	0	0	15	0	0	0
1,3-Dichlorobenzene	0	0	0	0	0	0	0	0	0
1,4-Dichlorobenzene	0	0	0	0	6	3	1	3	4
p-Isopropyl toluene	0	13	10	44	64	280	190	66	3
1,2-Dichlorobenzene	0	0	0	0	0	0	0	0	0
n-Butylbenzene	0	47	35	17	36	27	11	13	1
1,2,4-Trichlorobenzene	0	0	0	0	0	0	1	0	0
Naphthalene	0	22	340	55	29	92	98	45	8
Hexachlorobutadiene	0	0	0	0	0	0	0.1	0	0
1,2,3-Trichlorobenzene	0	0	0	0	0	0	0.1	0	0

Bkg. air = background air; tip floor = tipping floor; shred = air by shredder; indoor = indoor air at facility; dig. = back end of digester; fresh = fresh from newly formed active compost; mid-age = midway in composting, 1/5 to 4/5 into the composting process; old = end of active composting process; cure = on-site but outside.
Source: Eitzer, 1995.

reported in MSW facility air. The latter two compounds are a product of the wood chips or wood waste used in composting biosolids. One sample high in toluene and xylene at the tipping floor was the result of recent spillage of a load of paint.

Table 10.10 shows the maximum observed concentrations for all the facilities. These data are compared to the threshold limit value (TLV) in

TABLE 10.10. Maximum Observed Concentration of Volatile Organic Compounds in Comparison to Threshold Limit Values in Workplace Air.

Compound	Maximum Observed Concentration (μg/m³)	TLV[1] (μg/m³)
Trichlorofluoromethane	915,000	5,620,000
Acetone	166,000	1,800,000
Carbon disulfide	150	31,000
Methylene chloride	260	174,000
1,1-Dichloroethane	1	400,000
2-Butanone	320,000	590,000
Chloroform	54	49,000
1,1,1-Trichloroethane	15,000	1,900,000
Carbon tetrachloride	290	31,000
Benzene	700	32,000
Trichloroethane	1300	270,000
2-Hexanone	6600	20,000
Toluene	66,000	188,000
Tetrachloroethane	5600	339,000
4-Methyl-pentanone	16,000	205,000
Chlorobenzene	29	46,000
Ethylbenzene	178,000	434,000
m,o-Xylene	15,000	434,000[2]
p-Xylene	6900	434,000[2]
Styrene	6100	213,000

TABLE 10.10. (continued).

Compound	Maximum Observed Concentration ($\mu g/m^3$)	TLV[1] ($\mu g/m^3$)
Isopropyl benzene	370	246,000
n-Propyl benzene	1200	not listed
4-Chlorotoluene	240	not listed
1,3,5-Trimethylbenzene	2200	123,000[2]
1,2,4-Trimethylbenzene	1000	123,000[2]
sec-Butylbenzene	220	not listed
1,3-Dichlorobenzene	2	not listed
1,4-Dichlorobenzene	90	451,000
p-Isopropyl toluene	4800	not listed
1,2-Dichlorobenzene	1	150,000
n-Butylbenzene	210	not listed
1,2,4-Trichlorobenzene	9	123,000[2]
Naphthalene	1400	52,000
Hexachlorobutadiene	4	210
1,2,3-Trichlorobenzene	6	not listed

[1]American Conference of Governmental Industrial Hygienists—Threshold Limit Value: The time weighted average concentration for a normal 8-hour work day and a 40-hour work week, to which nearly all workers are exposed, day after day, without adverse effect.
[2]Value for sum of all isomers of each compound.
Source: Eitzer, 1995.

workplace air as referenced by the American Conference of Governmental Industrial Hygienists (ACGIH, 1992). Values indicate that the concentrations are considerably below the threshold values for workers in these facilities. The data also suggest that there is no potential health hazard from VOCs in outdoor air to the general public near similar composting facilities.

Heida et al. (1995) measured air samples collected from exhausts of two ventilation shafts from an organic waste composting facility in the Netherlands (Table 10.11). Alkylbenzenes were the most predominant compounds. Lemonene, a compound present in woody material, was found in concentrations exceeding 140,000 $\mu g/m^3$. However, it is not very toxic to humans. Comparisons were provided for some Dutch TLVs.

TABLE 10.11. *Concentrations of Volatile Organic Compounds in the Air of Exhausts from an Organic Waste Composting Facility.*

Compound	Left Vent Shaft Concentration ($\mu g/m^3$)	Right Vent Shaft Concentration ($\mu g/m^3$)	Netherlands and Dutch TLV ($\mu g/m^3$)
Aromatic Hydrocarbons			
Benzene	5	<1	30,000
Toluene	43	76	375,000
Ethylbenzene	21	19	435,000
Xylene	67	53	400,000
Naphthalene	36	44	50,000
Other alkylbenzenes	11,000[1]	13,000[1]	—
Chlorinated Hydro-carbons			
Dichloromethane	<1	<1	—
Dichloroethane	<1	<1	—
1,1-Dichloroethane	<1	<1	—
1,2-Dichloropropane	<1	<1	—
Trichloromethane	<1	<1	—
Tetrachloromethane	<1	<1	12,600
Trichloroethane	<1	<1	45,000
Tetrachloroethane	<1	<1	240,000
1,1,1-Trichloroethane	<1	<1	—
1,1,2-Trichloroethane	<1	<1	—

TABLE 10.11. (continued).

Compound	Left Vent Shaft Concentration (μg/m³)	Right Vent Shaft Concentration (μg/m³)	Netherlands and Dutch TLV (μg/m³)
Chlorinated Hydrocarbons *(continued)*			
1,1,2,2-Tetrachloroethane	< 1	< 1	—
Monochlorobenzene	< 1	< 1	350,000
Dichlorobenzene	< 1	< 1	300,000
Aliphatic Hydrocarbons			
Sum aliphatic	57,000[1]	52,000[1]	—
C6-Fraction	8000	10,000	360,000
C7-Fraction	3800	2300	1,600,000
C8-Fraction	940	3100	1,450,000
C9-Fraction	7700	4000	1,060,000
C10-Fraction	4600	5600	—
C11-Fraction	26,000	22,000	—
C12-Fraction	5000	5100	—
Other Hydrocarbons			
Limonene	140,000	120,000	560,000
Other Compounds			
Hydrogen sulfide ppm	0.4	0.5	—

[1]The high concentration of alkylbenzenes and aliphatic hydrocarbons is caused by limonene.
Source: Heida et al., 1995.

AIR DISPERSION MODELING FOR COMPOSTING FACILITIES

Air dispersion models are used to analyze and project air-quality impacts from proposed and exisiting composting facilities. Models can be used to demonstrate compliance with air quality standards, to assess strategies to achieve acceptable emissions levels, and to predict the impact of a proposed site on surrounding communities. As regulating agencies debate siting and buffer zone issues, modeling allows regulators and facility planners to evaluate the movement of odor and specific compounds under a diverse range of conditions. Modeling is less costly than extensive field sampling and provides more quantitative information on facility impacts on neighboring communities, which in turn allows planners to optimize siting.

Regulatory Models

Atmospheric dispersion modeling is a technique for estimating pollutant concentrations caused by source emissions. While it is impossible to fully model the complexities of the atmosphere and the exact transport and dispersion of pollutants, a series of mathematical formulae have been developed from empirical and theoretical studies to reliably estimate pollutant concentrations.

Incorporating these formulae into computer-based models greatly increases the ability to model numerous sources and receptors, as well as to compare various pollution-control strategies and their impact on receptor concentrations. In particular, dispersion models evaluate changes in receptor concentrations due to changes in stack height and exit velocity.

The USEPA and various state agencies have identified preferred models for regulatory use. In addition, numerous other public domain and proprietary models are available for specific modeling situations. The *U.S. EPA Guideline on Air Quality Models* (Revised) provides an overview of models and model use (EPA-450/2-78-027R). The Industrial Source Complex (ISC) model is widely considered to be the industry standard due to its broad applicability and flexibility in handling multiple sources, multiple receptors, terrain, and historical weather data.

Model Parameters

Models used to estimate pollutant concentrations consider meteorological conditions, emission rates, and design parameters of the source(s). This information is used to determine rise, direction, and spreading of the plume and the dispersion of pollutant concentrations within that plume. The models then calculate receptor concentrations based on their location (distance, bearing, and elevation) in relation to the plume.

Meteorological parameters that affect pollutant dispersion include wind speed, wind direction, amount of turbulence (mechanical and thermal), vertical thermal structure, and height to which the plume mixes in the atmosphere. Source parameters include location, type of source (point, area, or volume source), height of release, emission rate, emission velocity, and size of the source.

Gaussian Dispersion and Dispersion Parameters

Most regulatory air models utilize Gaussian, or normal, distribution to estimate the size of a plume and the distribution of pollutants within it. A plume spreads in both horizontal and vertical directions as it travels away from the point of emission (Figure 10.13). The models assume that the plume is essentially symmetrical. The magnitude of vertical and horizontal spreading is a function of the meteorological parameters listed above as well as the type of terrain over which the plume passes.

Treatment of Terrain

For the purposes of dispersion modeling, the elevation of terrain in relation to a source is classified as follows:

- simple terrain: elevation at or below the height of the pollutant release (e.g., stack tip)
- intermediate terrain: elevation between the height of release and the final plume rise
- complex terrain: elevation above the final plume rise

The current complex terrain regulatory model (CTDMPLUS) calculates a dividing line in the atmosphere based on the meteorological conditions and the terrain features (Figure 10.14). The portion of the plume above the dividing stream line tends to seek a path over the terrain, while the portion below the line generally seeks a path around the terrain.

Reliability of Model Results

In order to properly interpret modeling results, it is necessary to understand the level of accuracy and reliability of the regulatory air models. The *U.S. EPA Guideline on Air Quality Models* provides guidance in the proper use of computer models and interpretation of the results. The guideline states that "... models are reasonably reliable in estimating the magnitude of highest concentrations occurring sometime, somewhere within an area." Comparative studies of modeled versus actual concentrations have found that (1) the regulatory air models tend to be overpredictive, (2) errors in the

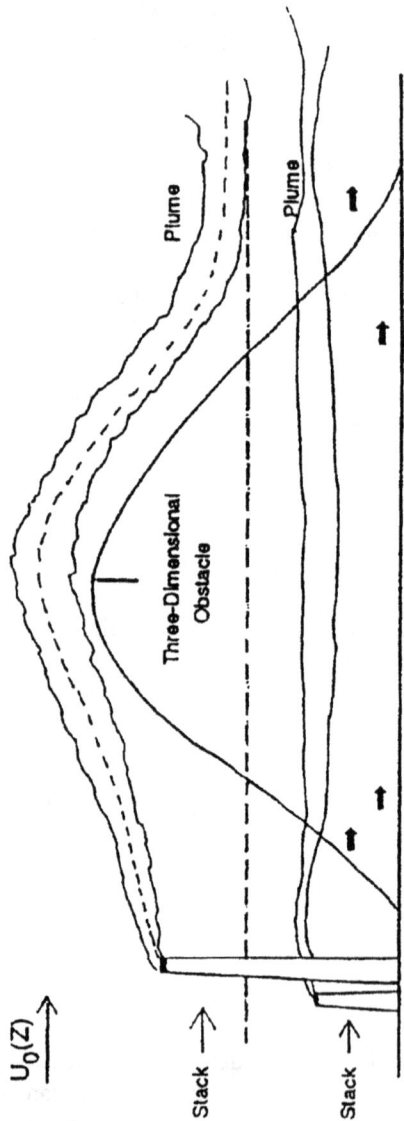

FIGURE 10.13. Spreading of a plume emitted from a stack. (Data from Schultze, 1993.)

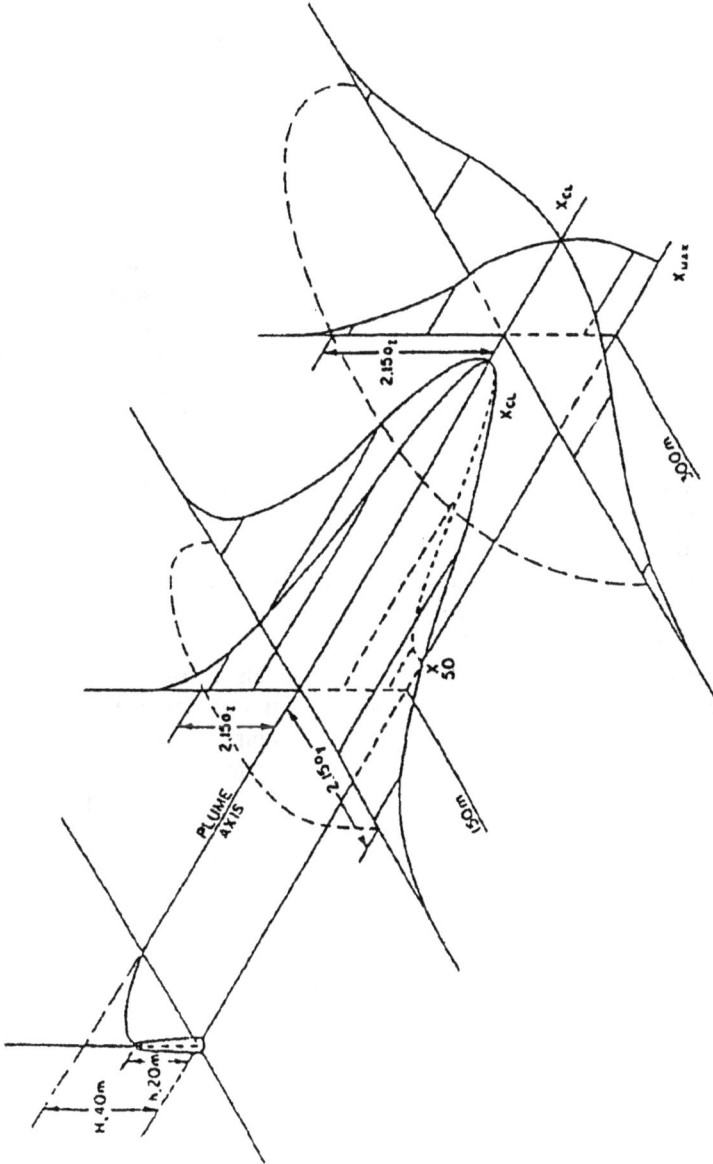

FIGURE 10.14. Spreading of a plume downwind from a stack showing Gaussian dispersion. (Data from Schultze, 1993.)

range of plus or minus 10% to 40% can be expected, and (3) the modeled concentration at a specific receptor for certain meteorological conditions does not always correlate well with actual measured concentrations. The USEPA guideline states that "uncertainties do not indicate that an estimated concentration does not occur, only that the precise time and locations are in doubt."

Peak-to-Mean Conversion for Assessing Odor Impacts

Air models are most commonly used to predict the concentrations of specific pollutants for regulatory purposes. One hour is the shortest averaging period considered by air regulations and as output by the models. However, odor impacts are frequently the cause of concern at composting facilities. An odor impact can be realized over a comparatively short time period, as short as it takes for one breath. Therefore, it is widely acknowledged that one-hour average concentrations calculated by models need to be converted to project the peak short-term concentration that may occur.

Two approaches to quantifying short averaging-period odor impacts are typically used. The first utilizes dispersion parameters modified for instantaneous releases (Slade, 1968) models the odor release as a series of quasi-instantaneous releases. This type of model is commonly referred to as a "fluctuating plume" or "puff" model.

The second approach is to utilize preferred models such as ISC and convert the one-hour concentration to a peak concentration for a much shorter averaging period utilizing a combination of empirical and measured data relating long-term and short-term averages. This latter approach is explained in the following paragraphs.

Examples of Dispersion Modeling for Composting Facilities

An air-dispersion modeling study performed for a private composting facility in the Northeast United States offers good insight into the use of dispersion models. Odor complaints about the facility had become so acute that the state regulatory agency had shut it down pending major capital modifications (E&A Environmental Consultants, Inc., 1994).

The purpose of the assessment was threefold: (1) to predict maximum downwind odor concentrations during most adverse weather conditions, (2) to compare predicted odors to an agency-imposed standard of "no detectable malodor," and (3) to determine whether the existing design was sufficient or whether additional equipment was needed to provide better odor control and/or atmospheric dispersion.

The facility retrofit considered included increased aeration capacity, improved process air handling, and exhaust treatment with a biofilter. It was also necessary to determine whether elevated release through rooftop ventilators was needed to improve dispersion.

The standard EPA-approved hierarchy for modeling was used. The first step involved using a screening model that identified most adverse weather conditions and maximum receptor concentrations. If the results indicated that it might be problematic to meet regulatory standards, refined modeling was employed.

The results of the screening model are summarized in Figure 10.15. Release of untreated compost exhaust through roof vents and open side walls led to unacceptable maximum odors at the property line and nearby residences. Increased aeration capacity, improved process air handling, and exhaust treatment with biofilters reduced odor impacts but still resulted in unacceptable odor levels under adverse weather conditions.

Further retrofit was necessary given the size of the operation and the close proximity to the property line and residences. Enclosure of the biofilters and exhaust through rooftop ventilators provided sufficient velocity, height, and mixing to increase atmospheric dispersion before the odorous exhaust plume impacted the ground. A value of 7 D/T at a property line is generally considered protective of surrounding receptor.

This study provided the facility with the information necessary to select the appropriate retrofit. It also gave the regulators the necessary information based on an accepted modeling protocol.

A second example of how models can help a compost facility assess odor involves an enclosed biosolids composting facility equipped with a chemical scrubber. Nearby neighbors complained of frequent nuisance odors generated by the facility. On-site odor sampling determined that the scrubber stack was the primary source of odors. The facility was committed to solving the odor problem and needed to determine the most effective way to do so. Given that the scrubber performance was optimal and odor emissions were low, control measures focused on increasing dispersion by (1) increasing stack height and/or (2) adding dilution air and increasing the stack exit velocity.

Source parameters, terrain elevations, receptor locations, and local historical weather data were prepared for modeling using ISCST. Model results were converted to project peak short-term concentrations. A nuisance odor threshold of 4 D/T was established by correlation to the standard butanol intensity scale. The model results for existing conditions confirmed that nuisance odors occurred with some frequency at nearby receptors (Figure 10.16). The model also indicated that a 7.6 m (25 ft) increased stack height did little to reduce the frequency of nuisance conditions (Figure 10.17). However, the addition of an additional fan that nearly doubled the stack flow

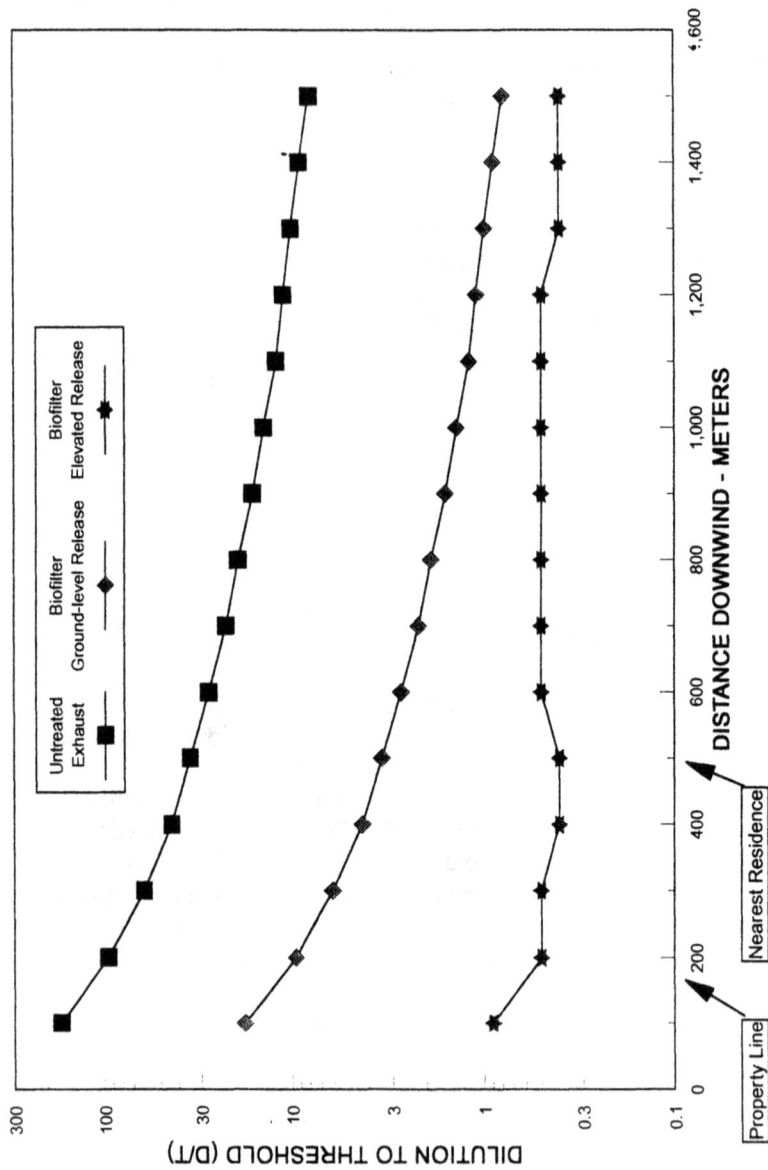

FIGURE 10.15. Effect of treatments on predicted odor levels for a biosolids composting facility.

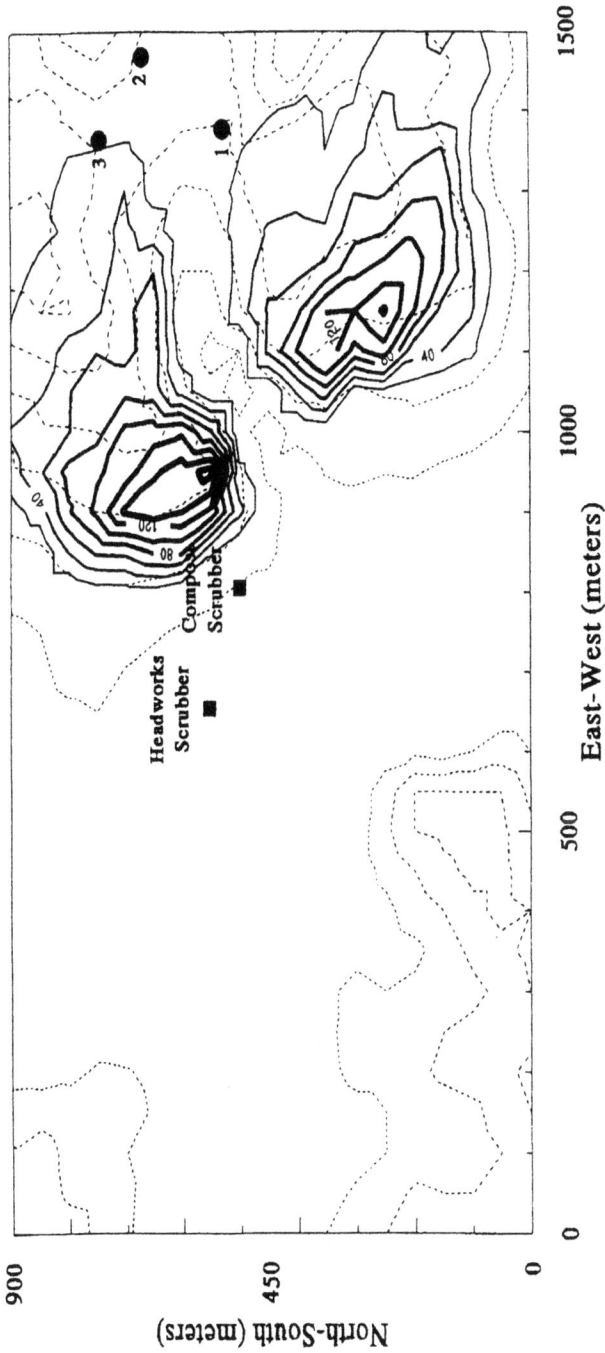

* Number of Hours per Year with a 30-Second Peak Odor Exceeding 4 D/T (3 Butanol Equivalents).

—— 160   —— 120   —— 80   —— 40   —— 0

FIGURE 10.16.   Occurrence of odors greater than 4 D/T* under current operations.

337

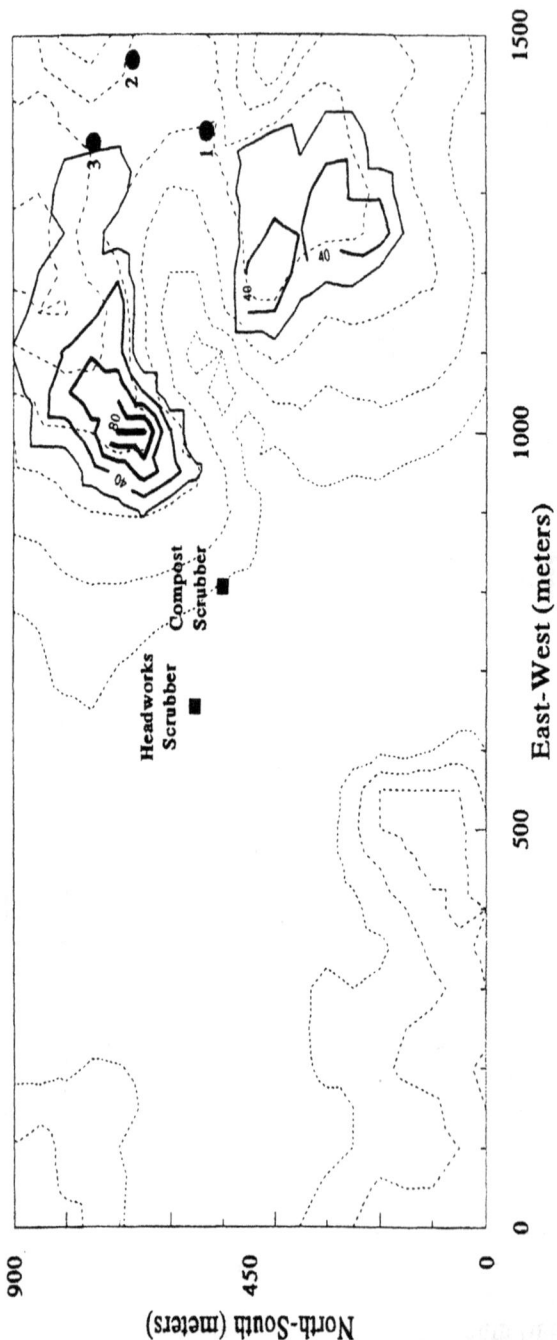

FIGURE 10.17. Occurrence of odors greater than 4 D/T* if compost stack height is increased by 7.6 m (25 ft).

* Number of Hours per Year with a 30-Second Peak Odor Exceeding 4 D/T (3 Butanol Equivalents).

—— 120   —— 80   ·—· 40   ···· 0

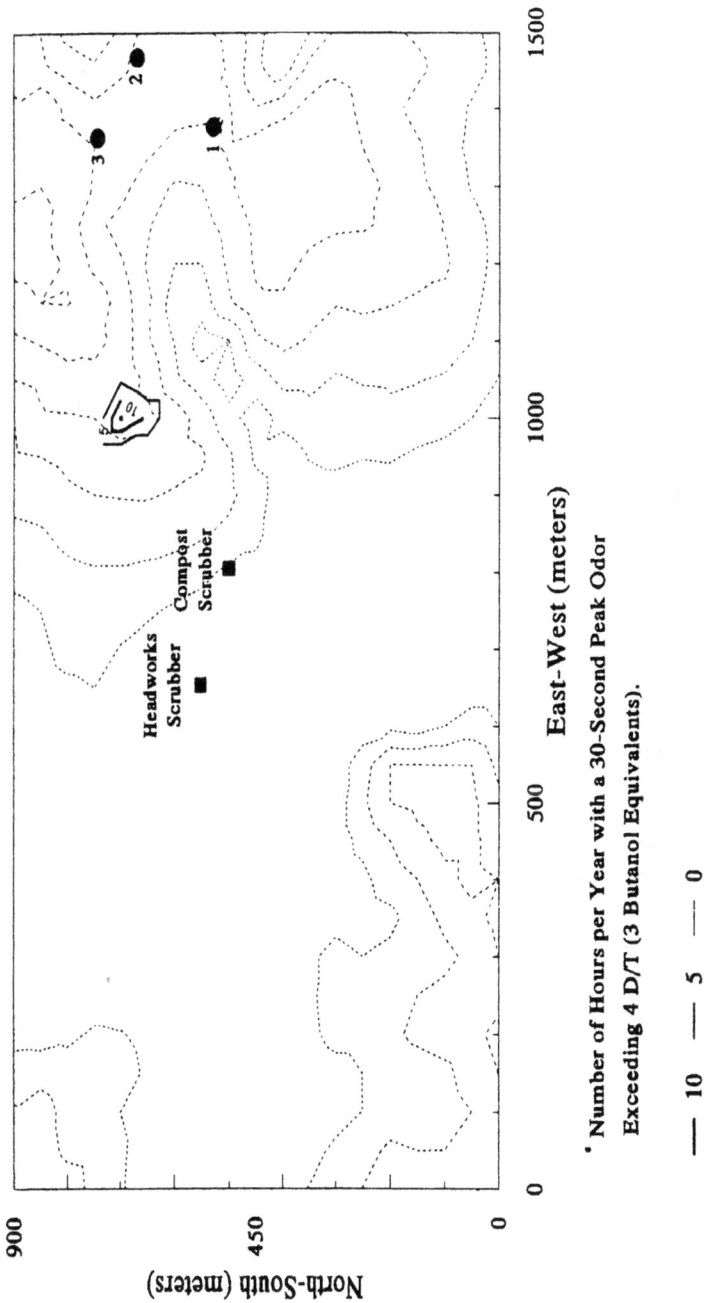

FIGURE 10.18. Occurrence of odors greater than 4 D/T* by increasing compost stack height to 7.6 cm (25 ft) and doubling the flow and velocity.

* Number of Hours per Year with a 30-Second Peak Odor Exceeding 4 D/T (3 Butanol Equivalents).

— 10 — 5 — 0

339

rate and diluted the stack tip odor concentrations dramatically reduced nuisance odors at critical receptors according to the model (Figure 10.18). These results provided the facility operator a basis for assessing the various options to control off-site impacts.

CONCLUSION

Odors and VOCs are emitted during the composting process. Intensity of odor emissions is often a function of the methodology used, process management, and the ability to capture and treat the emissions. Windrow composting can result in the highest level of odors. The exhaust from negatively aerated static pile can be scrubbed in biofilters or chemical scrubbers. Managing aeration, moisture and temperature can minimize odor problems.

Volatile organic compound emissions, like odors, can be removed from point-source process elements. The levels reported from MSW and biosolids composting operations have been found to be low and do not represent hazard to workers. Volatile organic compounds are major odor sources and need to be treated, except at remote facilities where odors do not impact nearby receptors.

More information on odor and VOC data is needed. No published data on odors or VOCs were found for yard waste facilities. Such information would enable facilities to modify their design and manage operations to reduce potential negative impacts on the environment.

Odor modeling is a valuable tool for identifying and modifying management and operations of existing composting facilities. Modeling can also assist in siting and design of new facilities.

REFERENCES

ACGIH. 1992. *American Conference of Governmental Industrial Hygienists, 1992–1993 Threshold Limit Values for Chemical Substances and Physical Agents and Biological Exposure Indices.* ACGIH. Cincinnati, OH.

E&A Environmental Consultants, Inc. 1993. *Food Waste Collection and Composting Demonstration Project.* Final Report for City of Seattle Solid Waste Utility, Bothell, WA.

E&A Environmental Consultants, Inc. (1994). In-house data.

Eitzer, B. D. 1995. Emissions of volatile organic chemicals from municipal solid waste composting facilities. *Environ. Sci. & Tech.* 29(4):896–902.

Heida, H., F. Bartman, and S. C. van der Zee. 1995. Occupational exposure and indoor air quality monitoring in a composting facility. *Am. Ind. Hyg. Assoc. J.* 55(1):39–43.

Hentz, L. H., C. M. Murray, J. L. Thompson, L. L. Gasner, and J. B. Dunson, Jr. 1992. Odor control research at the Montgomery County regional composting facility. *Water Environ. Res.* 64(1):13–18.

Hooper, J. E. and S. Cha. 1988. *Odor Perception and Its Measurement: An Approach to Solving Community Odor Problems.* TRC Environ. Consultants, East Hartford, CT.

Hovsenius, G. 1987. Odour investigations at two composting plants in Sweden, pp. 501–510. In M. de Bertoldi et al. (eds.). *Compost: Production, Quality and Use.* Elsevier Applied Sci. London, England.

Iacoboni, M. D., T. J. LeBrun, and J. Livingston. 1980. Deep windrow composting of dewatered sewage sludge, pp. 88–108. In *Proc. Nat'l. Conf. Municipal & Industrial Sludge Composting.* Philadelphia, PA.

Kissel, J. C. and C. L. Henry. 1992. *Emissions of Volatile and Odorous Organic Compounds from Municipal Waste Composting Facilities: A Literature Review.* The National Composting Council, Alexandria, VA.

Lees, P. S. J. and M. S. Tockman. 1987. *Evaluation of Possible Health Impact of WSSC Site II Sewage Sludge Composting Operations.* Johns Hopkins Univ. School of Hygiene and Public Health. Report to Maryland Dept. of Health and Mental Hygiene, Baltimore, MD.

Miller, F. C. 1993. Minimizing odor generation, pp. 220–241. In H. A. J. Hoitink and H. M. Keener (eds.). *Science and Engineering of Composting and Design, Environmental, Microbiological and Utilization Aspects.* The Ohio State University, Wooster, OH.

Schultze, R. H. 1993. *Practical Guide to Atmospheric Dispersion Modeling.* Trinity Consultants, Inc., Dallas, Texas.

Slade, D. 1968. *Meteorology and atomic energy.* TID-24190, U.S. Atomic Energy Commission, Oak Ridge, Tennessee.

Toffey, W. E., L. H. Hentz, Jr., and M. Haibach. 1995. Control of odor and VOC emissions from the largest aerated-static-pile biosolids composting facility in the U.S. Paper presented at *WEFTEC '95.* Water Environ. Fed., Miami Beach, FL.

USEPA. 1992. *Draft Guidelines for Controlling Sewage Sludge Composting Odors.* OWEC, Washington, DC.

Van Durme, G. P., B. F. McNamara, and C. McGinley. 1992. Bench-scale removal of odor and volatile organic compounds at a composting facility. *Water Environ. Res.* 64(1):19.

Van Durme, G., B. McNamara, and C. McGinley. 1990. Characterization and treatment of composting emissions at Hampton Roads Sanitation District. Paper presented at the *63rd Annual Water Pollution Control Federation Conf.* Washington, DC.

Walker, J. M. 1993. Control of composting odors, pp. 185–218. In H. A. J. Hoitink and H. M. Keener (eds.). *Science and Engineering of Composting and Design, Environmental, Microbiological and Utilization Aspects.* The Ohio State University, Wooster, OH.

Wilber, C. and C. Murray. 1990. Odor source evaluation. *BioCycle.* 31(3):68–72.

Williams, T. O. and F. C. Miller. 1993. Composting facility odor control using biofilters, pp. 262–281. In H. A. J. Hoitink and H. M. Keener (eds.). *Science and Engineering of Composting: Design, Environmental, Microbiological and Utilization Aspects.* The Ohio State University, Wooster, OH.

CHAPTER 11

Soil Physical and Chemical Manifestations

INTRODUCTION

Compost is an organic matter resource. The feedstock, its characteristics, and level of contaminants determine the amount of organic matter in relation to other constituents. As a source of organic matter, its primary impact is on the soil physical properties. However, several important soil chemical characteristics are also affected by the addition of organic matter. These physical and chemical properties include:

(1) Soil physical properties
  - soil structure: bulk density, porosity and aeration, and soil strength
  - water relations: water retention, available water to plants, and soil water content
  - infiltration and permeability
  - erosion and runoff
  - soil temperature
(2) Soil chemical properties
  - cation exchange capacity
  - soil pH
  - electrical conductivity (EC)
  - macronutrients

Compost is often referred to as a "soil conditioner." As such, it improves the soil physical properties that play a significant role in crop production. Water retention is increased, and more water becomes available to plants.

Figure 11.1a shows a cornfield with and without the addition of compost. The dark soil color indicates where the compost was applied. As a result of

343

FIGURE 11.1a. Effect of compost application on corn growth.

droughty conditions, corn in the fertilized control plots was more stressed than in places where compost was applied. This was evident by the wilted condition of the corn leaves in the fertilized control plot versus the compost-amended plot (Figure 11.1b). Examination of the root structure (Figure 11.1c) showed much smaller root development in the control. By comparison, the roots on plants grown in compost-amended soil were much more extensive with many more root hairs, which are necessary for water and nutrient uptake. Compost amendment to soil reduces compaction, thus enabling plant roots to develop better. Soil physical properties can also affect runoff and erosion through improvement of soil structure, resulting in greater infiltration. Runoff and erosion are the two most important parameters affecting non-point source pollution from agricultural lands.

The soil organic matter also changes several important soil chemical characteristics. These include cation exchange capacity, pH, and exchangeable bases. Moreover, since most feedstocks contain some amount of plant macro- and micronutrients, the use of compost at rates considerably higher than fertilizers often provides sufficient levels of these elements for plant growth. Soil salinity can also be affected depending on the characteristics of the compost. An increase in salinity can prevent seed germination and reduce plant growth or crop yields.

Khaleel et al. (1981) and Shiralipour et al. (1992) reviewed some of the

FIGURE 11.1b. Corn leaf on fertilized control plot during drought (left); corn leaf from compost-amended soil during drought (right).

CONTROL   COMPOST

FIGURE 11.1c. Effect of compost on corn root development.

literature on the effect of compost and other organic wastes application on soil properties. Prior to 1970 there was little published information about the effect of compost on soil properties. Often it takes several years before changes in soil physical properties are observed when compost or organic matter is added. Unlike chemical properties, which are relatively easy to measure, physical properties are much harder to quantify. As a result, there are considerably fewer publications on the changes to soil physical characteristics due to compost application.

Organic matter in compost varies from as low as 30% to a high of 70%. Some variability is due to different methods of analysis. The Walkley Black (Nelson and Sommers, 1982) and the Mebious (Mebious, 1960) methods utilize large samples that attempt to measure the available carbon, whereas the dry combustion method uses small samples to measure the total C (Nelson and Sommers, 1982). Because of the small sample, the latter method is much more prone to errors when assessing nonhomogeneous compost samples. This is particularly true with biosolids compost, which has wood chips, or MSW compost, which contains plastics and other contaminants.

The way compost is prepared affects the amount of organic matter or organic C in the compost. The effect of compost on soil organic matter levels has been reported by numerous researchers (Tester, 1990; Jacobowitz and Steenhuis, 1984; Hortenstine and Rothwell, 1973b; Terman and Mays, 1973; Mays et al., 1973; Epstein and Wu, 1994; Mays and Giordano, 1989; Avnimelech et al., 1990; Epstein et al., 1976).

A literature review by McConnell et al. (1993) found that compost applied at rates varying from 18 to 146 t/ha produced a 6% to 163% increase in organic matter. Mays and Giordano (1989) reported on long-term effects of MSW-sludge compost application during 1968–1972. During that period, a total of 90 to 2,240 t/ha of compost had been applied. In the 0–15 cm soil depth, organic matter increased from 1.6% in the control plots to 4.9% in the 2,240 t/ha plots. At the 15–30 cm soil depth, the percent organic matter ranged from 1.0% for the control plots to 2.3% for the 224 t/ha plots.

Tester (1990) found that the application of compost affected the organic matter content of a loamy sand soil several centimeters below the zone of application (Figure 11.2). This effect at the lower soil depths is probably due to movement of soluble organic fraction in the compost. Similar results were reported by Avnimelech et al. (1990). Specifically, organic carbon in the 0–10 cm depth increased from 6.1 mg/g in the control to 10.1, 15.6, and 22.6 mg/g for the 80, 160, and 400 m^3/ha treatments, respectively. In the 10–30 cm zone, the values were 7.6, 9.2, 11.2, and 16.2 mg/g for the control, 80, 160, and 400 m^3/ha treatments, respectively.

Epstein et al. (1976) found no increase in organic C when applications of

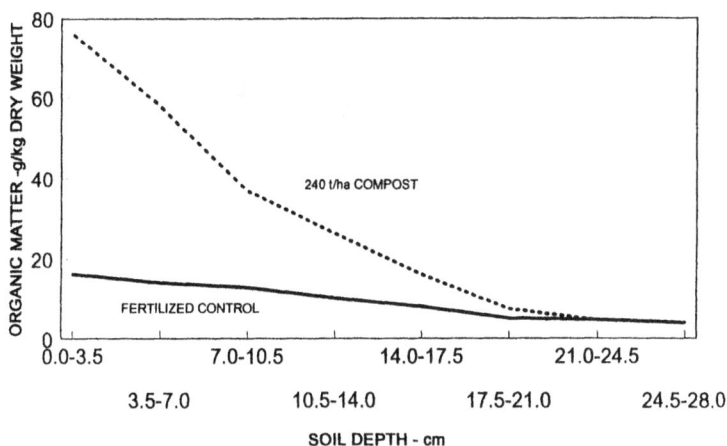

FIGURE 11.2. Comparison of organic matter in a loamy sand with a compost application of 240 t/ha and a fertilized control. (Data from Tester, 1990.)

40 and 80 t/ha were applied to silt loam soil (Aquic Hapludult), but did obtain a measurable increase by 1.5% to 2% from the 160 and 240 t/ha treatments.

Although the primary use of compost is as a soil conditioner, relatively large quantities are often needed to achieve improvements in the soil physical conditions. Consequently, the compost can provide significant levels of plant nutrients, especially nitrogen and phosphorus. Also, the addition of compost provides many of the minor elements or plant micronutrients (see Chapter 6).

This chapter discusses the soil changes that occur as a result of application of compost to soils. In some cases where data are limited, information as the result of organic matter application is presented. Finally, the nitrogen reactions in the soil are presented including the nitrogen cycle.

EFFECT OF COMPOST APPLICATION ON SOIL PHYSICAL PROPERTIES

Soil Structure

Soil structure refers to the arrangement of the soil particles; that is, the arrangement and stability of soil aggregates and pore space and pore size distribution. Organic matter improves soil structure through soil aggregation. Figure 11.3 shows a schematic model of soil particles forming an aggregate by binding organic matter (Emerson, 1959). Soils with poor structure such as sands and clays benefit much more from organic matter application than loams, which generally have a good soil structure. Good soil structure provides for better root growth. With improved root development, plants are better able to utilize water and nutrients.

Soil structure is difficult to evaluate. Soil physicists have attempted to measure changes in soil structure by measuring aggregate stability and percent aggregates. These parameters indicate the binding effect of organic matter on soil particles. The benefits of compost on soil aggregation are more evident and pronounced in fine-textured soils (e.g., clays, clay loams, silty clay loams) and in very coarse-textured soils such as sands, and sandy loams. Many soil properties such as bulk density, porosity, aeration, erosion and runoff, and hydraulic conductivity are either directly or indirectly affected by soil structure.

Figure 11.4 shows the effect of compost application on aggregation and improved soil structure. The soil crusted in the unamended control. Crusting increased runoff and reduced the amount of water that infiltrated the soil, which reduced the amount of water available for plant growth. Little data are available on direct measurements of aggregate stability or aggregation. Manifestations due to organic matter application may take several years and, therefore, are difficult to study.

Avnimelech and Cohen (1988) showed that modest amounts of compost (70 t/ha) increased aggregate stability. Pagliai et al. (1981) reported that only high rates (230 t/ha) added to a sandy loam resulted in observable changes in aggregate stability. Hernando et al. (1989) found that aggregate stability increased in plots amended with 30 and 60 t/ha. They also noted that the C content at 90 and 180 days of incubation statistically correlated with aggregate stability, indicating a formation of cationic bridges.

Avnimelech et al. (1993) presented the results of several studies on the effect of MSW compost on aggregate density and stability. Soil aggregate density was significantly reduced when soils were incubated under aerobic conditions. No effect was found under anaerobic conditions, however. Similarly, turbidity of the soil suspension, a measure of soil structural instability, was lower under aerobic conditions but not under anaerobic conditions. The authors attributed this phenomenon to the production of polysaccharides, polyurinides, and humic acid. Polysaccharides and uronic acid concentrations also increased under aerobic but not under anaerobic conditions.

Soil porosity is a measure of the size and arrangement of the voids in the soil matrix. Soil porosity affects aeration and water movement. A clay soil can have many more small pores and thus a greater total pore space than a sandy soil, but the individual pores are extremely small, restricting water

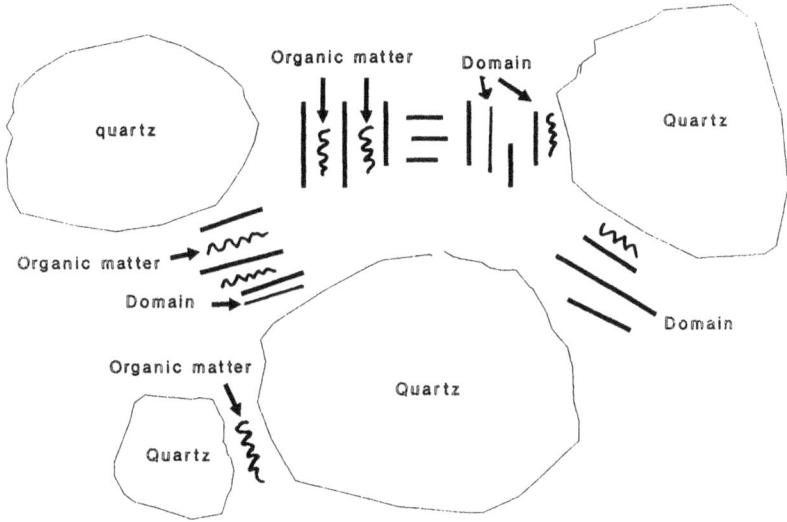

FIGURE 11.3. Model of soil structure showing binding of aggregates by organic matter. (After Emerson, 1959.)

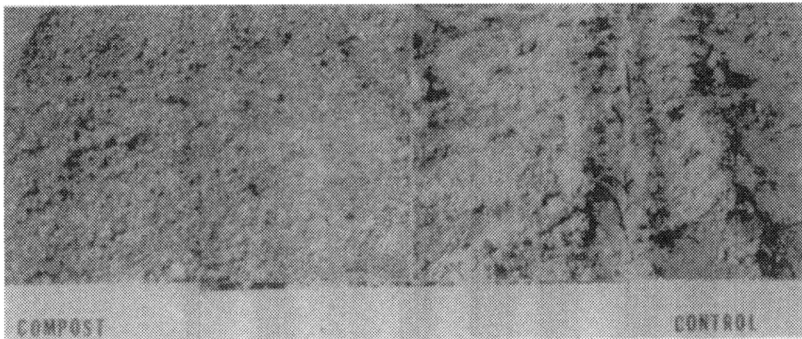

FIGURE 11.4. Effect of compost on soil surface conditions. The compost-amended soil is highly aggregated whereas the fertilized control shows crusting.

movement. Indirect measurements of porosity involve the determination of bulk density and relating this parameter to the specific gravity of a soil. If the specific gravity of a soil is known, then soil porosity can be calculated as follows:

Percent porosity = 1 − bulk density/specific gravity × 100

Normally, a specific gravity of 2.65 g/cc is used since it represents an average density of soil minerals. The measurement of specific gravity is difficult.

Several investigators have reported that composted MSW and composted sewage sludge increased the total porosity of soils ranging from loamy sand to clays (Pagliai et al., 1981; Tester, 1990; Schrader, 1967).

One of the most comprehensive studies on the effect of compost on soil porosity was conducted by Pagliai et al. (1981). The effect on total porosity is shown in Figure 11.5. Composts were prepared from aerobically and anaerobically digested biosolids combined with the organic fraction of urban waste. Two rates of compost were applied. Total soil porosity was significantly higher in plots amended with compost. However, no difference was found between the two application rates.

The effect on pore size distribution is shown in Figure 11.6. Small pores in the range of 0.5–50 μm are the pores that retain water necessary for plant growth and microorganisms; medium-sized pores, 50–500 μm, are necessary for water transmission and aeration; finally large pores, >500 μm, are fissures and play a relatively small part in water movement and retention.

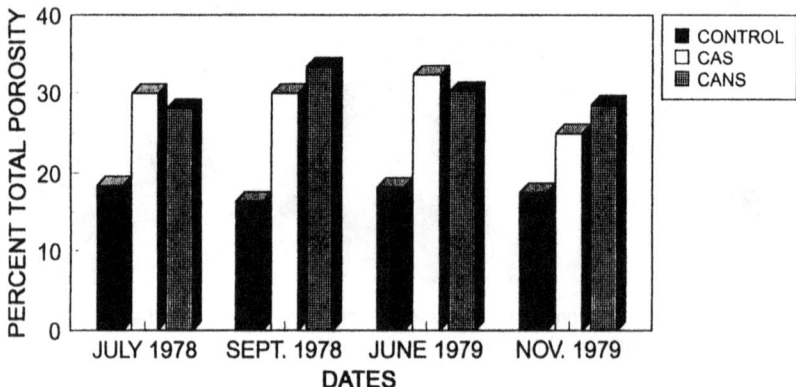

Average for the low and high rates of application
No significant differences for the two rates
CAS = compost from aerobic sludge; CANS = compost from anaerobic sludge.

FIGURE 11.5. Effect of compost incorporated into a soil on total porosity. (Data from Pagliai et al., 1981.)

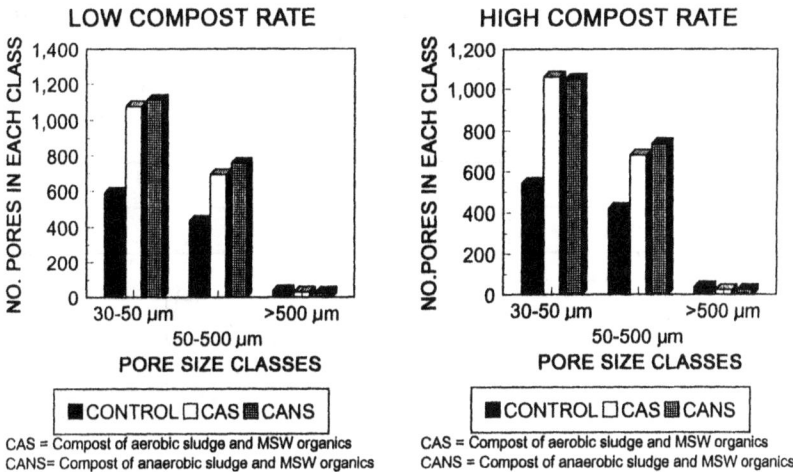

FIGURE 11.6. Effect of compost application on soil pore size distribution. (Data from Pagliai et al., 1981.)

Compost increased the pore size distribution by increasing the number of pores in the small and medium-sized classes while reducing the number of large pores. This is an indication of better soil structure and, therefore, a more favorable environment for plant growth.

Giusquaiani et al. (1995) found that total porosity (pores > 50 μm) increased linearly with compost rates to field plots regardless of soil depth and sampling time (Table 11.1). The greater porosity was due to an increase in elongated pores. Data were also obtained for pore distribution. In the plots with the 90 t/ha rate, the increase in elongated pores involved both the water transmission pores of 50 to 500 μm and pores >500 μm.

"Soil bulk density" is the weight per unit volume of a soil mass including the pore space. Thus, bulk density is an indication of soil compactness. The higher the bulk density, the more difficult it is for plant roots to proliferate. Consequently, the potential for plants to extract water and nutrients is reduced. The energy required to till soil also increases with compactness.

Bulk Density

The addition of compost usually results in a decrease of soil bulk density. The greatest impacts are on clays and other dense soils (Duggan and Wiles, 1976; Mays et al., 1973; Mays and Giordano, 1989; Guidi and Petruzzelli, 1989). Jacobowitz and Steenhuis (1984) found that at high rates of sludge-compost application, the soil bulk density was lowered significantly. At a

TABLE 11.1. Changes in Total Porosity with Depth as Affected by MSW Compost Applied at 10 and 90 t/ha.

	Total Porosity (percent)		
	Control	10 t/ha	90 t/ha
November 1991			
0–10 cm depth	10.1	11.2	16.9
20–30 cm depth	7.2	9.4	13.5
June 1992			
0–10 cm depth	17.7	18.7	24.6
20–30 cm depth	10.0	12.1	16.6
November 1992			
0–10 cm depth	10.8	12.1	18.2
20–30 cm depth	8.9	10.8	13.5

Source: Giusquiani et al., 1995.

rate of 50 t/ha, the bulk density was lower than the control but not to a statistically significant extent. Application of MSW composted with biosolids also resulted in a decrease in soil bulk density from 1.37 g/cc for the control to 1.32, 1.27, 1.22, and 1.12 g/cc for 46, 82, 164, and 327 t/ha, respectively.

The effect of compost application on the soil bulk density for two different soil types is shown in Figure 11.7. In both soils, bulk density decreased with increased application depths. The greatest impact was at the 3–6 cm depth. This effect was more pronounced for the silt loam soil than for the sandy loam soil.

Duggan and Wiles (1976) observed similar results with a loamy soil. Bulk density decreased from 1.43 g/cc for the control plots to 1.31, 1.26, 1.24, and 0.99 g/cc when 8, 16, 50 and 200 t/ha MSW continuing sewage sludge was applied.

Tester (1990) showed that the addition of biosolids compost significantly reduced the bulk density of an Evesboro sandy loam soil. After five years of compost application for all treatments, 60, 120, and 240 t/ha, bulk density was reduced compared to the fertilized control plot. This reduction occurred to a depth of 30 cm, which was considerably lower than the zone of compost application. This phenomenon suggests that organic carbon was leached into deeper depths.

In a second study, Tester (1990) evaluated bulk density and surface area approximately one month after four years of continuous application of 268 t/ha of compost (Figure 11.8). Bulk density was lower for all treatments while surface area increased. The increase in surface area was linear and was believed to contribute to greater water retention.

Epstein and Wu (1994) evaluated the effect of MSW compost application on bulk density as related to moisture tension (Figure 11.9). Bulk density was much lower for the two compost treatments at all moisture tensions.

A decrease in bulk density was also reported by Giusquiani et al. (1995) with application of 10 t/ha/yr and 90 t/ha/yr MSW compost to a clay soil. Bulk density changed from 1.55 g.cm to 1.46 g/cm for the 10 t/ha and 1.38 g/cc for the 90 t/ha application rates.

Khaleel et al. (1981) related bulk density to soil organic carbon due to waste application. They derived a linear regression equation relating bulk density to organic carbon. A significant linear relationship was found between increases in organic carbon due to waste incorporated into soil and percent reduction in bulk density.

Soil Strength

Another measure of soil structure is soil strength, which is usually measured with a penetrometer. For agricultural application, low soil penetrome-

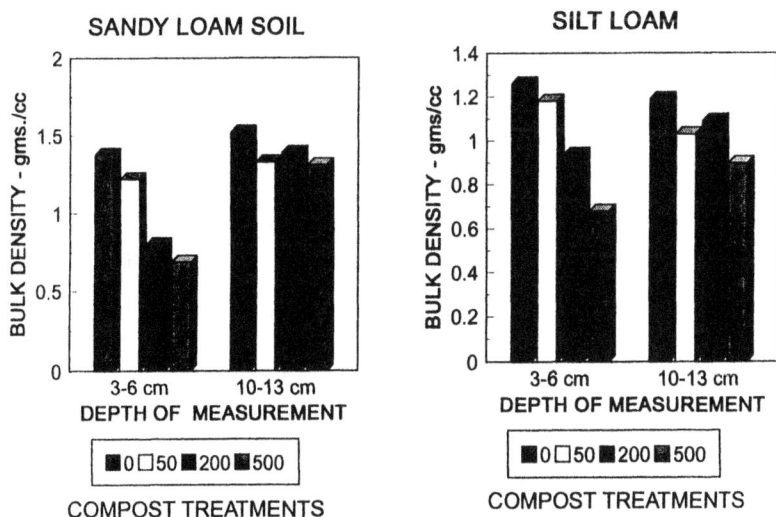

FIGURE 11.7. Effect of compost application on soil bulk density. (Reprinted with permission from Jacobowitz and Steenhuis, 1984.)

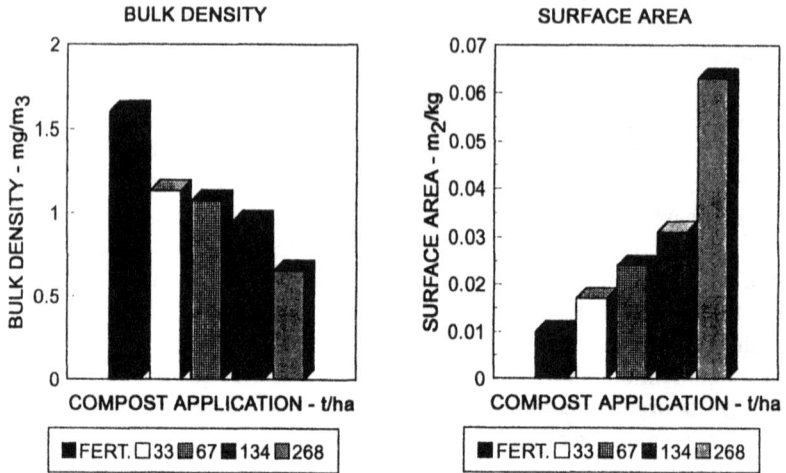

FIGURE 11.8. Effect of fertilizer and compost application on bulk density and surface area approximately one month after four years. (Data from Tester, 1990.)

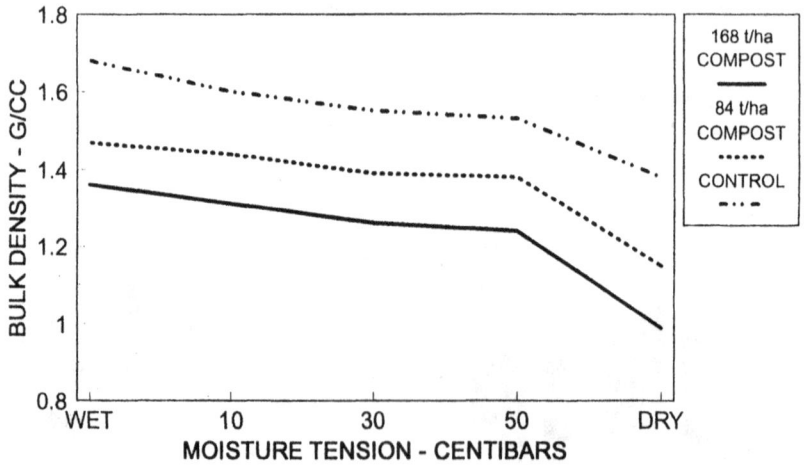

FIGURE 11.9. Effect of MSW compost application on soil bulk density as a function of soil moisture. (Data from Epstein and Wu, 1994.)

ter readings indicate a more favorable environment for root penetration and less resistance and energy requirements for tillage implements. In engineering applications, such as highway embankment construction and engineering fills, lower values are often desirable. Shear strength is also used in engineering to determine the value of soil material for highway construction.

Several studies have shown that soil strength, measured with a penetrometer, decreased with the addition of compost (Mays et al., 1973; Hornick, 1988; Tester, 1990; McCullough et al., 1993; Avnimelech et al., 1990).

For example, Mays et al. (1973) measured unconfined compression strength using a penetrometer and found a significant decrease at high compost application rates. At rates exceeding 164 t/ha, a significant reduction in penetrometer resistance compared to the control was noted.

In testing a sandy soil for engineering applications, McCullough et al. (1993), found that as the percent of compost increased in relation to the amount of sand, proctor penetrometer readings decreased. Terman and Mays (1973) measured unconfined compression strength, also using a penetrometer. Application of 0, 44.8, 82.9, 163.6, and 327.0 t/ha of MSW compost resulted in unconfined compression strengths of 0.276, 0.276, 0.248, 0.228 and 0.145 MPa, respectively. Only the two higher rates were statistically significant.

Tester (1990) reported that biosolid compost reduced penetrometer readings in an Evesboro loamy sand. Data for two depths are shown in Figure 11.10. After five years following compost application, penetrometer readings were significantly lower where compost had been applied. Even at a 24

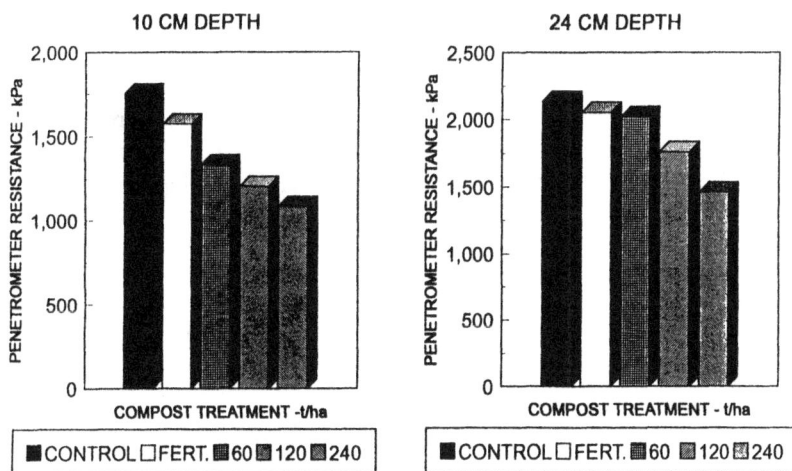

FIGURE 11.10. Effect of compost application on penetrometer resistance at two different soil depths. (Data from Tester, 1990.)

cm depth, considerably below the depth of compost application, penetrometer readings were statistically significantly lower for the 120 and 240 t/ha
application plots than for the 60 t/ha and fertilized control plots.

Avnimelech et al. (1990) evaluated the effect of placement and soil
manipulation of MSW compost on cone penetration. Penetrometer resistance was affected both by the method of soil mechanical treatment (mode
of tillage) and compost. Compost application provided some reduction in
penetrometer resistance (Figure 11.11). Roto-tilled and surface-applied
compost resulted in a greater reduction of resistance than plowing in the
compost. Increased application of MSW compost reduced penetration resistance (Figure 11.11). Even at depths of 10 to 20 cm, which was below the
zone of application, resistance was reduced 20% to 25% compared to the
control. This could be the result of organic carbon leaching into lower layers.
Penetrometer resistance effects were significant only during the season in
which application took place.

Water Relations—Soil Water Retention and Available Water to Plants

Water is held in soils under a force in capillaries due to surface tension.
Energy is required for plants to extract water. Less energy to remove water
is needed at high moisture contents than when a soil is dry. Further, more
energy is required to extract water from clay soils than from sands, since the
former have very small capillaries that do not readily transmit water.

The relation of the energy required to extract water to the soil water content
is expressed as "desorption curves." The expression for the force or energy
by which water is held is termed "matric potential" and is expressed
numerically as bars or kPa. The plant available water is considered to be the
difference in moisture contents at field capacity and the permanent wilting
point. Field capacity is the moisture held between 0.1 (10 kPa) and 0.33 bars
(33 kPa); the wilting point is the moisture held at 15 bars (1500 kPa).

Epstein et al. (1976) showed that application of 240 t/ha of compost
increased the water retained at specific water potentials. When compost was
added to soils, it increased the amount of retained water. Figure 11.12 from
Jacobowitz and Steenhuis (1984) shows the effect of adding biosolids
compost to a sandy soil on water retention. As the proportion of compost to
sand increased, the retention of water increased.

Compost increases the water-holding capacity of soils (Mays et al., 1973;
Epstein et al., 1976; Jacobowitz and Steenhuis, 1984). The increased water-
holding capacity provides for higher water availability to plants. Thus, under
drought conditions plants can survive longer. Furthermore, where irrigation

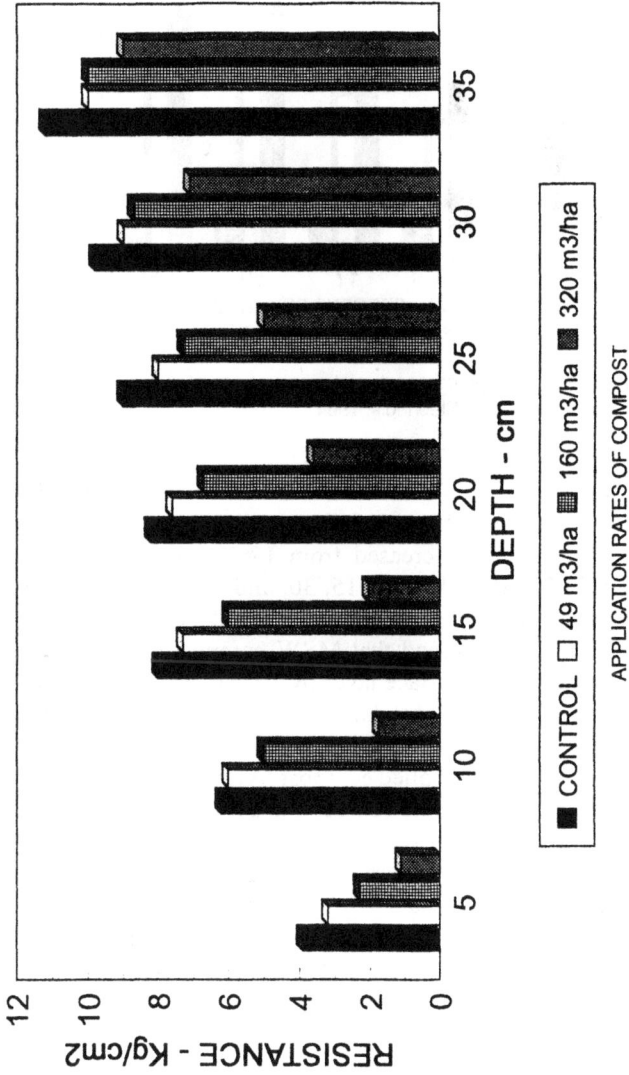

FIGURE 11.11. Effect of compost application rates on cone penetrometer resistance. (Data from Avnimelech et al., 1990.)

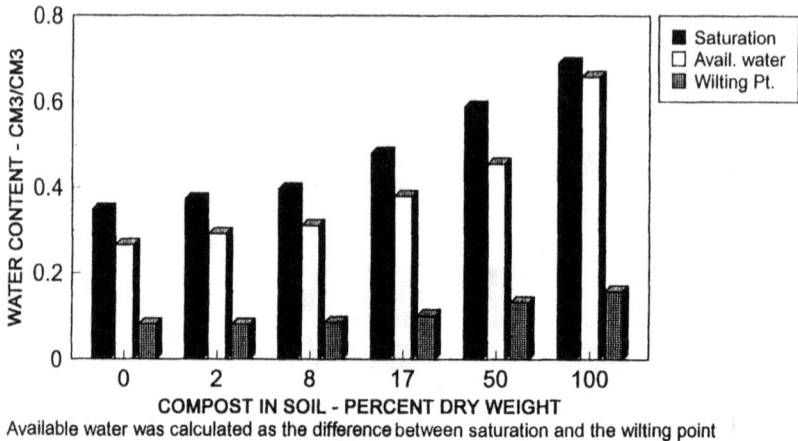

Available water was calculated as the difference between saturation and the wilting point

FIGURE 11.12. Effect of sludge compost additions on soil water retention. (Reprinted with permission from Jacobowitz and Steenhuis, 1984.)

is used, less water is needed. Water retention varies with the type of soil. For example, sandy soils retain less water than loams or clays.

Water-holding capacity increased from 17.23 where no compost was added to 18.6, 20.33, and 24.77 for 15, 30, and 60 t/ha of compost added, following 180 days of incubation (Hernando et al., 1989). In a field study, Avnimelech et al. (1990) found that MSW compost increased the water-holding capacity in the subsurface layer by 825% compared to 341% in the surface layer to which compost had been applied. They attributed this difference to the leaching of organic carbon in the surface layer, which consisted of the residues of applied compost (Avnimelech et al., 1990). This finding suggests that the soluble organic carbon fractions are the ones that are principally responsible for water retention.

Epstein et al. (1976) showed that the available water, the difference between -0.33 bar and -15 bar, was 12.5% for the control and 14.5% for soil amended with 240 t/ha biosolids compost. Application of an MSW compost to a sandy clay loam soil increased the water retention, which resulted in an increase of the available water to plants (Epstein and Wu, 1994). Specifically, a 38% and 63% increase in the available water resulted from the application of 168 and 336 t/ha of MSW compost.

Further, applying compost to soils increases water infiltration and improves the permeability in soils. Increased infiltration, in turn, increases the soil moisture content while also reducing the potential for runoff. Many clayey soils have very small pores so that water movement is restricted. The permeability or hydraulic conductivity of these soils is low. As a result, water does not penetrate into the soil but results in runoff. Avnimelech et al. (1990) found that the water content increased with increasing compost application rates. A positive linear relationship was found between organic carbon and moisture at the 0–10, and 10–30 cm depth. This relationship was expressed by the following equations:

Water content = 25.4 + 0.341 × organic carbon (0–10 cm depth)

Water content = 26.7 + 0.825 × organic carbon (10–30 cm depth)

This increase in water content is the result of greater infiltration. No direct measurements of infiltration were found in the literature.

Jacobowitz and Steenhuis (1984) reported that the saturated hydraulic conductivity of sand-sludge compost mixtures increased with increasing amount of compost. The hydraulic conductivity of a soil is a measure of permeability. With 100% compost, the maximum saturated hydraulic conductivity was 3.74×10^4 cm/day compared to 3.44×10^2 cm/day for the unamended sand. With a 50% compost and sand mixture, the hydraulic conductivity was 1.01×10^4 cm/day. This indicated a high increase in permeability for the 50% and 100% compost compared to the control.

Tester (1990) reported that increasing rates of compost increased the soil water content of a loamy sand soil (Figure 11.13). At the 5 cm depth, the

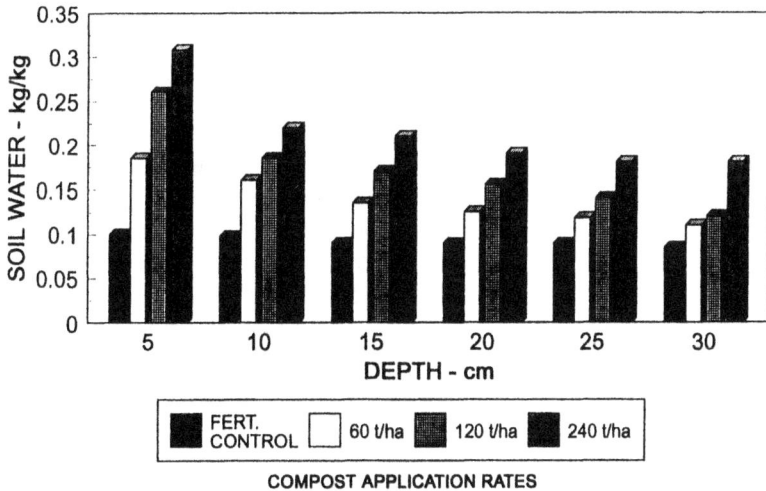

FIGURE 11.13. Effect of compost application on soil water for fescue-plot treatments five years after a single application. (Data from Tester, 1990.)

increase for the 240 t/ha was over three times the amount of water in the control. Even at the 30 cm depth, the soil water for the 240 t/ha plots was twice that of the control. Serra-Wittling et al. (1996) showed that the application of biowaste compost increased the water content 2.4- to 4.5-fold at all matric potentials. The water content of the compost was very high, 0.432 g/g, at the wilting point (-1580 kPa) compared to 0.095 g/g for the soil. Soil water retention increased significantly when amended with compost.

MSW compost-amended sandy soil retained more water over a range of matric potentials (Turner et al., 1994). The compost applied was immature and continued to decompose in the field. Table 11.2 provides the data for the three compost application rates. The water-retention characteristic curves at the lower water potentials were similar. Toward higher water potentials approaching the wilting point, the curves should be very different.

Runoff and Soil Erosion

Runoff occurs when the rate of water infiltration into a soil is less than the amount of precipitation or irrigation. Erosion is the result of water movement over a soil surface and the removal of soil particles. In addition to soil movement, chemicals and fertilizers are carried out in the water stream, resulting in pollution. Eutrophication often results from the enrichment of water bodies by fertilizers as a result of runoff. Runoff and erosion are the major causes of non-point source pollution.

Compost tilled into soils eliminates soil crusting and increases infiltration, and permeability, thus reducing the potential for runoff and erosion. Soil surface crusting is a major source of runoff as water cannot infiltrate the soil. However, compost applied as a mulch intercepts raindrops, reducing their energy and potential for dislodging soil particles.

Relatively little data are available on the effect of compost on runoff and erosion. Byrd (1987) conducted a study evaluating the effect of sludge-compost application on runoff, soil waterborne pollutants, and erosion. The soil was a silty clay and the plots were seeded to fescue grass. Figure 11.14 shows the total suspended solids or erosion and runoff for several storms in 1986. As illustrated, runoff and erosion were much higher for the control plot. The fertilized compost plot was slightly better than the unfertilized compost plot. This could have been the result of better grass coverage over the plot.

Soil Temperature

The effect of compost application on soil temperature has been shown to be rather modest (Jacobowitz and Steenhuis, 1984). Both for a sandy loam

TABLE 11.2. Volumetric Water Content for Three Applications of MSW Compost to a Sandy Soil over a Two-Year Period.

Matric Potential (kPa)	Year	Soil	MSW Compost Application Rate (t/ha)		Statist. Signif.[1]
			67	134	
		Volumetric Water Content (%)			
12	1992	8.50	13.5	20.0	L*
12	1993	7.25	10.6	12.1	L**
	t-test[1]	*	NS	NS	
50	1992	7.50	12.3	18.5	L*
50	1993	6.15	9.40	11.4	L**
	t-test[2]	**	NS	NS	
100	1992	6.50	12.3	17.3	L*
100	1993	5.35	8.48	10.4	L**
	t-test	**	NS	NS	

[1]Means among MSW compost rates are linear at the 1% (**) or 5% (*) level of statistical significance using SAS GLM procedures.
[2]Means between years are different at the 1% (**) or 5% (*) level of statistical significance using SAS t-test procedures.
NS = not statistically significant.
Source: Turner et al., 1994.

EFFECT ON EROSION
(TOTAL SUSPENDED SOLIDS)

EFFECT OF COMPOST ON RUNOFF

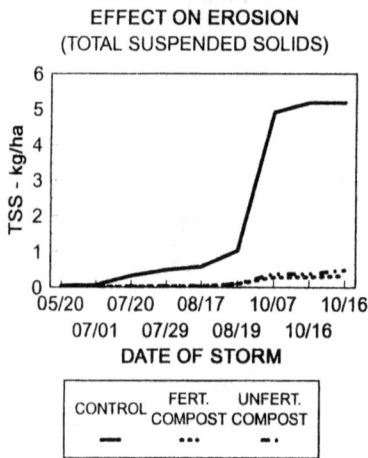

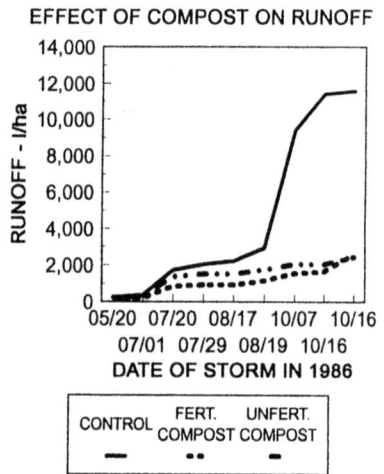

FIGURE 11.14. Effect of compost on erosion and runoff. (Data from Byrd, 1987.)

and silt loam soils, the daytime temperatures were slightly higher when 500 t/ha was applied. Nighttime soil temperatures were not affected, however. Overall, compost application resulted 1–2°C cooler temperatures for the 200 and 500 t/ha compost treatments.

EFFECT OF COMPOST APPLICATION ON SOIL CHEMICAL PROPERTIES

The effect of compost on soil chemical properties is a result of the interaction of the organic matter with soil minerals and the chemical characteristics of the compost. The latter is often a function of the feedstock characteristics and the method of preparation. The single most important effect of organic matter on soil chemical properties involves cation exchange capacity (CEC) since this property governs the potential fertility of a soil. An increase in organic C results in an increase in the CEC (Kononova, 1975).

Cation Exchange Capacity

Cation exchange capacity (CEC) is a measure of a soil's capacity to retain cations on soil colloids as a result of negative charges. The Soil Science Society of America defines CEC as "the sum total of exchangeable cations that a soil, soil constituent, or other material can adsorb at specific pH. It is usually expressed in centimoles of charge per kilogram of exchanger ($cmole_c/kg$)."

The negative charges of the soil matrix are obtained from isomorphic substitutions within the structure of layer silicate minerals, broken bond at mineral edges and external surfaces, dissociation of acidic functional groups in organic compounds, and the adsorption of certain ions on particle surfaces. Positively charged cations (e.g., K^+, Ca^{++}) are attracted to the negative surfaces of clay particles. Thus, CEC provides for the retention of plant nutrients and makes them available to plants. This property also precludes the potential for leaching of cations into groundwater or lower soil layers where they cannot be available to plant roots.

The two most important soil constituents that impact CEC are clay minerals and organic matter. The CEC of organic matter is often much higher than that of the clay minerals. The CEC of soil clay minerals generally ranges from 3 to 5 $cmol_c/kg$ (100 me/100 g) for kaolinite; 30 to 40 $cmol_c/kg$ for illite; and 80 to 150 $cmol_c/kg$ for montmorillonite (Stevenson, 1994). Soil organic matter can have CECs ranging from 100 to 200 $cmol_c/kg$ (MacCarthy, 1990). Stevenson (1994) stated that total acidities of humic acids range from 485 to 870 cmol$_c$/kg. Values as high as 1400 cmol$_c$/kg for fulvic acid explain the significant contribution of humus to the soil's CEC.

Composition occurs during composting transformation of the organic matter. Grebus et al. (1994) measured CEC as a function of time of composting (i.e., in relation to decomposition of yard waste compost). During the first 25 days, the CEC increased from less than 15 cmol$_c$/kg to over 27 cmol$_c$/kg. Over the next 100 days, the rate of increase was lower (Figure 11.15). Since the CEC of organic matter is a function of the ionization of COOH groups and phenolic OH, little change in CEC would be expected when the compost is stable.

Hortenstine and Rothwell (1973a) reported that a significant increase in CEC of an Arredondo sand occurred only at the high MSW compost application rates. In another experiment with a Leon fine sand, Hortenstine and Rothwell (1973a) found a significant increase in the CEC with applications from 3.67 cmol$_c$/kg for the control to 4.81 and 7.14 cmol$_c$/kg for the 128 and 512 t/ha treatments, respectively. However, the increase was not significant at the lower rates of 2, 8 and 32 t/ha. Finally, applying compost to phosphate mine tailings increased the CEC from 0.65 cmol$_c$/kg to 1.20 cmol$_c$/kg with an application of 70 t/ha.

Epstein et al. (1976) found that the CEC of a silt loam soil (Aquic Hapludult) increased using a biosolids compost. The CEC of the untreated soil ranged from 5.5 to 6.4 cmol$_c$/kg. No difference in the CEC was noted at the 40 t/ha treatment but a very significant increase was noted with the 240 t/ha application. After two years' application of the CEC, there was a decrease in CEC for the higher rate. The difference between the control and the high rate of application was still significant.

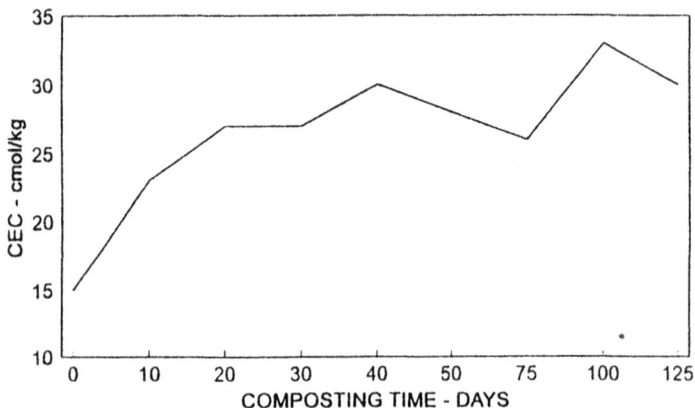

FIGURE 11.15. Changes in cation exchange capacity as a function of yard waste composting time. (Reprinted with permission from Grebus et al., 1994.)

Epstein and Wu (1994) evaluated the effect of an MSW compost on several soil chemical properties for a silty clay loam. Compost applications of 168 and 336 t/h increased the CEC by 10% and 24%, respectively, compared to the fertilized control.

Soil pH

The acidity of a soil (pH) impacts metal solubility, plant uptake and movement, plant growth, soil microorganisms, and many other soil attributes and reactions. Most plants grow best near a neutral pH.

The pH of most stable compost products ranges from 6.5 to 7.5. Exceptions can be found with either very acidic feedstocks such as grape pomace or in composts prepared with lime. Hortenstine and Rothwell (1973b) found no effect of MSW compost application on the pH of a sandy soil. Similarly, in another experiment with phosphate mine tailings, no significant change in pH was noted (Hortenstine and Rothwell, 1973a).

Mays et al. (1973) reported that application of MSW and biosolids compost increased pH from 5.4 for the control to 6.8 with an application for 327 t/ha. In a laboratory incubation study, Hernando et al. (1989) found that pH increased from 6.1 for the soil alone to 7.6, 7.7, and 7.8 with soil amended with MSW compost at 15, 30, and 60 t/ha, respectively. Increases in soil pH were also reported by Jacobowitz and Steenhuis (1984), Epstein et al. (1976), and Tester (1990) for a biosolids compost, and Terman and Mays (1973) for an MSW and biosolids compost.

Tester (1990) found that the effect of pH extended well below the zone of compost application, using a compost prepared from biosolids conditioned with lime during the dewatering process. The pH profile was reportedly similar to the profile of Ca movement reported by Tan et al. (1985) for mobilization of surface-applied Ca by biosolids. The pH of the loamy sand soil remained above a pH of 7.5 to a depth of 50 cm and then decreased rapidly. At 70 cm the pH was approximately 5.5.

Electrical Conductivity (EC)

Electrical conductivity is a measure of the salt content of the soil solution. Salinity affects germination and plant growth both directly and indirectly. Plants differ in their tolerance to salts. Low salt-tolerance crops can tolerate EC values of 2 to 4 mmhos/cm; medium salt-tolerance crops can tolerate EC values 4 to 10 mmhos/cm; and high salt-tolerance crops can tolerate EC values of 10 to 18 (Bernstein, 1964). Although the concern over high salinity most often relates to plant growth, Tester and Parr (1983) reported that leaching of salts from a biosolids compost-amended Evesboro loamy sand increased respiration rate as measured by CO_2 evolution. Increase in decomposition rate as a result of leaching of salts increased the rate of nitrogen immobilization. Presumably microbial activity was inhibited by the high salt content.

Madariaga and Angle (1992) reported that soluble salts inhibited the soil population of *Bradyrhizobium japonicum*, the microsymbiont of soybeans. The population of this nitrogen-fixing bacteria, which is present in root nodules of soybeans, was reduced to nearly undetectable levels when a biosolids compost high in soluble salts was added to a soil. Biosolids compost that was leached with water reduced the salt content but did not affect the heavy metal content. The greatest population of *Bradyrhizobium japonicum* was observed when the biosolids were leached with six consecutive leachings. This reduced the soluble salt content from 18 dS/m to 2.80 dS/m. When the number of leachings increased, there was no significant difference in population.

Compost prepared from different feedstocks or prepared with water containing salts may have different EC values. Grebus et al. (1994) found that EC increased with composting time. Manios and Syminis (1988) attributed the increase in EC to extensive decomposition of organic matter resulting in a high salt concentration. If decomposition is reduced due to high salt content, then the EC should reach a plateau where any increase in EC as a result of decomposition begins to inhibit microbial activity and the rate of decomposition.

The use of MSW compost can result in an increase in soil EC (Shiralipour et al., 1992). McConnell et al. (1993) reported that on occasion soluble salt levels in MSW compost have resulted in phytotoxicity to plants grown in the greenhouse. Hortenstine and Rothwell (1973a) showed an increase in salinity from 0.50 mmhos/cm in the control to 0.46, 0.65, 0.87, and 1.18 mmhos/cm for compost applications of 16, 32, 64, and 128 t/ha, respectively.

Epstein and Wu (1994) found that preparing MSW compost irrigation water increased the salt levels in the compost. As a result, plant germination and growth were suppressed until the salts were leached below the root zone. Compost added to a desert soil high in soluble salts increased the EC from 5.8 mmhos/cm to 11.4 and 13.1 mmhos/cm for 168 and 336 t/ha, respectively.

Epstein et al. (1976) showed that EC increased with increasing application rates of biosolid compost. Specifically, EC values ranged from a low of 0.41 mmhos/cm for a control to 4.15 mmhos/cm with an application of 240 t/ha. At the end of the growing season, EC values dropped considerably as a result of leaching and by the following growing season EC values were at levels tolerable to most plants.

Salinity levels in compost materials need to be assessed prior to utilization of the compost. Where salts can be a problem for plant growth such as in horticultural or agricultural applications, leaching of the compost may be necessary prior to seeding or transplanting. In field applications, rainfall or irrigation will leach out salts.

Further research is needed to identify the effects of the various components within the compost matrix on specific soil physical and chemical properties.

Nitrogen Availability in Soil

The nitrogen (N) in compost is predominantly in the organic form (see Chapter 2). Organic N is not soluble and does not leach through the soil; nor is it available to plants. For N to be available to plants, it needs to be mineralized (i.e., converted to the inorganic soluble form). The inorganic forms of N are ammonium (NH_4^+) and nitrate (NO_3^-).

Knowledge of the availability of N from compost is important in order to estimate the amount of N for crop growth. If the amount of available N from the compost is insufficient, then supplement N is needed. If too much N is available from the compost, excess N beyond the crop's needs can leach to groundwater.

The N cycle in the soil depicted in Figure 11.16 shows the various reactions and transformations that can occur when compost is applied to soil. For example, on incorporation into the soil, organic residues or compost

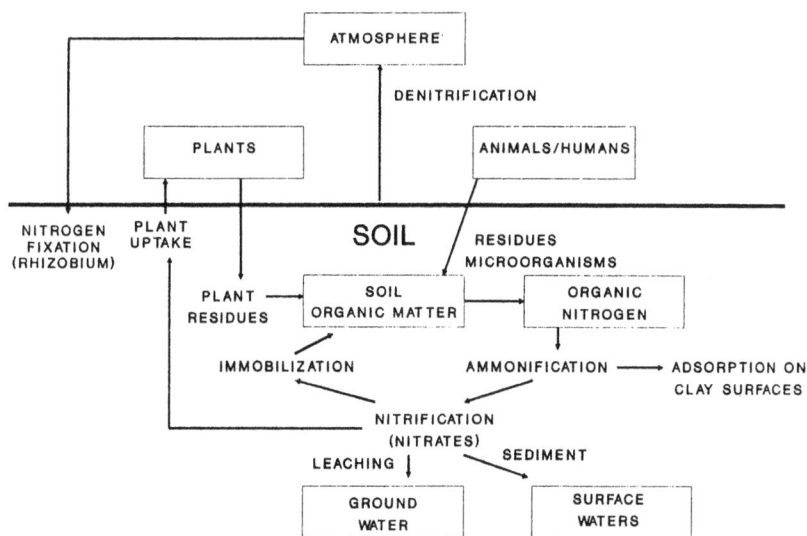

FIGURE 11.16. The nitrogen cycle in the soil.

from plants, animals and humans become part of the soil organic matter pool. The organic N undergoes mineralization by microorganisms. Ammonification converts the organic N to ammonia (NH_3). Some of the NH_3 is converted to ammonium ions (NH_4^+) and is adsorbed on clay surfaces where it is exchangeable with other cations or fixed in the clay structure. A major portion of NH_3 is nitrified by bacteria to nitrite (NO_2^-) which is very rapidly converted to NO_3^-. Nitrate N is soluble and is available to plants or if in excess of plant needs, it can leach with soil water into groundwater. Denitrification under anaerobic soil conditions can convert NO_2^- and NO_3^- to N_2 gas and be lost to the atmosphere.

As early as 1948, Bould (1948) evaluated the availability of N in waste materials. Similar results were obtained by four methods used to estimate the available N from straw-biosolids and MSW-biosolids compost at approximately 7% for mature compost (Table 11.3). Immature compost had higher values. During composting nitrogen is lost as ammonia. The remaining nitrogen is in the organic form and is more resistant to microbial attack.

Tester et al. (1977) found a linear relationship between the amount of municipal biosolids applied and the quantity of N mineralized. These researchers also found that in different soils vastly different amounts of N were mineralized. They attributed some of these differences to NH_4-N fixed on the surfaces of vermiculitic clay in one soil and indicated that this N is not available for nitrification. Maximum N mineralization of 6% of the compost N occurred with an application rate of 4% compost-loamy sand mixtures.

Epstein et al. (1978) determined the mineralization rates of compost from digested and raw biosolids compost. Cumulative N mineralized over a

TABLE 11.3. *Percent of Available Nitrogen in Straw-Biosolids Compost and MSW-Biosolids Compost as Determined by Four Different Methods.*

Method of Analysis	Straw-Biosolids Compost (% available N)		MSW-Biosolids Compost (% available N)	
	Mean	Range	Mean	Range
Nitrification pot method	7.4	4.3–10.2	6.4	4.4–7.9
Nitrification perfusion	7.8	7.8	8.4	7.3–9.5
Fermentation	14.6	7.7–22.0	11.2	5.7–14.3
Pot experiment by analysis	7.8	7.8	8.6	5.6–11.5

Source: Bould, 1948.

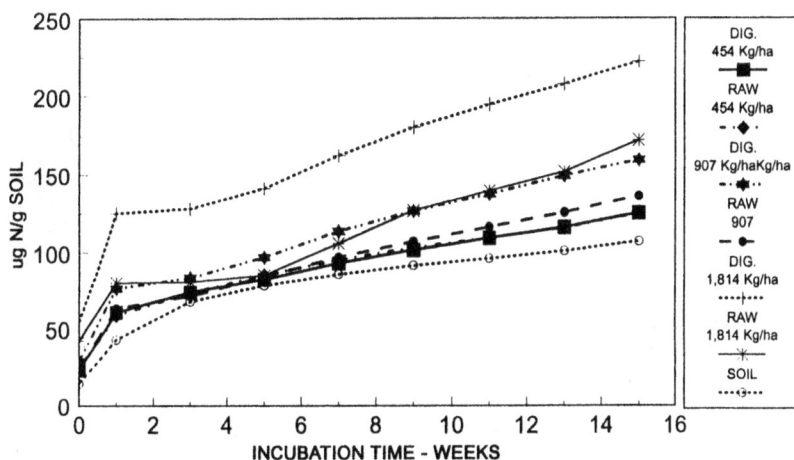

FIGURE 11.17. Cumulative total N mineralized for 454-kg N/ha, 907-kg N/ha, and 1814-kg/ha rate from soil (S), composted raw biosolids (CRB), and composted digested biosolids (CDB). (Data from Epstein et al., 1978.)

15-week incubation period is shown in Figure 11.17. The highest amount of mineralized N occurred with the high application rate of 1814 kg N/ha treatment of the digested biosolids compost. For all three treatment rates, less N was mineralized with raw biosolids composts than from the digested biosolids compost. This difference was not statistically significant but may indicate that the raw biosolids compost was less stable and immobilized more N.

For the two composts and for the three rates, the rate of N mineralization (Figure 11.18) showed that the majority of the mineralization occurred during the first three weeks of incubation. Thereafter the rate was fairly constant. Total N mineralized was linear with compost N application rates (Figure 11.19). The percentage mineralized over a 15-week period was in the range of 7% to 10% for the digested biosolids compost and 4% to 5% for the raw biosolids compost. Douglas and Magdoff (1991) studied the N mineralization of various organic residues and found that much of the compost had immobilized N (negative mineralization) or was similar to the control. This may indicate that the compost was unstable and that microorganisms were still utilizing N in their synthesis.

In an attempt to provide information about the transformations that occur when biosolids compost is added to soil, Epstein et al. (1978) conducted ^{15}N studies (Table 11.4). These studies showed that the raw biosolids compost promoted greater immobilization than the digested biosolids compost. The data also indicated that amendments high in available C are very active

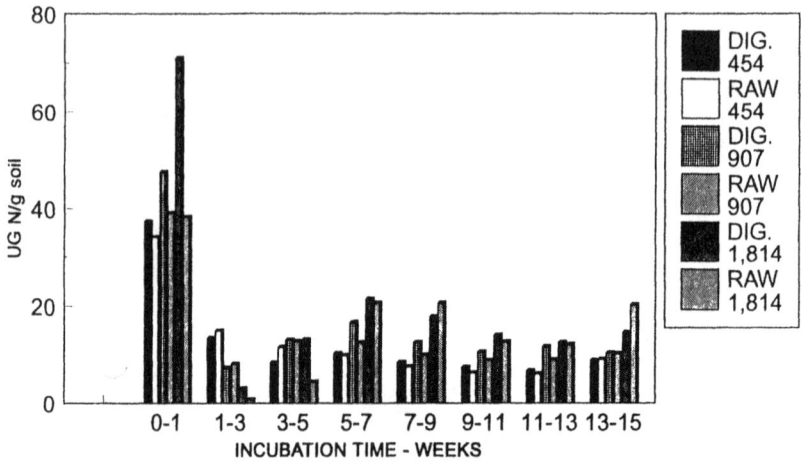

FIGURE 11.18. Inorganic nitrogen mineralized from biosolids compost during incubation applied at 454 kg/ha, 907 kg/ha, and 1814 kg/ha. (Data from Epstein et al., 1976.)

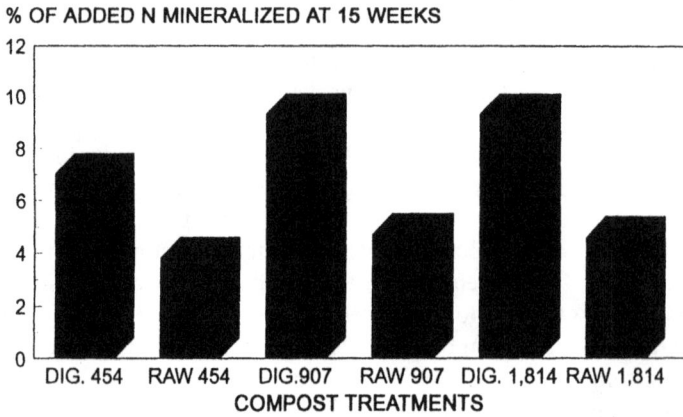

FIGURE 11.19. Percent of added nitrogen mineralized at 15 weeks. Values represent net mineralization. (Data from Epstein et al., 1976.)

TABLE 11.4. Nitrogen Transformations from Biosolids Compost Applied to Soil after 1.3 and 5 Weeks of Incubation.

Treatment	Incubation Time (weeks)	% 15N Recovered				
		NO_3^-	NH_4^+	Soluble Organic N	Organic N	Total
Soil	1	91.2	0.6	1.6	2.6	95.9
	2	84.5	0.1	5.6	2.4	92.5
	3	44.9	0.1	5.6	2.9	53.5
	L.S.D. (0.05)	19.4	0.3	7.7	0.9	16.4
Soil and compost from digested biosolids	1	84.6	1.5	3.2	4.9	94.0
	2	72.8	0.1	3.1	17.4	93.4
	3	28.9	0.0	3.9	28.3	61.0
	L.S.D. (0.05)	17.2	0.1	4.8	13.7	7.8
Soil and compost from raw biosolids	1	87.0	1.6	1.4	6.9	96.8
	2	65.1	0.2	3.3	25.9	90.8
	3	13.1	0.2	2.0	44.5	59.8
	L.S.D. (0.05)	9.4	0.5	0.3	5.4	8.8
	Pooled L.S.D.	10.2	1.8	3.7	5.1	8.3

Source: Epstein et al., 1978.

biologically and that significant quantities of NO_3-N can be lost by denitrification and immobilization. The N in compost produced from biosolids and a carbonaceous bulking agent (e.g., wood chips) is not easily mineralized. Evidently, composting may tie up large quantities of N in relatively unavailable forms and the N, when released, can be incorporated into the organic fraction as the C is oxidized.

Ama (1985) determined the distribution of various N compounds in four fractions of four composts. Three of the composts were from animal wastes; the fourth was a MSW compost. In the MSW compost, the highest amount of organic N was in the water-soluble fraction; the least amount of organic N was in the coarse solid fraction. This fraction represented the sediment after shaking a 100 g sample with water and allowing it to settle for 16 hours. Based on Stokes' law even this fraction would contain a significant amount of fine particles. Mineralized N ranged from 2.0 to 15.7% of the organic N for the various fractions, with an average of 7% to 8%. Total organic N was 1712 mg N/g dry matter. It was evident that different organic wastes and the different amounts of carbonaceous amendment affected the N mineralization rate.

In order to evaluate residual effects of compost application, an index capable of estimating available N would be useful. O'Keefe et al. (1986) evaluated N availability indexes for biosolids compost-amended soil. A long-term increase in N availability was found when biosolids compost was applied to soils. The autoclave-extractable N chemical index could be used to assess N availability. All the N availability indexes studied were linearly related to biosolids compost application.

Any N index needs to be calibrated to local soil-crop-climate conditions before it can be expressed in quantitative terms (O'Keefe et al., 1986). Two distinct periods of N mineralization were noted regardless of the rate of application. In the first period, lasting about 28 weeks, mineralization was rapid. The second period, from 28 to 72 weeks, was much slower and constant. The first period indicated easily mineralized organic N (Figure 11.20).

Hadas and Portnoy (1994) found that over a 32-week period inorganic N released by composted manure applied to soils ranged from 11% to 29% of their total N. Two percent to 12% was initially inorganic N and 1% to 5% was soluble organic N. Net mineralization rates of insoluble N did not exceed 10% of total N.

Based on all the studies reviewed, the rate of N mineralization from cured compost usually ranged between 7% and 10% over a 20- to 35-week period. Nitrogen mineralization is rapid initially and then decreases and remains fairly constant. The type of feedstock used does not appear to greatly affect the amount of N mineralized, provided the compost is mature or cured.

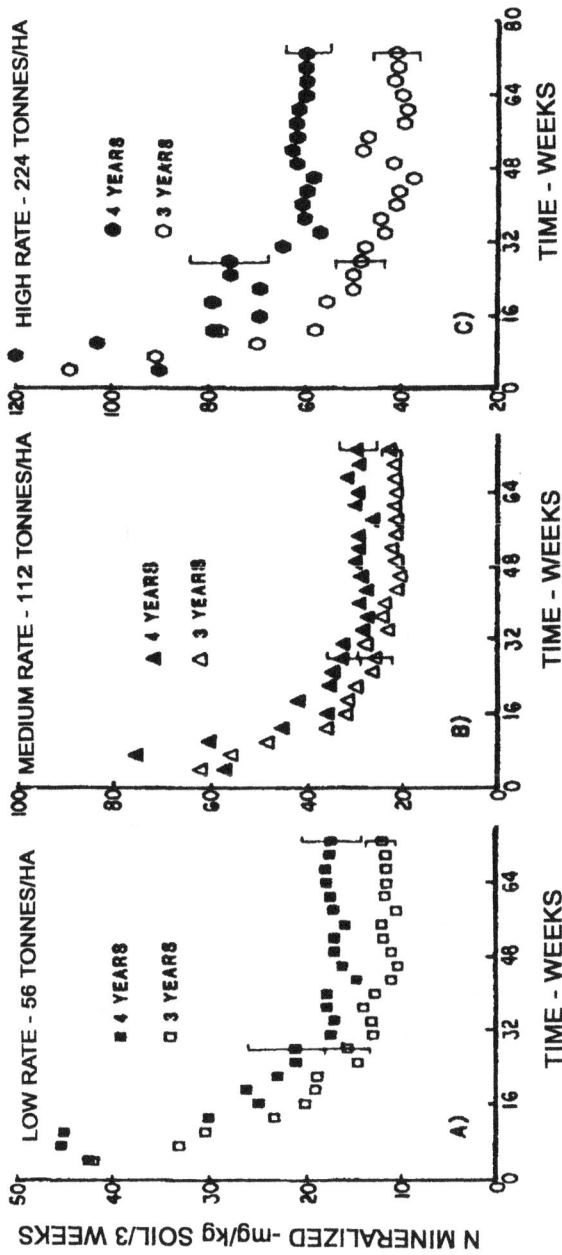

FIGURE 11.20. Net nitrogen mineralization in the zone of biosolids compost incorporation that received three and four yearly treatments of biosolids compost at three different rates. (Data from O'Keefe et al., 1986.)

Nitrogen Leaching

Soluble N compounds applied to soils for crop growth either in fertilizers or wastes can leach through the soil into groundwater if the amount of N exceeds the crop requirements. Nitrate-N in excess of 10 ppm in drinking water is considered a health hazard (National Research Council, 1978).

Maynard (1993) applied spent mushroom compost (SMC) and chicken manure compost (CMC) at rates of 56 and 112 tonnes/ha to a Merrimac fine sandy loam. The nitrogen content of the CMC was 2.5% (25,000 mg/kg) and 0.58% (5800 mg/kg) for the SMC. No nitrogen was added to the compost plots and the fertilizer control received nitrogen. Wells were installed in the groundwater for sampling. Depth of groundwater fluctuated from a low of 1.67 m to a high of 2.58 m. Table 11.5 shows the mean NO_3-N concentrations in groundwater beneath plots amended with compost. Figures 11.21a, 11.21b, and 11.21c list the monthly NO_3-N concentrations of groundwater during the time of the study.

During 1989–1990 (Figure 11.21a) generally higher levels of NO_3-N were found in groundwater beneath the fertilizer control plots than beneath the compost plots. In the control plots, levels above 10 ppm occurred from early May through early June, with a high level of 14.8 ppm. In 1990–1991 the highest levels occurred most frequently under the control plots. The highest level in the control plot was 7.0 ppm (Figure 11.21b). In 1991, higher levels

TABLE 11.5. *Mean Nitrate Concentrations in Groundwater beneath Plots Amended with Chicken Manure Compost (CMC) and Spent Mushroom Compost (SMC) at 56 and 112 tonnes/ha.*

	Compost Application Rate (tonnes/ha)				
	CMC 56	CMC 112	SMC 56	SMC 112	Control
Year	Mean Concentration of Nitrate				
1989–1990	1.2	1.8	2.2	2.6	4.4
1990–1991	2.3	2.9	3.1	2.0	4.0
1991	2.6	5.3	6.0	4.4	4.0
Overall mean	1.9	3.1	3.4	2.8	4.2
Significance[1]	a	bc	bcd	ab	d

[1]Same letter in row indicates values not significantly different at the 5% level according to the Newman-Keuls comparison test.
Source: Maynard, 1993.

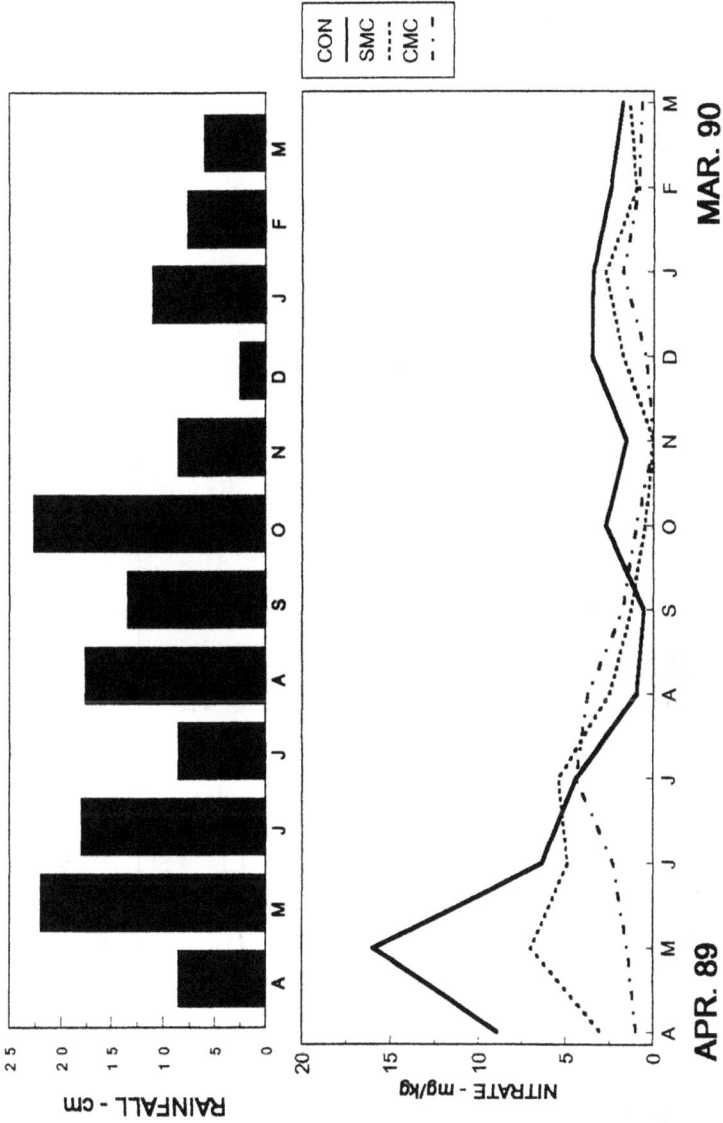

FIGURE 11.21a. Concentration of nitrate in groundwater beneath plots amended with one application of 112 tonnes/ha of spent mushroom compost (SMC) or chicken manure compost (CMC) or fertilized control (CON). (Reprinted with permission from Maynard, 1993.)

375

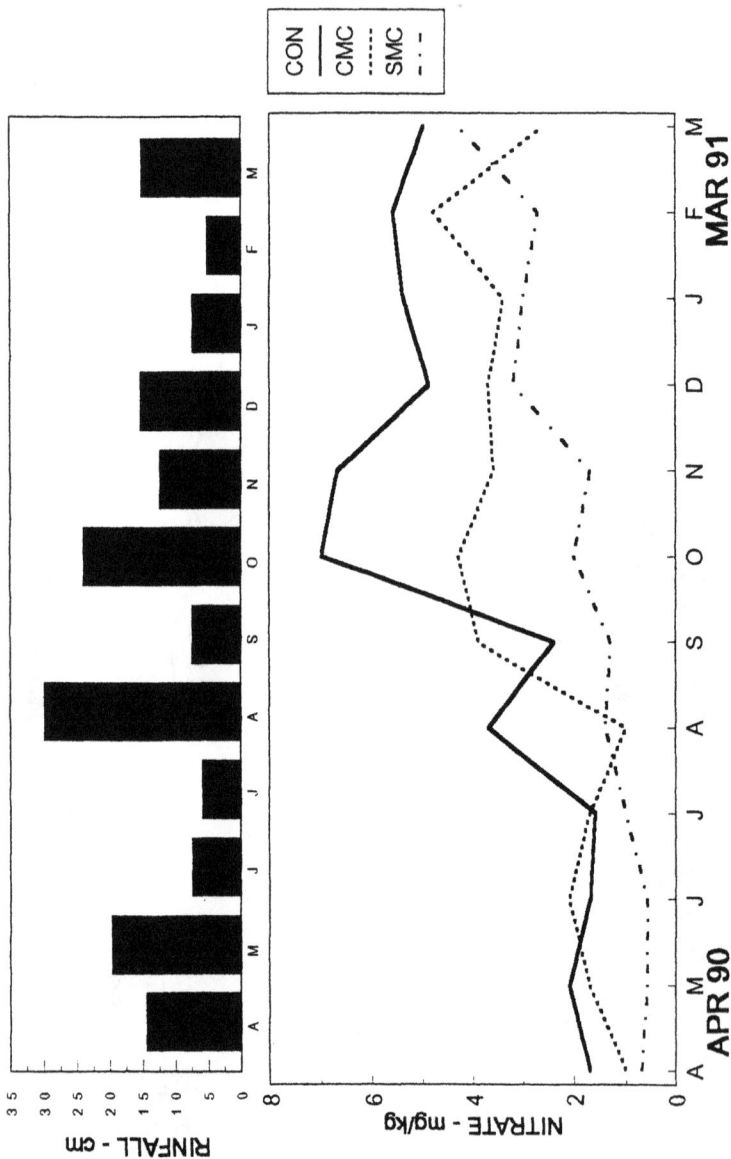

FIGURE 11.21b. Concentration of nitrate in groundwater in 1990–1991 beneath plots amended with two applications of 112 tonnes/ha of spent mushroom compost (SMC) or chicken manure compost (CMC) or fertilized control (CON). (Reprinted with permission from Maynard, 1993.)

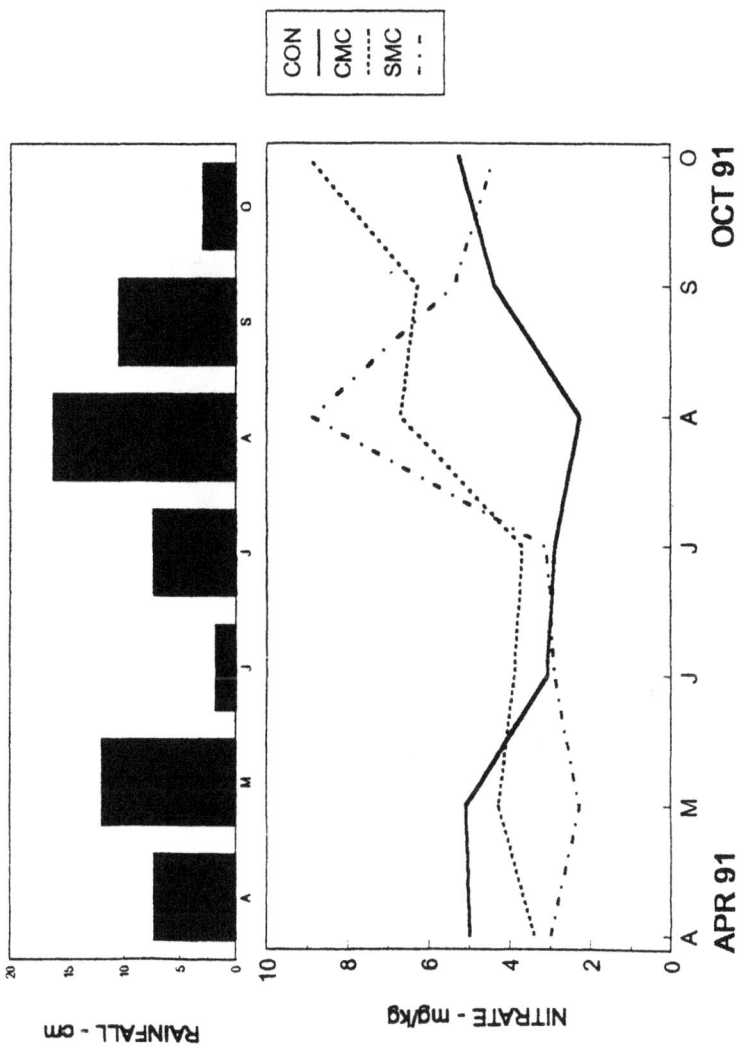

FIGURE 11.21c. Concentration of nitrate in groundwater in 1991 beneath plots amended with three applications of 112 tonnes/ha of spent mushroom compost (SMC) or chicken manure compost (CMC) or fertilized control (CON). (Reprinted with permission from Maynard, 1993.)

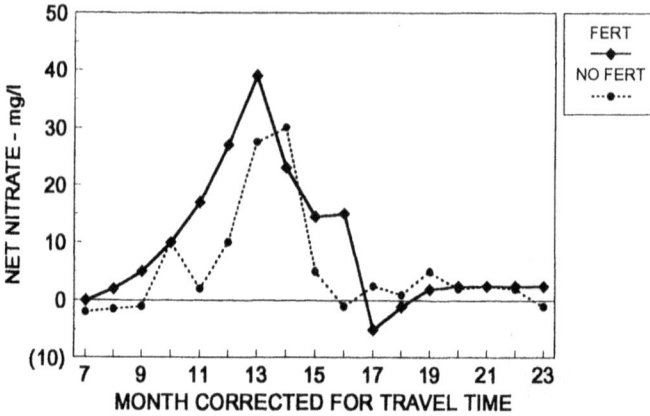

FIGURE 11.22. Net concentration of nitrate-nitrogen in turf plot wells corrected for ground-water travel time. (Data from Frink and Sawhney, 1994.)

of NO_3-N were found in groundwater beneath the compost plots (Figure 11.21c).

Frink and Sawhney (1994) studied the movement of N after the application of biosolids compost through lysimeters and turf plots. In the absence of plants, monitoring for N on the turf plots took place through wells that reached the shallow (1.5 to 2.4 m) perched water table of the Merrimac sandy loam. Compost was applied at a rate of 293 t/ha (6000 lbs/1000 ft^2) and harrowed into the top 10 to 15 cm of soil. The biosolids compost contained over 2% total-N in addition to some NO_3-N and NH_4-N. Apparently, it contained more available N than anticipated with the movement of significant amounts to groundwater. The actual contribution of NO_3-N from the compost was on the average 5.4 mg/l after correcting for the water travel time between the upgrade and downgrade wells.

Figure 11.22 shows the NO_3-N data. During two months in the compost plots without fertilizer, NO_3-N exceeded the 10 mg/l drinking water standard. Assuming 10% of the total-N became available, the amount of available N would be 586 kg/ha (523 lbs/ac). This is a high amount and accounts for the excessive leaching of N. This could have been avoided by first conducting an N mineralization study.

SUMMARY

Compost is primarily a soil-conditioning material that contains various levels of plant macro- (major) and micronutrients. Since all plants and soils contain trace elements, regardless of the feedstock, compost will contain trace elements.

Since most compost products have a high organic matter content, the effect on the soil physical and chemical properties is primarily the result of the addition of organic matter.

As a soil conditioner, compost improves the soil physical properties and, therefore, provides for a better root environment. As a result, plants are better able to utilize water and nutrients. Improvements in soil structure provides for increased water infiltration and permeability, which reduces the potential for runoff and erosion while at the same time increasing the water content of the soil. This effect is more evident in clayey soils, which have poor infiltration and low permeability to water. Under droughty conditions, compost-amended soils provide more water to plants. In irrigated areas, compost could reduce water application. Application of compost to sandy soils increases the water-retention properties and provides more water to plants.

Compost generally contains low levels of the major elements, N, P, and K. It also contains many of the micronutrients required for crop growth. In order to improve many soils, the quantities of compost needed also provide the necessary nitrogen. Knowledge of N mineralization is important in order to ensure sufficient N for plant growth and avoiding excess N, which could leach to groundwater.

REFERENCES

Ama, M. 1985. Properties of fine and water-soluble fractions of several composts. *Soil Water Plant Nutr.* 31(2):189–198.

Avnimelech, Y. and A. Cohen. 1988. On the use of organic manures for amendment of compacted clay soils: Effects of aerobic and anaerobic conditions. *Biological Wastes.* 26:331–339.

Avnimelech, Y., A. Cohen, and D. Shkedi. 1990. The effect of municipal solid waste compost on the fertility of clay soils. *Soil Technology.* 3:275–284.

Avnimelech, Y., A. Cohen, and D. Shkedi. 1993. Can we expect a consistent efficiency of municipal waste compost application? *Compost Sci. & Util.* 1:7–14.

Bernstein, L. 1964. *Salt Tolerance of Plants.* USDA Agric. Inform. Bull. No. 283. 23 pp.

Bould, C. 1948. Availability of nitrogen in composts prepared from waste materials. *Empire J. Exp. Agr.* 16:103–110.

Byrd, R. E. 1987. Stormwater-borne pollutant export from turfgrass established on soils amended with composted domestic wastewater sludges. M.S. Thesis, Virginia Polytechnic Institute and State U. Blacksburg, VA.

Douglas, B. F. and F. R. Magdoff. 1991. An evaluation of nitrogen mineralization indices for organic residues. *J. Environ. Qual.* 20:368–372.

Duggan, J. C. and C. C. Wiles. 1976. Effect of municipal compost and nitrogen fertilizer on selected soils and plants. *Compost Sci.* 17(5):24–31.

Emerson, W. W. 1959. The structure of soil crumbs. *J. Soil Sci.* 10:215.

Epstein, E., D. B. Keane, J. J. Meisinger, and J. O. Legg. 1978. Mineralization of nitrogen from sewage sludge and sludge compost. *J. Environ. Qual.* 7(2):217-221.

Epstein, E., J. M. Taylor, and R. L. Chaney. 1976. Effect of sewage sludge and sludge compost applied to soil on some soil physical and chemical properties. *J. Environ. Qual.* 5:422–426.

Epstein, E. and N. Wu. 1994. *Internal Document. The SAMM Compost Pilot Project Agricultural Study.* Final Report. E&A Environmental Consultants, Inc., Canton, MA.

Frink, C. R. and B. J. Sawhney. 1994. *Leaching of Metals and Nitrate from Composted Sewage Sludge.* 46 pp. Bull. 923. The Connecticut Agr. Expt. Sta., New Haven, CT.

Giusquiani, P. L., M. Pagliai, G. Gigliotti, D. Businelli, and A. Benetti. 1995. Urban waste compost: Effects on physical, chemical, and biological soil properties. *J. Environ. Qual.* 24:175–182.

Grebus, M. E., M. E. Watson, and H. A. J. Hoitink. 1994. Biological, chemical and physical properties of composted yard trimmings as indicators of maturity and plant disease suppression. *Compost Sci. & Util.* 2(1):57–71.

Guidi, G. V. and G. Petruzzelli. 1989. Effects of compost on chemical and physical characteristics of soil, pp. 53–56. In *Compost Production and Use: Technology Management, Application, and Legislation. Proc. Int. Symp. on Compost.*

Hadas, A. and R. Portnoy. 1994. Nitrogen and carbon mineralization rates of composted manures incubated in soil. *J. Environ. Qual.* 23:1184–1189.

Hernando, S., M. C. Lobo, and A. Polo. 1989. Effect of the application of municipal refuse compost on the physical and chemical properties of a soil. *The Sci. of the Total Environ.* 81/82:589–596.

Hornick, S. B. 1988. Use of organic amendments to increase the productivity of sand and gravel spoils: Effects on yield and composition of sweet corn. *Annals Agr. Res.* 7:346–352.

Hortenstine, C. C. and D. F. Rothwell. 1973a. *Composted Municipal Refuse as a Soil Amendment.* USEPA PB-222 422. Gainesville, FL.

Hortenstine, C. C. and D. F. Rothwell. 1973b. Pelletized municipal refuse compost as a soil amendment and nutrient source for sorghum. *J. Environ. Qual.* 2:234–245.

Jacobowitz, L. A. and T. S. Steenhuis. 1984. Compost impact on soil moisture and temperature. *BioCycle.* 25(1):56–60.

Khaleel, R., K. R. Reddy, and M. R. Overcash. 1981. Changes in soil physical properties due to organic waste applications: A review. *J. Environ. Qual.* 10:133–141.

Kononova, M. M. 1975. Humus of virgin and cultivated soils. In J. E. Giescking (ed.). *Soil Components. Vol. 1. Organic Components.* Springer-Verlag, New York.

MacCarthy, P., R. L. Malcolm, C. E. Clapp, and P. R. Bloom. 1990. An introduction to soil humic substances, pp. 1–12. In P. MacCarthy et al. (ed.). *Humic Substances in Soil and Crop Sciences: Selected Readings.* Amer. Soc. of Agronomy, Inc., Soil Sci. Soc. of Amer., Inc., Madison, WI.

Madariaga, G. M. and J. S. Angle. 1992. Sludge-borne salt effects on survival of *Bradyrhizobium japonicum. J. Environ. Qual.* 21:276–280.

Manios, V. L. and H. L. Syminis. 1988. Town refuse compost of Heraklio. *BioCycle.* 29(6):44–47.

Maynard, A. A. (1993). Nitrate leaching from compost-amended soils. *Compost Sci. & Util.* 1(2):65–72.

Mays, D. A. and P. M. Giordano. 1989. Landscaping municipal waste compost. *BioCycle.* 30(3):37–39.

Mays, D. A., G. L. Terman, and J. C. Duggan. 1973. Municipal compost: Effects on crop yield and soil properties. *J. Environ. Qual.* 2:89–92.

McConnell, D. B., A. Shiralipour, and W. H. Smith. 1993. Compost application improves soil properties. *BioCycle.* 33(1):61–63.

McCullough, E., R. Wetzel, and K. Buttry. 1993. Physical properties of composted materials mixed with sand. *Compost Sci. & Util.* 2:54–57.

Mebius, L. J. 1960. A rapid method for the determination of organic carbon in soil. *Anal. Chim. Acta.* 22:120–124.

National Research Council. 1978. *Nitrates: An Environmental Assessment.* National Academy of Sciences, Washington, DC.

Nelson, D. W. and L. E. Sommers. 1982. Total carbon, organic carbon and organic matter, pp. 539–579. In A. L. Page (ed.). *Methods of Soil Analysis. Part 2. Chemical and Microbiological Properties.* 2nd edition. Agronomy No. 9. Amer. Soc. of Agronomy, Madison, WI.

O'Keefe, B., J. Axley, and J. J. Misinger. 1986. Evaluation of nitrogen availability indexes for a sludge compost amended soil. *J. Environ. Qual.* 15(2):121–128.

Pagliai, M., G. Guidi, M. La Marca, M. Giachetti, and G. Lucamante. 1981. Effects of sewage sludge and composts on soil porosity and aggregation. *J. Environ. Qual.* 4:556–561.

Schrader, T. 1967. Composted town refuse and sewage sludge in viticulture. *Weiberg Keller.* 12:531–537.

Serra-Whittling, C., S. Houot, and E. Barriuso. 1996. Modification of soil water retention and biological properties by municipal solid waste compost. *Compost Sci. & Util.* 4(1):44–52.

Shiralipour, A., D. M. McConnell, and W. H. Smith. 1992. Physical and chemical properties of soils as affected by municipal solid waste compost application. *Biomass and Bioenergy.* 3:261–266.

Stevenson, F. J. 1994. *Humus Chemistry—Genesis, Composition, Reactions.* 2nd Ed. John Wiley & Sons, Inc., New York.

Tan, K. H., J. H. Edwards, and O. L. Bennett. 1985. Effect of sewage sludge on mobilization of surface-applied calcium in a Greenville soil. *Soil Sci.* 139:262–269.

Terman, G. L. and D. A. Mays. 1973. Utilization of municipal solid waste compost: research results at Muscle Shoals, Alabama. *Compost Sci.* 14(1):18–21.

Tester, C. F. 1990. Organic amendment effects on physical and chemical properties of a sandy soil. *Soil Sci. Soc. Amer. J.* 54:827–831.

Tester, C. F. and J. F. Parr. 1983. Decomposition of sewage sludge compost in soil: IV. Effect of indigenous salinity. *J. Environ. Qual.* 12:123–126.

Tester, C. F., L. J. Sikora, J. M. Taylor, and J. F. Parr. 1977. Decomposition of sewage sludge compost in soil: 1. Carbon and nitrogen transformations. *J. Environ. Qual.* 6(4):459–463.

Turner, M. S., G. A. Clark, C. D. Stanley, and A. G. Smajstrla. 1994. Physical characteristics of a sandy soil amended with municipal solid waste. *Soil Crop Sci. Soc. Florida Proc.* 53:24–26.

CHAPTER 12

Utilization of Compost

INTRODUCTION

The predominant use of compost has traditionally been in horticulture and agriculture to improve soil conditions and thereby enhance plant growth. In more recent years, compost has also been used as a medium for biofiltration, as a suppressant of plant diseases, and a filtering medium for contaminants found in storm water, as well as other uses not related to improving soil physical properties. This chapter focuses on the use of compost to enhance plant growth. The following chapter will provide information on the non-agronomic/horticultural uses.

Compost as an organic matter resource improves the soil's physical properties as discussed in Chapter 4. In addition, it provides various levels of macronutrients such as N, P, K, Ca and Mg, and many of the micronutrients required for plant growth. The level of macro- and micronutrients depends to some extent on the feedstock. Nitrogen is the one element that is reduced during the composting operation. The ultimate objective of compost utilization in agriculture/horticulture is increased plant growth or improved plant quality. However, another important function for compost is to improve soil physical conditions in order to reduce runoff and erosion, thereby reducing non-point source pollution.

Two important characteristics of the compost that must be considered are stability and soluble salts. An unstable compost can deplete soil nitrogen unless it is properly supplemented. Soluble salts can result in phytotoxicity. Leaching of the salts prior to seeding or planting alleviates these problems.

One of the major concerns of compost producers is the potential market. Compost in bulk is light weight (approximately 450 kg/m^3, or 1000 lb/cubic yard); consequently, it is costly to transport it great distances, which often limits the market. If the compost is bagged, however, it has a much higher value on a weight or volume basis, and can be shipped much greater distances.

Slivka et al. (1992) estimated that a total of 46 million tonnes or 78 cubic meters (51 million tons or 102 million cubic yards) of compost are produced in the United States in this century. They also indicated that the potential demand for compost in the United States could be 795 million cubic meters or 472 million tonnes (1040 million cubic yards or 520 million tons).

Table 12.1 shows the projected demand by application segments. Agricultural demand, by far the largest potential market, was based on the assumption that farms within the maximum economic hauling distance of 80 km from large cities (population greater than 100,000) would utilize large volumes of compost.

The predominant current use of compost is in horticulture, agriculture, and reclamation of disturbed soils. Shiralipour et al. (1992) subdivided the agricultural uses into subcategories, agronomic and horticultural crops. The

TABLE 12.1. Estimated Potential Demand for
Compost in the United States.

Application Segment	Potential Compost Demand, million cubic meters (million cubic yards)	
Landscaping	1.53	(2.0)
Delivered topsoil	2.83	(3.7)
Bagged/retail	6.12	(8.0)
Landfill final cover	0.46	(0.6)
Surface mine reclamation	0.15	(0.2)
Container nurseries	0.69	(0.9)
Field nurseries	3.06	(4.0)
Sod production	15.29	(20)
Silviculture	79.52	(104)
Agriculture	684	(895)
Total	795	(1040)

Source: Slivka et al., 1992.

listing below indicates the various subcategories that will be discussed in this chapter:

- horticulture: nurseries, landscaping, greenhouse, and sod production and turf
- agriculture and forestry: agronomic or field crops, forest and forest nurseries, vegetable crops, fruit trees, and viticulture
- disturbed soils: mine reclamation, public works, and roadside revegetation

Utilization of compost depends on product quality and consistency. The parameters of concern vary depending on the compost feedstock source and the intended use. The most important characteristics for different types of compost are:

- biosolids compost: pathogens, heavy metals, soluble salts, stability, odor, pH, and particle size
- MSW compost: pathogens, heavy metals, soluble salts, boron, stability, odor, maturity, pH, inerts (plastics, glass, metals, etc.) and particle size
- yard waste compost: stability, odor, maturity, inerts (plastics, metals, stones, etc.), particle size, and pH

Pathogens and heavy metals are less of a concern in yard waste than in compost containing biosolids or MSW. However, as indicated in previous chapters, yard waste can contain significant levels of pathogens and lead. Lead is not readily taken up by plants and does not present a risk to humans or plants.

In addition to these characteristics, the user generally wants the material to have a moisture content in the 35% to 50% range. At lower moisture contents, the compost is dry and dusty. Above 50% moisture, the compost is too wet and can form clumps. In addition, wet compost is expensive to transport and difficult to handle.

The data on the effect of compost on plant growth are quite extensive and more are being generated (Shiralipour et al., 1992a, 1992b). One objective of this chapter is to present current data and trends. The interested reader is encouraged to seek the original publication to obtain greater detail.

The effect of compost products on soil physical properties changes little with the different feedstocks, assuming the compost reaches the same level of stability. Chemical effects such as phytotoxicity, impacts on levels of trace elements or heavy metals, and addition of organic compounds vary depending on feedstock characteristics. The characteristics of the feedstocks, in turn, change with consumer habits (extent of recycling) and technology (e.g., separation). Finally, regulations affect the feedstock characteristics through restrictions on inputs.

When reviewing the literature it is important to recognize the changes that have occurred over the past few years. This is particularly true of chemical changes as a result of the composting of cleaner feedstocks and the changes resulting from regulations.

A good example is boron (B) in MSW. Compost produced prior to the 1980s usually contained fairly high levels of *B. purvis* and MacKenzie (1974) reported on B phytotoxicity from MSW compost that often contained high levels due to B in paper. This is not true today.

Another example is chlorinated pesticides, which are no longer used and hence are often found only in very small amounts. The Clean Water Act had a significant impact on reducing industrial discharges into the municipal wastewater system, resulting in large reductions of heavy metals in sewage biosolids.

The literature on the use of compost for agriculture, horticulture, forestry and disturbed lands is extensive. Readers who desire additional information on specific plant growth trials in particular media may refer to Table 12.2.

TABLE 12.2. Literature Cited on Uses of Compost in Agriculture, Horticulture, Forestry, and Disturbed Lands.

Plant Species	Compost	Soil/ Media	Study	Reference
Field Crops				
Corn, tall fescue	MSW	sil, cl	G	Terman et al., 1973
Tobacco	MSW	cl	F	Duggan, 1973
Sorghum	MSW	s	G	Hortenstine and Rothwell, 1973
Corn	MSW & bio	l	F	Duggan and Wiles, 1976
Fescue	Bio	ls	G	Sikora et al., 1980
Ryegrass	Bio	sil	G	Hileman, 1982
Corn	MSW	NA	F	Mays and Giordano, 1989
Wheat	Ind	l	G	Sikora and Azad, 1993

TABLE 12.2. (continued).

Plant Species	Compost	Soil/ Media	Study	Reference
Field Crops (continued)				
Corn	MSW	ls, sil	F	Wolkowski, 1994
Vegetables				
Tomato	Ind	media	G	Gouin and Shanks, 1981
Komatsuna	MSW	s	G	Chanyasak et al., 1983
Tomato, pepper, cucumber	Agr	media	G	Inbar et al., 1986
Bush bean, collard, squash, tomato	Mush	s	F	Stephens et al., 1989
Kohlrabi	Biogen	NA	F	Vogtmann and Fricke, 1989
Broccoli, cauliflower	CMC, mush	sl	F	Maynard, 1994
Ornamental				
Chrysanthemum	MSW	media	G	Conover and Joiner, 1966
Tulip poplar, dogwood	Bio	sl	F	Gouin, 1976
Ficus elastica, Dracaena terminalis, Monstera deliciosa, Fatshedera, Dieffenbachia	Bark	NA	G	Cappaert et al., 1973, 1976
Ilex cornuta, Thuja occidentalis, Viburnum burkwoodii	MSW	media	G	Sanderson and Martin, 1974
Cotoneaster	Bark	media	G	Sterrett and Fretz, 1977
Petunia, zinnia, marigold	Bio	media	G	Wootton, 1977

(continued)

387

TABLE 12.2. (continued).

Plant Species	Compost	Soil/ Media	Study	Reference
Ornamental *(continued)*				
Japanese holly, cherry laurel	Bio	media	G	Gouin, 1978a
Chrysanthemum	Bio	media	G	Cambell and Pirani, 1978
Ilex cornuta, Juniperus conferta, rhododendron "Evensong" *Viburnum burkwoodii,* snapdragon, easter lilies, geraniums	MSW/Bio	media	G	Sanderson, 1980
Monstera deliciosa, Cordyline terminalis	Bark	media	beds	Verdonck, 1980
Marigold	Bio	media	G	Chaney et al., 1980
Marigold, zinnia petunia	Bio	media	G	Wootton et al., 1981
Dwarf oleander, jasmine, ligustrum	Bio	media	G	Fitzpatrick, 1981
Ficus, Schefflera	Bio	media	G	Fitzpatrick, 1981
Pansy, snapdragon	Bio	media	G	Hemphill et al., 1984
West Indian mahogony, key lime, *Schefflera,* pink tabebuia, pigeon plum	Bio	media	G	Fitzpatrick, 1985
Orange jasmine, dwarf oleander, Cuban royal palm	MSW, MSW & bio	media	G	Fitzpatrick, 1989
Dieffenbachia, gardenia	Ind	media	G	Poole and Conover, 1989

TABLE 12.2. (continued).

Plant Species	Compost	Soil/ Media	Study	Reference
Ornamental *(continued)*				
Dracaena fragrans, Pontederia cordata, Quercus shumardii	MSW, yw	media	G	McConnell et al., 1991
Rudbeckia hirta	Bio, MSW, Ind	media	G	Bugbee, 1994
Pickerelweed	MSW, yw	media	G	McConnell et al., 1990
Pickerelweed	MSW, yw	media	G	McConnell and Shiralipour, 1991
Dracaena fragrans, Peperomia obatusifolia, Schefflera arboricola	yw	media	G	McConnell and Shiralipour, 1991
Ageratum	yw	media	N	MacCubbin and Henley, 1993
Pickerelweed	MSW, yw	media	G	McConnell et al., 1991

HORTICULTURE

The horticultural industry utilizes vast quantities of organic materials in the production of marketable plants. The high value of the crops grown enables the user to invest in high-quality products. In recent years the cost of natural organic materials such as peat has increased. The container-grown plant industry uses a soilless potting medium consisting of 75% or more of peat and pine bark (Shiralipour et al., 1992b). Many different media including rice hulls, pine bark, hardwood bark, and other material have been utilized successfully as replacements for peat moss. Compost can be an excellent substitute for peat or other media. The two most important requirements by the industry are quality and consistency.

Producers of greenhouse and nursery crops are ideal users of waste composts. In these situations, there is less potential health risk from pathogens or heavy metals (Gouin, 1978b). However, persons working in the container industry have expressed concern about health when using compost prepared with biosolids and other feedstocks. For example, the presence of glass shards in MSW compost poses a concern to the user's safety. It is important for the compost industry to maintain high standards for compost quality.

Ornamental

An early study by Conover and Joiner (1967) found that a composted MSW product could be substituted for peat moss in potted chrysanthemum production. A second study by Gogue and Sanderson (1970), however, reported that MSW compost caused salt injury to cut chrysanthemum.

Sanderson and Martin (1974) utilized an MSW and biosolids compost to evaluate the growth of several ornamental plants. The compost was produced in windrows during a 12- to 16-week period by the city of Mobile, Alabama. A significant increase in plant dry weight and height was found for *Ilex cornuta* and *Thuja occidentalis*, but not for *Viburnum burkwoodii* (Figure 12.1). The plants were grown in full sun under field conditions. *Ilex* and *Thuja* grown in the compost medium were heavier and taller than those grown in the sphagnum peat medium.

Gouin (1976) evaluated the effect of utilizing a biosolids compost on the growth of several woody ornamentals. The growth response (mean stem length) of *Liriodendron tulipifera* (tulip poplar) and *Cornus florida* (dogwood) was significantly higher with compost applications. The 224

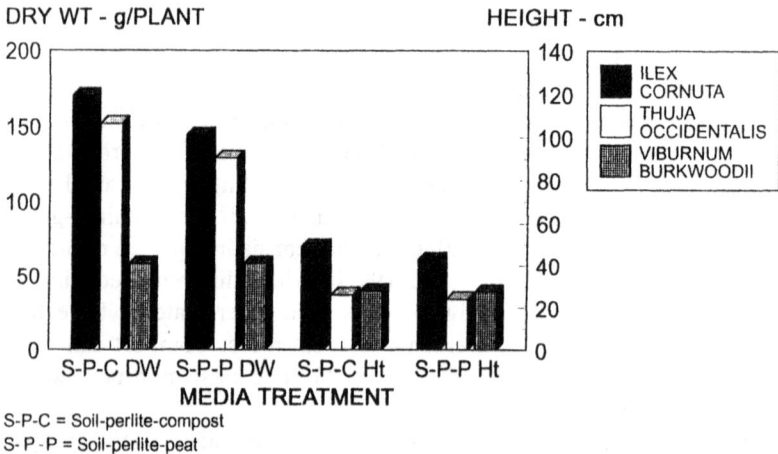

S-P-C = Soil-perlite-compost
S-P-P = Soil-perlite-peat

FIGURE 12.1. Comparison of plant dry weight and plant height for three woody ornamentals in MSW/biosolids compost-amended medium and a sphagnum peat moss-amended medium. (Data from Sanderson and Martin, 1974.)

tonnes/ha (100 t/a) treatment gave the best response. By comparison, increasing the compost rate to 448 tonnes/ha (200 t/a) resulted in a decreased growth. The higher compost rate also resulted in a lower mean number of seedlings. There was no significant difference in mean number of seedlings between the 0, 112, and 224 tonnes/ha (0, 50, 100 t/a) treatments. Additional studies on potted root cuttings of *Cotoneaster congestus* and *Jasminum nudiflorum* found that these grew best in a potting medium containing three parts of screened biosolids compost and two parts of composted municipal leaves.

In a subsequent study, Gouin and Walker (1977) showed that screened biosolids compost provided better growth for both *Liriodendron tulipifera* and *Cornus florida* than unscreened compost under field conditions (Figure 12.2). Seedlings of *L. tulipifera* grown in screened compost suffered less winter die-back than all other treatments, including seedlings growing in adjoining nursery beds. Specifically, there was a noticeable difference in root growth and proliferation. Seedlings grown in soils amended with 112 and 224 tonnes/ha (50 and 100 t/a) appeared to develop a more fibrous root system than other treatments including fertilized nursery beds. *Cornus florida* seedlings had better root growth in soils amended with 0 and 112 tonnes/ha (0 and 50 t/a). Finally, screened compost applications increased water retention, pH, N, P, K, Ca, and Mg of a loamy sand soil in the nursery.

A mixture of subsoil or sand with a leaf biosolids compost at a ratio of 25:75 provided the best growth for *Ilex crenata* (Japanese holly) with and without fertilizer. *Prunus laurocerasus* (cherry laurel) did not show much response to compost. Optimum levels of compost in both sand- and subsoil-composted biosolids blends for both species appeared to be between 50% and 75% in both fertilized and unfertilized treatments (Gouin, 1978a).

Flowering and House Plants

The texture of compost affects water movement, water retention, porosity and, therefore, aeration and bulk density. Plants differ in their requirements for these variables. Wootton (1977) evaluated the effect of different particle-size composted digested biosolids on the growth of three flowering annuals, petunia, zinnia and marigold. As compost particle size increased, growth of plants in both the unfertilized and the fertilized treatments decreased. The best growth occurred with compost particles between 0–2.4 mm when mixed with 50% vermiculite; however, there was no significant difference between the compost-vermiculite mixture and 100% compost.

Cambell and Pirani (1978) evaluated the effect of biosolids compost as a growing medium for chrysanthemum cultivar Bright Golden Anne. Compost was used in a series of mixes of soil, peat moss, sand, and perlite. The highest

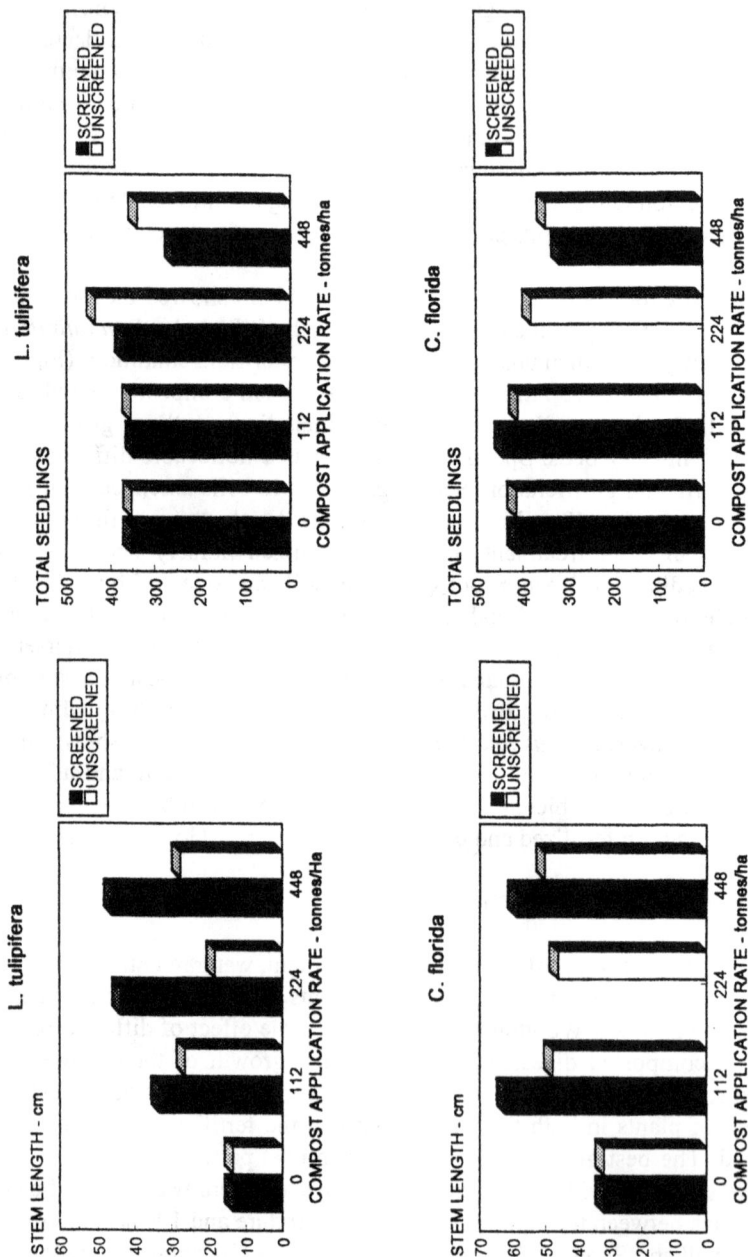

FIGURE 12.2. The effect of screened and unscreened biosolids compost on stem length and total number of *Lirodendron tulipifera* and *Cornus florida* seedlings harvested. (Data from Gouin and Walker, 1977.)

average plant height and greatest spread occurred with a compost and peat moss. All the compost treatments were equal to or better than the control.

Response to compost used as a container medium varies with the type of compost. MSW-biosolids compost produced in the 1970s in Mobile, Alabama, resulted in marginal necrosis on the lower leaves of chrysanthemums (Sanderson, 1980). Although this does not affect the quality of cut flowers, it would affect the marketability of a potted plant. Foliar analysis showed excessive levels of P, K, B, Mn, and Zn. In an earlier study, Gogue and Sanderson (1973) showed that B was the phytotoxic element. Today the B content of MSW and other composts is low.

Verdonck (1980) studied the effect of "wet" and "dry" bark compost on the growth of *Monstera deliciosa* var. *Borgsigiana* and *Cordyline terminalis,* two common house plants. The "wet" and "dry" barks were the result of two different debarking methods at paper mills. Verdonck found that the two bark composts reacted differently. The "wet" bark produced better growth in both species over a longer period of time. This difference was attributed to the physical properties of the bark. The "dry" bark had finer particle sizes resulting in faster degradation, which Verdonck believed resulted in unfavorable growth conditions.

Measuring the respiration rates of the two media would have confirmed this hypothesis. The "dry bark" may not have decomposed sufficiently; upon wetting, therefore, carbon for microbial metabolism became available. Proper curing would have alleviated this problem. This study points out that if a feedstock results in a compost that either inhibits or fails to promote growth compared to other compost products, it is important to find the reason and determine what corrective action is needed.

Interaction between elements can cause or alleviate phytotoxic effects. Chaney et al. (1980) evaluated four media with 0%, 33%, 67%, and 100% by volume of biosolids compost additions on the growth of marigolds. Trace elements, limestone, potassium nitrate and superphosphate were added or deleted from the compost and the Cornell mix (equal volumes of sphagnum peat moss and vermiculite) control. Data were collected on elemental composition, conductivity (soluble salts), plant growth parameters, pH, and several elemental (P, B, Cu, Zn, Mn, Fe) analysis of plant tissues.

Soluble salts limited yield at the 67% and 100% compost media. The conductivity at these levels ranged from 0.93 to 1.68 mmho/cm. The 33% compost mix with KNO_3 provided equal growth compared to the complete Cornell mix. The compost supplied considerably more Fe, Zn, and Cu than is normally added to media and corrected an apparent marginal Cu deficiency stress of the complete Cornell mix.

For data to be valuable to users, they should determine or associate a plant's growth response to prevailing conditions so that principles can be established and the data can provide insight into what could happen to similar species. Often data indicates the benefit or detriment of a particular medium with no explanation of the cause of the condition observed or measured. Results of this nature are limited in use as they may apply only to specific conditions.

Fitzpatrick (1981) evaluated biosolids compost potting mixes in conjunction with incinerator ash and sand compared to a peat-sawdust-sand potting mix. The study was conducted using three plants, dwarf oleander (*Nerium oleander*), ligustrum (*Ligustrum japonicum* var. *rotundifolium*), and jasmine (*Jasminum volubile*).

For jasmine, the poorest growth was obtained with a compost-incinerator ash-sand mixture at a ratio of 16:4:5. The best growth was obtained with the peat-sawdust-sand (6:4:1) and compost-incinerator ash-sand (4:1:5). With ligustrum, the best growth occurred with compost-incinerator ash (4:1) and compost-incinerator ash-sand (8:2:5). Finally, with oleander the best growth was obtained with compost-incinerator-ash (4:1) and compost-incinerator ash-sand (16:4:5).

No explanation was given for the difference in growth response, especially between jasmine and the other two species. Also, no information was given on the composition of the mixes, foliage analysis, or physical properties of the mixes. This makes it difficult to determine if the effect was due to the compost or the incinerator ash. Different amounts of ash and sand could greatly affect the soil-water relationships, particularly drainage. Drainage affects the soluble salt content as well as the chemical state of several elements, which could result in different sensitivities by the various plants.

Wootton et. al. (1981) evaluated the effect of biosolids compost on the growth of three flowering annuals, *Tagetes erecta* L. cv. Golden Jubilee ('Golden Jubilee' Marigold), *Zinnia elegans* Jacq. cv. Fire Cracker ('Fire Cracker' Zinnia), and *Petunia hybrida* Hort. cv. Sugar Plum ('Sugar Plum' Petunia). The study showed that as the particle size of the compost increased, growth of plants in both fertilized and unfertilized treatments decreased. Growth determined by shoot dry weight was greatest in media containing a greater portion of fine compost particles. No nutrient deficiency or toxicity symptoms were observed. Differences were attributed to porosity and available water.

Another study on the effect of particle size using a compost derived from a completely different feedstock was reported by Gouin and Shanks (1981). They conducted several studies using a gelatin (food processing) waste on the growth of zinnias, geraniums, marigold, chrysanthemum, and poinsettia. Different-sized particles, pH, and soluble salts were evaluated to determine their effect on zinnias and geraniums.

The best growth was obtained with 20% compost and the 6.3 mm mesh particle size. The authors indicated that this food processing waste compost could be used for growing poinsettias and chrysanthemums and most annual bedding plants. Leaching of soluble salts before fertilizer is applied was recommended.

Hemphill et al. (1984) found that composted biosolids increased growth of pansy (*Viola tricolor* 'Super Swiss Mix') and snapdragon (*Antirrhinum majus* 'Floral Carpet red'). Increasing the proportion of biosolids compost from 25% to 50% increased media pH, soluble salts, air-filled pore space, and nutrient levels. The authors noted that bark, sawdust, and biosolids compost could be substituted completely or in part for more expensive components such as sphagnum peat moss. Plant growth in biosolids compost media was equal to or better than growth in several commercial media. No phytotoxicity from heavy metals was noted. Problems that could arise from extreme pH or soluble salts can be solved inexpensively. For example, the acidity or alkalinity of compost could be adjusted with lime or gypsum and soluble salts could be leached from the compost.

In more recent work, Fitzpatrick (1989) evaluated the growth of Cuban royal palm (*Roystonea regia*), dwarf oleander (*Nerium oleander*), and orange jasmine (*Murraya paniculata*) in compost from pulp and paper mill, MSW and MSW-biosolids compost. Plant height of orange jasmine and Cuban royal palm was not significantly affected by any of the three compost products compared to the control medium. Oleander growth was significantly greater in the pulp and paper mill compost than the control; however, growth was significantly reduced in the MSW-biosolids compost compared to the pulp and paper mill and MSW composts. Data on the chemical and physical parameters showed that the MSW-biosolids had the highest electrical conductivity (soluble salts). Fitzpatrick and Verkade (1991) believed that optimum growth was impeded by settling, oxidation, and compaction of MSW composts.

It is clear from these and previous studies that individual species may respond very differently to different types of compost. Differences in response appear to be due to sensitivity to soluble salts, moisture conditions, bulk density, and sensitivity to specific elements such as B. Many of these conditions can be easily modified prior to using the compost, however. For example, soluble salt levels as well as B and other soluble elements can be leached prior to planting. Further, moisture, bulk density and other physical parameters can be modified during preparation of the planting media.

An evaluation of the growth of pickerelweed (*Pontederia cordata* L.), a perennial Florida native wetland plant, on several different compost products showed that MSW and yard waste compost can be used as a potting medium

(McConnell et al., 1990). Pickerelweed produced the greatest number of attractive purple inflorescence in a commercial mixture, Metro 500 (a formulation with pine bark, ash, composted pine bark, peat moss, sand, and vermiculite), which was mixed with yard waste compost in a 1:1 ratio. The greatest amount of biomass was produced with Metro 500 under 0% shade. Plant growth in 0% shade in all mixes comprised of 50% MSW compost, fine yard waste compost, coarse yard waste compost and 50% Metro 500 compared favorably to the Metro 500 alone. Neither potting medium particle size, bulk density, soluble salts, or pH was associated with observed differences in plant growth.

McConnell et. al. (1991) evaluated nine different potting media on the growth of *Dracaena fragrans* L. 'Massangeana' (corn plant) and *Pontederia cordata* L. The amount of municipal yard waste compost that could be used to formulate potting media depended on the methodology of production and other components used to formulate the potting mix. *D. fragrans* grew best in a commercial mix and yard waste at a 7:1 ratio. *P. cordata* grew best in the commercial mix and fine yard waste compost on a 1:1 ratio. Different fertilizer rates were used. It appeared that high fertilizer rates overcame some of the inhibitory effects when MSW and yard waste were used at high rates.

For another experiment (McConnell and Shiralipour, 1991), 13 different potting mixes were used to grow *Dracaena fragans* L. Ker-Gawl (corn plant), *Peperomia obtusifolia* A. Eieter (Peperomia), and *Schefflera arboricola* H. Ayata (Dwarf Schefflera). The tallest plants of *D. fragans* were those grown in a potting mix containing pine bark, ash, composted pine bark, peat moss, sand and vermiculite. However, plants grown in 100% yard waste were as tall as those grown in a commonly used potting mix by South Florida growers for this species.

The best growth of *P. obtusifolia* was in a sand (33%) and composted yard waste (67%). *S. arboricola* grew best in a peat:pine bark:sand with yard waste compost (87.5%:12.5%) and sand (33%) yard waste compost (67%) mixture. The authors concluded that composted yard waste can be used successfully as a potting mix for these three plants.

Zachary and Shiralipour (1994) published the results of the growth of salvia using 67 tonnes/ha (30t/a) of compost and nitrogen. Three types of compost, MSW, yard waste and biosolids, were used along with varying rates of N. They reported a significant increase in yield over the control (no compost). Specifically, for all six N treatments, the yield increased significantly at N applications of 67 kg/ha (60 lb/a) compared to the no nitrogen control. No significant interaction was found between compost and nitrogen applications. Some of the data are shown in Table 12.3. The highest yield was obtained with a 67 tonnes/ha (30 t/ha) of compost containing 53% MSW and 47% yard waste compost and 67 kg/ha (60 lb/a) of N.

TABLE 12.3. Effect of Municipal Solid Waste Compost (MSW), Yard Waste Compost Alone or with Biosolids Compost on Growth of Salvia.

Compost mg/ha (t/a)/ N kg/ha (lb/a)[1]	100% MSW (g·FW[2])	53% MSW, 47% YW (g·FW)	96% MSW, 4% BIO (g·FW)	52% MSW, 46% YW, 2% BIO (g·FW)	96% YW, 4% BIO (g·FW)	100% YW (g·FW)
0/0 (0/0)	149	191	149	156	141	85
0/34 (0/30)	227	269	184	198	177	124
0/67 (0/30)	225	295	273	226	248	138
67/0 (30/0)	355	385	350	325	311	262
67/34 (30/30)	354	425	363	354	375	276
67/67 (30/60)	369	449	380	389	386	304

[1]Numbers in parentheses are in tons/acre and lb/acre.
[2]Grams fresh weight, average of nine plants (three plots, three plants from each plot).
Source: Zachary and Shiralipour, 1994.

Sod Production and Turf Grass Establishment

The production of sod represents one of the best potential uses of compost. Harvested sod removes a layer of topsoil and eventually depletes the entire layer. The use of compost for turf grass production for home lawns, parks, athletic fields, cemeteries, or institutional grounds can result in improved soil physical properties while adding nutrients and organic matter. Compost can be used in turf grass production as (1) a soil amendment for the establishment of turf grass, (2) a fertilizer source for maintenance of established turf grass, and (3) a soil amendment or growth medium for commercial sod production (Hornick et al., 1979).

Neel et al. (1978) evaluated 10 media consisting of five waste products spread 10 cm deep on black plastic sheets. Two species, *Cynodon dactylon* L. ('Tifgreen' Bermuda grass) and *Paspalum notatu* Flugge ('Argentine' Bahia grass), were evaluated. The three composted products were biosolids, sugarcane processing by-products and yard waste compost.

Excellent-quality Bahia grass was produced in biosolids compost; the least acceptable was produced in yard waste with biosolids. Excellent rooting occurred in seven days. In 51 days, the Bahia grass had formed sod that was comparable in tear strength to commercially available sod. Bermuda grass developed sufficient coverage in 65 days. When transplanted, the Bermuda grass sod rooted much more quickly than commercial sod.

Angle et al. (1981) studied the establishment of turf grass from both seed and sod using biosolids compost. Turf grass quality increased with increased compost application. No significant difference was found between Kentucky bluegrass and tall fescue.

Quality also increased with compost applications as a result of the increased supply of plant nutrients and improvement in the soil physical properties. One of the most significant changes occurred with the soil bulk density, which decreased from 1.48 g/cm^3 with 0 compost to 0.84 g/cm^3 with 720 Mg/ha (320 t/a). The optimum growth rate appeared to be between 180 Mg/ha (80 t/a) and 360 Mg/ha (160 t/a). Establishment from seed was enhanced by the application of 260 Mg/ha (115 t/a). There was no difference between the species studied.

Another recent study evaluated sod production on MSW compost over plastic (Cisar and Snyder, 1992). Fertilized and non-fertilized treatments were compared. The species tested included *Paspalum notatum* Flugge ('Argentine' Bahia grass), *Cynodon transvaalensis* x *C. dactylon* ('Tifway' Bermuda grass), and *Stenotaphrum secundatum* (Walt.) Kuntze (St. Augustine grass). Six weeks after seeding or sprigging, both fertilized and non-fertilized compost had discolored leaf blade tissues and poor growth.

Bahia grass leaf tissue had low N concentration, which suggested that some of the fertilizer N was immobilized.

No data were given on the state of the compost's stability. The MSW compost came from the Agripost facility in Dade County, Florida, and was most likely not fully stabilized. High initial salinity (2.85 dS/m) may have caused some phytotoxicity. At 5 months, fertilized sod had sufficient coverage for harvest. By comparison, normal commercial sod production takes 9 to 24 months. One to three weeks after transplanting on a sandy soil, the compost-produced sod demonstrated better growth and longer roots than commercially grown sod. Table 12.4 shows the visual ratings from 1 to 10 (10 = best quality) and the percent cover. For two of the grasses, visual quality increased with time. Thus, there was considerable improvement in cover at later dates. This may have been the result of leaching of salts and greater stability of the compost in the soil with time.

The data clearly show that the use of compost can result in excellent turf grass production. Often during construction of new homes and commercial developments there is a need for soil improvement as the remaining local

TABLE 12.4. *Visual Quality and Coverage Rating for MSW Compost-Grown Turfgrass.*

	Variable			
	Visual Quality		Cover (percent)	
Grass	9 August	31 October	9 August	31 October
Bahia grass	6.5	6.2	51	72
Bermuda grass	6.3	6.8	51	77
St. Augustine	6.0	7.2	49	84
Significance	NS	<0.10	NS	NS
Fertilizer treatment				
Fertilizer	8.0	8.9	76	99
No fertilizer	4.6	4.7	25	57
Significance	<0.01	<0.01	<0.01	<0.01
Grass × fert. interaction	NS	NS	NS	NS

Source: Cisar and Snyder, 1992.

soil is poor and low in organic matter. The addition of compost can improve these marginal soils at considerably lower cost than bringing in topsoil. However, the quality of the compost needs to be high so as not to result in initial poor growth.

In the U.S. Department of Agriculture publication on the use of biosolids compost, Hornick et al. (1979) recommended the following rates of application (for complete recommendations refer to the publication):

- turf grass establishment:12 kg/m²–36 kg/m² (2000–6000 lbs/1000 ft²); incorporate with top 10 to 15 cm (4–6 in.) of soil
- surface mulch: 3.6 kg/m²–4.2 kg/m² (600–700 lb/ft²); broadcast uniformly on surface before seeding small seeded species or after seeding large seeded species
- maintenance: 2.4 kg/m²–4.7 kg/m² (400–800 lb/1000 ft²); broadcast uniformly on surface
- sod production: (1) incorporated with soil: 18 kg/m²–36 kg/m² (3000– 6000 lb/ft²); incorporate with top 10–15 cm (4–6 in.); (2) unincorporated with soil: 36 kg/m²–108 kg/m² (6000–18,000 lbs/100 ft²; apply uniformly to surface

AGRICULTURAL CROPS

During the past 26 years the use of compost in agriculture has been subject to considerable research. Table 12.5 summarizes some of the published literature. Since it is impossible to review all these studies, this table provides readers with the opportunity to seek data specific to their needs.

Field Crops

In 1969, the U.S. Public Health Service (now USEPA) and the Tennessee Valley Authority (TVA) jointly operated a solid waste/biosolids composting operation in Johnson City, Tennessee. The compost produced was evaluated as an amendment for several agricultural crops (Mays et al., 1973).

Figure 12.3 shows the forage sorghum and Bermuda grass yields obtained with compost and the addition of nitrogen. The forage sorghum data represent the total yield over a three-year period (1969, 1970, 1971). Compost was applied in 1969 and 1970. Compost increased yield with all three nitrogen application rates. The highest yield was obtained with 164 mt/ha of compost and 189 kg/ha of N. Bermuda grass yields increased with compost applications. The highest yield was obtained with 72 Mg/ha of compost and 180 kg/ha of nitrogen.

Tietzen and Hart (1969) compared the use of fresh and mature compost

TABLE 12.5. *Literature Published as Related to Specific Agronomic Crops.*

Crop	Literature Citation
Potatoes, rye, oats	Tietzen and Hart, 1969
Sorghum, Bermuda grass, corn	Mays et al., 1973 Wong and Chu, 1985
Sorghum	Hortenstine and Rothwell, 1973
Corn	Duggan and Wiles, 1976
Fescue	Sikora et al., 1980
Ryegrass	Hileman, 1982
Komatsuna	Chanyasak et al., 1983
Tomatoes, peppers, cucumbers	Inbar et al., 1986
Kohlrabi	Vogtmann and Fricke, 1989
Broccoli	Buchanan and Gliessman, 1991
Wheat	Sikora and Azad, 1993
Broccoli, cauliflower, eggplant, tomatoes, peppers	Maynard, 1994
Corn	Wolkowski, 1994
Lettuce, broccoli	Zachary and Shiralipour, 1994

to a fertilized control (no compost). The fresh compost reduced yields of rye and oats the first year. After one year, however, both compost yields exceeded the control plot. Fresh compost plots had higher yields and was better than the mature compost plots. For the total nine-year study, yields were 11.1% higher with the mature compost and 11.7% higher with fresh compost than those from the fertilizer control plots.

Increasing MSW-biosolids compost application to a silt loam soil increased forage sorghum yields over three years (Mays et al., 1973). Compost was applied annually. The study also evaluated corn and Bermuda grass. Without the addition of N, corn grain yields increased with increasing compost application rates up to 112 tonnes/ha annually and then decreased slightly. Applying N to Bermuda grass did not increase yield significantly when comparing the compost and the fertilized plots. The compost application without N produced significantly more forage at the high compost rate in 1969 than the no N plot. In 1970 all the compost rates were significantly higher than the no N control.

Hortenstine and Rothwell (1973) reported on a greenhouse study with sorghum where several rates of MSW-biosolids compost were applied to a sand. The application of 8 tonnes/ha of compost increased the yield of two sorghum (*Sorghum bicolor* L. Moench) crops compared to the control. The yield of sorghum was higher at the 64 tonnes/ha compost application rate than a fertilized control.

Sikora et al. (1980) found that a significantly greater fescue grass yield was obtained from biosolids compost amended loamy sand and silt loam soils. Yields were linearly related to compost amendment for both soils. The addition of N fertilizer substantially increased yields. After 100 days, the fertilizer nutrients were depleted and the 4% to 6% compost additions continued to provide nutrients to the fescue. In a greenhouse study using industrial wastes from jute and sugarcane mills, Sikora and Azad (1993) found that 50% compost and 50% fertilizer combinations were equal or better than 100% complete fertilizer in producing wheat grain yields.

Wolkowski (1994) studied the effect of compost of different stability and maturity levels on corn grain yields grown on two soils (Figure 12.4). Fresh compost (0 days) reduced yield as the rate of compost application increased. The highest yields for both were obtained with the older, more stable compost. However, the difference was greater at all three compost application rates with the loamy sand soil. Yields were higher when grown on the

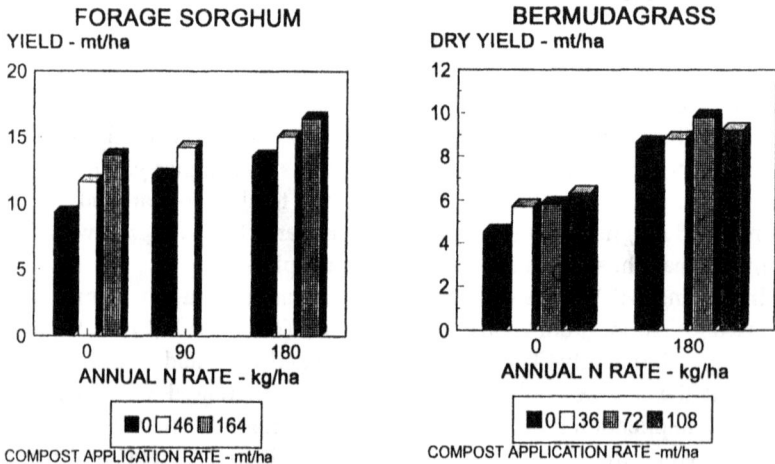

FORAGE SORGHUM
YIELD - mt/ha

BERMUDAGRASS
DRY YIELD - mt/ha

ANNUAL N RATE - kg/ha

ANNUAL N RATE - kg/ha

■0□46▦164
COMPOST APPLICATION RATE - mt/ha

■0□36▦72▧108
COMPOST APPLICATION RATE -mt/ha

FIGURE 12.3. Effect of MSW/biosolids compost and nitrogen addition on forage sorghum and Bermuda grass yields. (Data from Mays et al., 1973.)

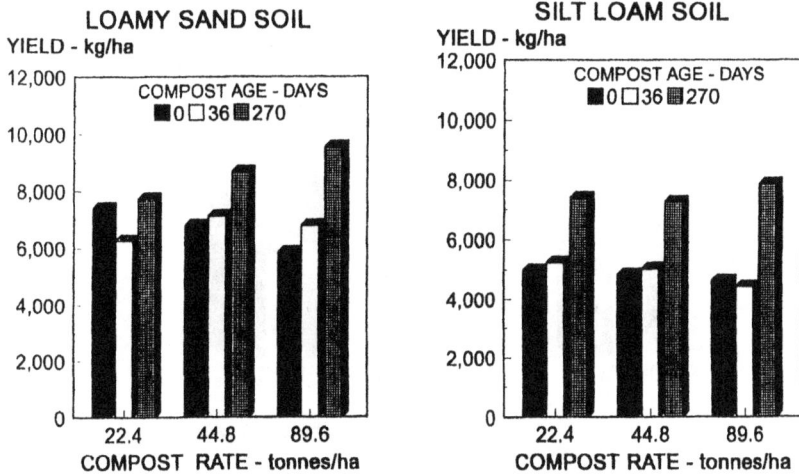

FIGURE 12.4. Effect of compost age and application rate on corn grain yields on two different soils. (Data from Wolkowski, 1994.)

silt loam soil. The author attributed this finding to better leaching of phytotoxic organic acids in the more coarse texture (better drained) soil.

Vegetable Crops

Improperly stabilized compost can reduce yields. For example, Tietzen and Hart (1969) found that fresh garbage compost reduced the yield of potatoes below the yields of a fertilized control plot. Mature compost produced higher yields, especially with low levels of supplemental N addition. Increasing the N above 45 kg/ha (40 t/a) in the fertilized plots produced yields similar to the mature compost plots.

Inbar et al. (1986) studied the response of tomatoes, peppers and cucumber to composted separated manure (CSM) and composted grape pomace. The composts were compared to peat and peat/vermiculite and perlite media with regard to germination and seedling growth. Pepper, tomato and cucumber seedlings developed faster in media containing compost than the peat/vermiculite media.

Vogtmann and Fricke (1989) studied the nutrient value and utilization of biogenic (source-separated) wastes on the yield of kohlrabi (*Brassica oleracea* Gongylods cultivar 'Trero'). Tuber yields with biogenic compost were comparable to a fertilizer control, and both were significantly higher than unfertilized control. However, there was no statistical difference in yield of tops for the three treatments (Figure 12.5).

Buchanan and Gliessman (1991) investigated the use of a compost derived

YIELD - dt/ha

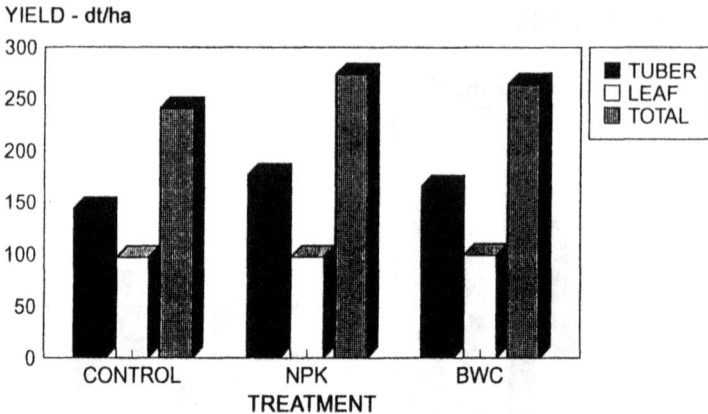

FIGURE 12.5. Yield of kohlrabi (*Brassica oleracea* Gongylods) grown with biogenic waste compost or NPK fertilizer. (Data from Vogtmann and Fricke, 1989.)

from spent mushroom bedding, horse manure, fava bean, and hay residues and vermicompost on the growth of broccoli. Fresh weight yields were greatest in the 3 tonnes/ha compost treatment with 75 kg N/ha; 30 tonnes/ha compost treatment; and 30 tonnes/ha vermicompost treatment.

Maynard (1994) evaluated spent mushroom (SMC) and chicken manure (CMC) compost on broccoli, cauliflower, eggplant, tomatoes, and peppers. The study was conducted over a three-year period on two soils, a fine sandy loam and a sandy loam. The compost was applied at 56 and 112 tonnes/ha and was the only source of nutrients.

The yields for the five crops are shown in Figures 12.6a and 12.6b. The yields of spring broccoli and cauliflower from the unfertilized CMC plots were equal to the fertilizer control during the three years on both soils. In 1990 cauliflower yields were significantly higher at the 112 tonne/ha compost rate than the control. In 1990 and 1991, spring broccoli yields were lower in the 56 tonnes/ha SMC plots for both soils and in the 112 tonnes/ha loamy sand soil. Fall broccoli yields were also lower in the SMC treatments. There was no difference in broccoli yield between the control and the CMC treatments for both soils. Fall cauliflower yields were generally lower than the control with the 56 SMC treatment. Yields for eggplant, tomatoes and peppers followed similar trends, with the CMC treatments outyielding the fertilized control and the SMC treatments having lower yield than the control. The difference in yields was attributed to the availability of N since the chicken manure had a higher total N, NO_3-N, and NH_4-N.

Zachary and Shiralipour (1994) reported on the effect of six compost products on the yield of lettuce and broccoli. The document provides

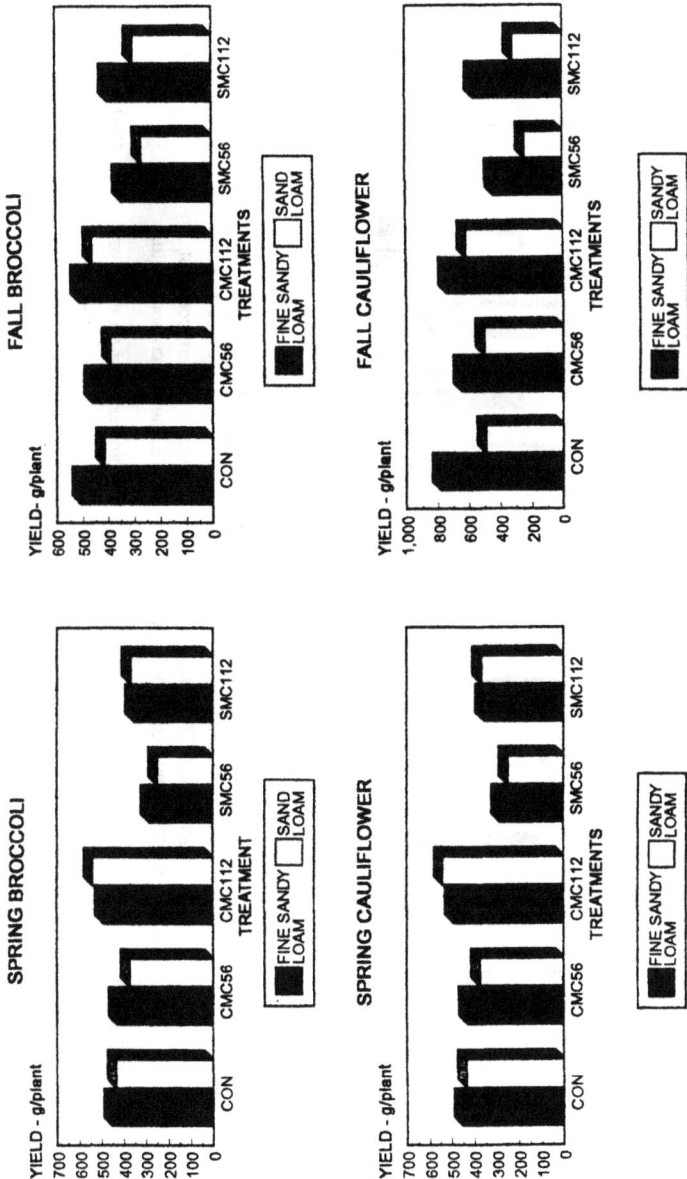

FIGURE 12.6a. The effect of chicken manure and spent mushroom compost on three-year (1989–1991) average yield of broccoli and cauliflower grown on two soils. (Reprinted with permission from Maynard, 1994.)

Treatments: CON - control; CMC - chicken manure compost;
1994   SMC - spent mushroom compost; 56 and 112 tonnes/ha compost. rate

405

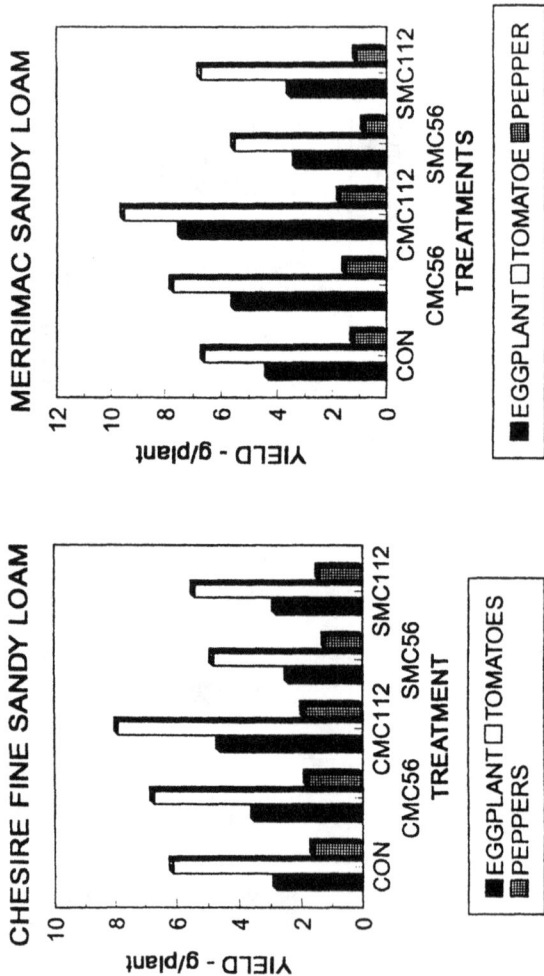

FIGURE 12.6b. The effect of chicken manure and spent mushroom compost on three-year (1989–1991) average yields of eggplant, tomatoes, and peppers grown on two soils. (Reprinted with permission from Maynard, 1994.)

considerable data on macro- and micronutrients, heavy metals and toxic organics. The composts included 100% MSW; 53% MSW, 47% yard waste (MSW/YW); 96% MSW, 4% biosolids (MSW/BIO); 52% MSW, 46% yard waste, and 2% biosolids (MSW/YW/BIO); 96% yard waste, 4% biosolids (YW/BIO); 100% yard waste (YW). Each of the compost treatments was applied with and without additional N.

Figures 12.7 and 12.8 show the yield data for lettuce and broccoli, respectively. For each of the compost treatments, the highest fertilizer control is shown. A no-fertilizer and a lower-fertilizer rate was reported, but since there was a yield response for the higher rate, only that rate is given.

Compost with nitrogen generally provided the best yields. With the MSW, MSW/YW, MSW/BIO and MSW/YW/BIO, yields exceeded the high fertilized control. All of the YW treatments had the lowest yield regardless of N addition. The high N rates in the YW compost treatment did improve lettuce yield over the control. With broccoli, the highest yields were obtained with the three biosolids compost treatments amended with 67 kg/ha N. Yard waste compost still produced the lowest yields. However, there was an N response. The data showed that significant yield increases for both lettuce and broccoli could be obtained with compost application rates of 34 kg · ha (15 t/a) and 67 kg/ha (30 t/a) depending on the type of compost used. Analysis of lettuce leaves and broccoli heads for uptake of heavy metals showed no significant relationship between compost application or type.

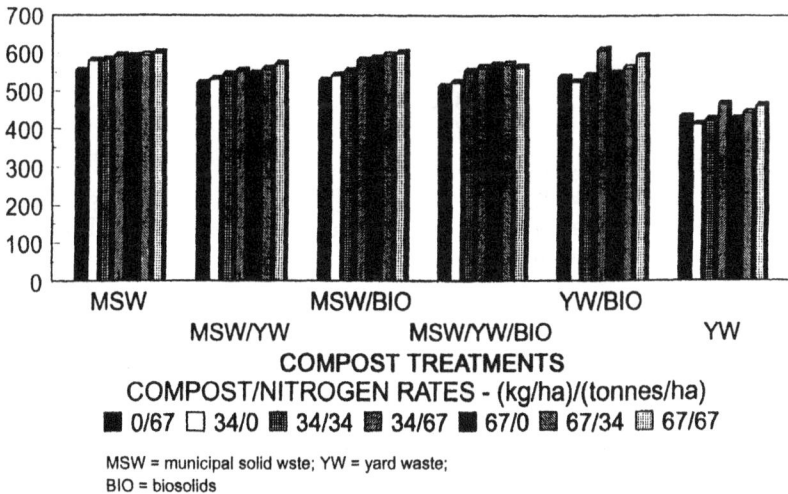

FIGURE 12.7. Effect of compost and nitrogen on yield of lettuce. (Data from Zachary and Shiralipour, 1994.)

YIELD - g/plant FRESH WEIGHT

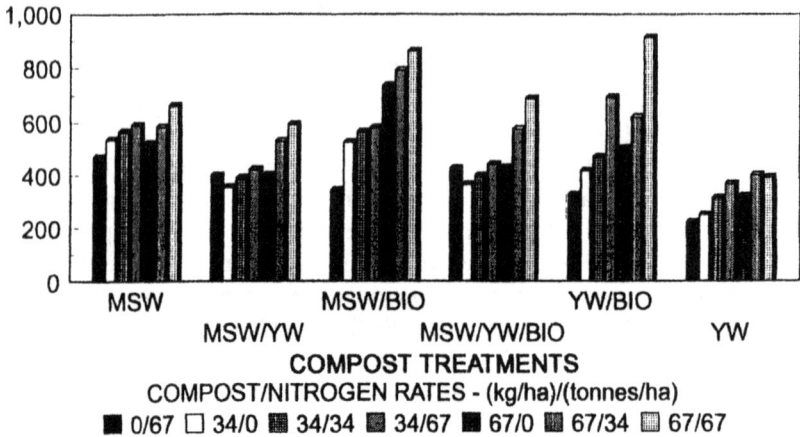

FIGURE 12.8. Effect of compost and nitrogen on yield of broccoli. (Data from Zachary and Shiralipour, 1994.)

SILVICULTURE

Forest nurseries are an excellent market for compost. The use of compost in established forests is most unlikely. Transportation and application of a low-density, low-nutrient product such as compost makes its use uneconomical. Because of these factors, the literature on forest application is sparse.

One of the earlier studies in the United States (Bengtson and Cornette, 1973) found that MSW compost application increased the amount of available moisture, decreased soil acidity, increased soil organic matter, increased cation exchange capacity, and increased the amount of exchangeable Ca, Mg, and K. It appears that the compost was of poor quality both from a physical and a chemical perspective. The authors stated: "No adverse effects of compost on soil or trees were observed, but aesthetics of the site were degraded somewhat by the residues of non-degradable particulates which persisted on the soil surface." The study also found that the N content of pine foliage was reduced following application and incorporation of the high rate of compost. Recovery was rapid, however.

Gouin (1977) found that sewage biosolids compost was a good amendment in seedbeds used to grow Norway spruce (*Picea abies* L.) and white pine (*Pinus strobus* L.) (Table 12.6). Norway spruce seedlings demonstrated comparable growth compared to an Osmocote control when top-dressed

TABLE 12.6. *The Effect of Screened and Unscreened Biosolids Compost on Mean Stem Length and Total Number of Norway Spruce Seedlings and Mean Grade Score and Total Number of White Pine Seedlings.*

Treatment	Tonnes/ha	Norway Spruce		White Pine	
		Total No. Seedlings	Mean Stem Length (cm)	Total No. Seedlings	Mean Grade Score[1]
Osmocote top-dressed + pine sawdust[2,3]	1 + 112	581a[4]	223a	448b	2.04a
Screened compost top-dressed	112	434abc	21.7a	489b	1.92
Screened compost incorporated	112[5]	436abc	19.8b	759ab	2.19a
Screened compost incorporated	224	399abc	20.6ab	756ab	2.19a
Screened compost incorporated	448	239c	18.4bc	481b	2.33a
Unscreened compost incorporated	112	530ab	17.2c	831a	2.33a
Unscreened compost incorporated	224	398abc	19.2c	721ab	2.16a
Unscreened compost incorporated	448	330bc	14.7d	576ab	2.15a

[1]Mean grade score: 1 = vigorous health seedlings; 2 = weak stems and stunted seedlings; 3 = chlorotic and dead seedlings.
[2]Osmocote 18-6-12.
[3]Applied November 1974, harvested March 1976.
[4]Mean separation in columns by Duncan's multiple range test, 5% level. Numbers in column with the same letter are not significantly different.
[5]Applied October 1973, harvested March 1976.
Source: Gouin, 1977.

409

with 112 tonnes/h sewage biosolids compost or grown in seedbeds amended with 224 dry tonnes/ha of screened compost. Screened compost at 112 and 448 tonnes/ha and unscreened compost at 112, 224, and 448 tonnes/ha resulted in smaller mean stem length.

More seedlings with white pine were produced in soils amended with compost; the highest amount was found with the 112 tonnes/ha unscreened compost. The lowest number of pine seedlings was obtained with the Osmocote treatment. It appears that considerable damage was done to seedlings by damping-off organisms. Since compost has been shown to be effective in reducing plant pathogens, its application may have been the reason for the higher number of seedlings in the compost treatments. Additional studies by Gouin et al. (1978a) also showed that red maple growth was greater in soil amended with compost in earlier years.

Jokela et al. (1990) assessed the effect of MSW compost application on growth and elemental concentrations of slash pine after 16 years of compost application. Compost increased tree growth and understory growth without long-term deleterious effects on the ecosystem.

The application of MSW mixed with biosolids to a poorly drained soil in Florida resulted in doubling the biomass of slash pine (*Pinus elliottii* Engelm.) the first year when 448 tonnes/ha was applied Bengtson and Cornette, 1973). Both tree height and biomass increased with increasing rates of compost, although the composting was done over a short period of time and, therefore, was not stable, delaying organic wastes. In addition to these areas gravel pits and other disturbed soils are much closer to urban populations, which generate wastes. A major advantage of using compost derived from waste is the low potential for negatively impacting human health and the environment. Furthermore, most of these soils are low in organic matter and fertility, making reclamation costly and difficult.

Many states now require that bonding be posted by the reclamation companies until the land is restored to aesthetic conditions. Compost applied to these soils adds organic matter, increases fertility, and improves soil physical properties, thereby assisting in revegetation. Most of the research on utilization of compost for mine and disturbed soil reclamation was conducted in the 1970s.

Hortenstine and Rothwell (1972) applied compost at rates of 35 and 70 tonnes/ha with and without fertilizer to plots on phosphate mine tailings in Central Florida. The compost increased cation exchange capacity, water-holding capacity, electrical conductivity, organic matter, and K, Ca, Mg contents of the sand tailings. Sorghum (*Sorghum bicolor* L. Moench) and oats (*Avena sativa* L.) were planted after the compost was disked. Although growth was poor compared to normal agricultural conditions, it was considerably better than the sparse vegetation on the surrounding area.

A very different approach to land reclamation was the use of MSW with 5% biosolids compost to stabilize an abandoned ash pond (Duggan and Scanlon (1974). The objective was to control dust and to improve the appearance of the site. Kentucky-31 fescue and white sweet clover were seeded on all the plots except the control. Virginia pine, black locust and eastern red cedar were broadcast seeded on several plots while others were planted to European alder and Virginia pine.

The best coverage with fescue and clover was obtained with the 224 tonnes/ha treatment. Broadcast seeding of Virginia pine, black locust and eastern red cedar was unsuccessful. Seventy-seven percent of the planted Virginia pine survived after two years. The best result was with European black alder, which also grew best on the 224 tonnes/ha compost-treated soil.

One of the most elaborate reclamation sites is Constitution Gardens in Washington, D.C., initiated prior to the 200th birthday of the United States (Patterson, 1975; Cook et al., 1979). Considerable topsoil would have been required for the 42-acre project. Instead, the Ecological Services Laboratory of the National Capital Parks, U.S. Department of Interior, felt that sewage biosolids compost produced at the U.S. Department of Agriculture's Beltsville project mixed with leaf mold could be used for soil modification using existing soil.

Today Constitution Gardens is thriving and is an excellent example of the use of urban wastes. No visual differences in turf quality were observed between soil modified by the sewage biosolids compost, soil modified with refuse compost or unmodified soil. The sewage biosolids compost increased by 50% water infiltration both in intake per hour and total intake over a three-hour period. On the other hand, infiltration was decreased with the refuse compost. The sewage bioslids contained a considerable amount of wood chips, which could have contributed to the lower compaction and greater water infiltration.

CONCLUSION

Considerable literature is available on the use of compost as a soil amendment. The response to compost is primarily successful with light (sandy) soils and heavy (clay) soils. Compost has been shown to be an excellent medium for plant growth, whether in containers or in the field.

Factors affecting the response to compost include compost type, particle size, soil type, and physical and chemical characteristics.

The impact of chemical properties such as soluble salts and pH is well known. There is a need to better identify the specific physical factors, stability, and maturity criteria that affect plant growth in order to predict response. Otherwise every user would need to experiment with specific crops.

Currently most of the compost produced in the United States is used in the horticultural field. The use of compost in agriculture predominantly takes place in Florida and California.

REFERENCES

Angle, J. S., D. C. Wolf, and J. R. Hall III. 1981. Turfgrass growth aided by sludge compost. *BioCycle.* 22(6):40–43.

Bengtson, G. W. and J. J. Cornette. 1973. Disposal of composted municipal waste in a plantation of young slash pine: Effects on soil and plants. *J. Environ. Qual.* 2(4):441–444.

Buchanan, M. and S. R. Gliessman. 1991. How compost fertilization affects soil nitrogen and crop yield, *BioCycle.* 32:72–77.

Bugbee, G. J. 1994. Growth of *Rudbeckia* and leaching of nitrates in potting media amended with composted coffee processing residue, municipal solid waste and sewage sludge. *Compost Sci. & Utilization.* 2(1):72–79.

Cambell, F. J. and A. M. Pirani. 1978. Composted sludge as a growing media for pot mums. *Florogram.* 11:15–17, G1–G3.

Cappaert, I., O. Verdonck, and M. De Boodt. 1973. Barkwaste as a growing medium for plants. *Proceedings Symposium Artificial Media in Horticulture. Acta Horticulturae* no. 37, 1974, p. 2041.

Cappaert, I., O. Verdonck, and M. De Boodt. 1976. Composting of bark from paper mills and the use of bark compost as a substrate for plant breeding. *Compost Sci.* 17(5):18–20.

Chaney, R. L., J. B. Munns, and H. M. Cathey. 1980. Effectiveness of digested sewage sludge compost in supplying nutrients for soilless potting media. *J. Amer. Hort. Sci.* 105(4):485–492.

Chanyasak, V., A. Katayama, M. F. Hirai, S. Mori, and H. Kubota. 1983. Effects of compost maturity on growth of komatsuna (*Brassica rapa* var *perfidis*) in Neubauer's pot. *Soil Sci. Plant Nutr.* 29(3):239–250.

Cisar, J. L. and G. H. Snyder. 1992. Sod production on a solid-waste compost over plastic. *HortScience.* 27(3):219–222.

Conover, C. A. and J. N. Joiner. 1966. Garbage compost as a potential soil component in production of *Chrysanthemum morifolium* 'Yellow Delaware' and 'Oregon.' *Proc. Fla. State Hort. Soc.* 79:424–429.

Cook, R. N., J. C. Patterson, and J. R. Short. 1979. Compost saves money in parkland restoration. *Compost Sci./Land Util.* 20(2):43–44.

Duggan, J. C. 1973. Utilization of municipal refuse compost. *Compost Sci.* 14(2):24–25.

Duggan, J. C. and D. H. Scanlon. 1974. Evaluation of municipal refuse compost for ash pond stabilization. *Compost Sci.* 15(1):26–30.

Duggan, J. C. and C. C. Wiles. 1976. Effects of municipal compost and nitrogen fertilizer on selected soils and plants. *Compost Sci.* 17(5): 24–31.

Fitzpatrick, G. 1981. Evaluation of potting mixes derived from urban waste products. *Proc. Fla. State Hort. Soc.* 94:95–97.

Fitzpatrick, G. 1985. Container production of tropical trees using sewage effluent, incinerator ash and sludge compost. *J. Environ. Hort.* 3(3):123–125.

Fitzpatrick, G. 1989. Solid waste compost as growing media. *BioCycle.* 30(9):62–64.

Fitzpatrick, G. and S. D. Verkade. 1991. Using solid waste compost for horticultural production, pp. 502–513. In A. B. Bottcher et al. (ed.). *Proc. Conf. Envir. Sound Ag.* Univ. Fl., Gainesville, FL.

Gogue, G. J. and K. C. Sanderson. 1970. Foliar analysis of *Chrysanthemum morifolium*, (cv.) 'Albatross' and 'CF #2 Good News' grown in processed garbage amended media. *HortScience.* 5:311.

Gogue, G. J. and K. C. Sanderson. 1973. Boron toxicity of chrysanthemums. *HortScience.* 8:473–475.

Gouin, F. R. 1976. Marked growth response of woody plants with screened composted sewage sludge. *Proceedings of the International Propagators' Soc.* 26:195–201.

Gouin, F. R. 1977. Conifer tree seedling response to nursery soil amended with composted sewage sludge. *HortScience.* 12:341–342.

Gouin, F. R. 1978a. Forest seedlings thrive on composted sludge. *Compost Sci./Land Util.* 19(4):28–30.

Gouin, F. R. 1978b. New organic matter source for potting mixes. *Compost Sci./Land Util.* 19(1):26–28.

Gouin, F. R. and J. B. Shanks. 1981. Composted gelatin waste aids crops. *BioCycle.* 22(4):41–45.

Gouin, F. R. and J. M. Walker. 1977. Deciduous tree seedling response to nursery soil amended with composted sewage sludge. *HortScience.* 12(1):45–47.

Hemphill, D. D., Jr., R. L. Ticknor, and D. J. Flower. 1984. Growth responses of annual transplants and physical and chemical properties of growing media as influenced by composted sewage sludge amended with organic and inorganic materials. *J. Environ. Hort.* 2(4):112–116.

Hileman, L. H. 1982. Fortified compost product shows promise as a fertilizer. *BioCycle.* 23(1):43–44.

Hornick, S. B., J. J. Murray, R. L. Chaney, L. J. Sikora, J. F. Parr, W. D. Burge, G. B. Willson, and C. F. Tester. 1979. *Use of Sewage Sludge Compost for Soil Improvement and Plant Growth.* U.S. Department of Agriculture, Sci. and Educ. Adm. Agr. Rev. and Manuals. ARM-NE-6. Beltsville, MD.

Hortenstine, C. C. and D. F. Rothwell. 1972. Use of municipal compost in reclamation of phosphate-mining sand tailings. *J. Environ. Qual.* 1(4):415–418.

Hortenstine, C. C. and D. F. Rothwell. 1973. Pelletized municipal refuse compost as a soil amendment and nutrient source for sorghum. *J. Environ. Qual.* 2(3):343–345.

Inbar, Y., Y. Chen, and Y. Hadar. 1986. The use of composted separated cattle manure and grape marc as peat substitute in horticulture. *Acta Horticulturae.* 178:147–154.

Jokela, E. J., W. H. Smith, and S. R. Corbet. 1990. Growth and elemental content of slash pine 16 years after treatment with garbage composted with sewage sludge. *J. Environ. Qual.* 19:146–150.

MacCubbin, T. J. and R. W. Henley. 1993. Evaluation of a yardwaste compost as a potting medium amendment for production of potted *Ageratum. Proc. Fla. State Hort. Soc.* 106:302–305.

Maynard, A. A. 1994. Sustained vegetable production for three years using composted animal manures. *Compost Sci. & Util.* 2(1):88–96.

Mays, D. A. and P. M. Giordano. 1989. Landspreading municipal waste compost. *BioCycle.* 30(3):37–39.

Mays, D. A., G. L. Terman, and J. C. Duggan. 1973. Municipal compost: Effects on crop yields and soil properties. *J. Environ. Qual.* 2(1):89–92.

McConnell, D. B., M. E. Kane, and A. Shiralipour. 1990. Growth of pickerelweed in municipal solid waste compost and yard trash compost. *Proc. Fla. State Hort. Soc.* 103:165–167.

McConnell, D. B., M. E. Kane, and A. Shiralipour. 1991. Influence of shade and fertilizer levels on pickerelweed growth in composted solid waste and yard trash. *Soil and Crop Sci. Soc. Proc.* 50:145–154.

McConnell, D. B. and A. Shiralipour. 1991. Foliage plant growth in potting mixes formulated with composted yard waste. *Proc. Fla. State Hort Soc.* 104:311–313.

McConnell, D. B., A. Shiralipour, and M. E. Kane. 1991. *Using Composted Waste Materials as Plant Growth Substrates.* Fl. Agr. Exp. Sta. J. Series No. N-00410.

Neel, P. L., E. O. Burt, and P. Busey. 1978. Sod production in shallow beds of waste material. *J. Amer. Soc. Hort. Sci.* 103(4):549–553.

Patterson, J. C. 1975. Enrichment of urban soil with composted sludge and leaf mold—Constitution Gardens. *Compost Sci.* 16(30):18–21.

Poole, R. T. and C. A. Conover. 1989. Growth of *Dieffenbachia* and gardenia in various potting ingredients. *Proc. Fla. State Hort. Soc.* 102:286–288.

Purvis, D. and E. J. MacKenzie. 1974. Phytotoxicity due to boron in municipal compost. *Plant and Soil.* 40:231–235.

Sanderson, K. C. 1980. Use of sewage-refuse compost in the production of ornamental plants. *HortScience.* 15(2):173–178.

Sanderson, K. C. and W. C. Martin, Jr. 1974. Performance of woody ornamental in municipal compost medium under nine fertilizer regimes. *HortScience.* 9(3):242–243.

Shiralipour, A., D. B. McConnell, and W. H. Smith. 1992a. *Uses and Benefits of Municipal Solid Waste Compost. A Database Review.* Center for Biomass Energy Systems and Dept. of Environ. Hort. U. of Fl., Gainesville, FL.

Shiralipour, A., D. B. McConnell, and W. H. Smith. 1992b. Uses and benefits of MSW compost: A review and an assessment. *Biomass and Bioenergy.* 3(3–4):267–279.

Sikora, L. J., and M. I. Azad. 1993. Effect of compost fertilizer combinations on wheat yields. *Compost Sci. & Util.* 1(2):93–96.

Sikora, L. J., C. F. Tester, J. M. Taylor, and J. F. Parr. 1980. Fescue yield response to sewage sludge compost amendments. *Agron. J.* 72:79–84.

Slivka, D. C., T. A. McClure, A. R. Buhr, and R. Albrecht. 1992. Compost: United States' supply and demand potential. *Biomass and Bioenergy.* 3(3–4):281–299.

Stephens, J. M., G. C. Henry, B. F. Castro, and D. L. Bennett. 1989. Mushroom compost as a soil amendment for vegetable gardens. *Proc. Fla. State Hort. Soc.* 102:108–111.

Sterrett, S. B. and T. A. Fretz. 1977. Effect of nitrogen source and rate on composted hardwood bark media and subsequent growth of *Cotoneaster. J. Amer. Soc. Hort. Sci.* 102(5):677–680.

Terman, G. L., J. M. Soileau, and S. E. Allen. 1973. Municipal waste compost: Effects on crop yields and nutrient content in greenhouse pot experiments. *J. Environ. Quality.* 2(1):84–89.

Tietzen, C. and S. A. Hart. 1969. Compost for agricultural land. *J. Sanitary Engin. Div.* SA 2. Proc. Paper 6506:269–287.

Verdonck, O. 1980. Utilization of pine bark compost in horticulture. *Compost Sci./Land Util.* 21(1):22–23.

Vogtmann, H. and K. Fricke. 1989. Nutrient value and utilization of biogenic compost in plant production. *Agric., Ecosystems and Environment.* 27:471–475.

Wolkowski, R. P. 1994. Landspreading MSW compost in Wisconsin: Effect on corn yield, nutrient and metal uptake, and soil nitrate-N. Paper presented at the *Seventeenth International Madison Waste Conf.,* Madison, WI.

Wong, M. H. and L. M. Chu. 1985. Yield and metal uptake of *Cynodon dactylon* (Bermuda grass) grown on refuse-compost-amended soil. *Agric., Ecosystems and Environ.* 14:41–52.

Wootton, R. D. 1977. Evaluation of composted, digested sludge as a medium for growing ornamental plants. Ph.D. Dissertation, Univ. of Maryland.

Wootton, R. D., F. R. Gouin, and F. C. Stark. 1981. Composted, digested sludge as a medium for growing flowering annuals. *J. Am. Soc. Hort. Sci.* 106(1):46–49.

Zachary, J. and A. Shiralipour. 1994. *Compost field experiment guide for California communities.* Community Environmental Council, Inc. Santa Barbara, CA.

CHAPTER 13

Compost Utilization II

INTRODUCTION

As indicated in the previous chapter, compost has traditionally been used as a soil amendment. This chapter provides information on non-traditional uses of compost. The two applications that have received the greatest attention are (1) the use of compost as a medium for plant disease suppression and (2) the use of compost as a medium for biofiltration.

Numerous studies have shown that the use of organic amendments can reduce or suppress plant pathogens. For example, Linford et al. (1938) reported that adding organic materials such as coarse grass, chopped pineapple, and sugarcane resulted in the reduction in the population of root-knot nematode. Duddington and Duhoit (1960) showed that green manuring affected populations of *Heterodera major*, cereal root eelworm. Lear (1959) reported that castor bean pomace applied to soil reduced nematode populations.

Control of plant diseases through composting can be achieved in two different ways. First, the thermophilic temperatures achieved through composting can destroy many plant disease organisms. Second, some compost products have been shown to suppress plant disease organisms (Hoitink and Fahy, 1986; Hoitink and Kuter, 1986). Finally, specific physical, chemical and biological properties of compost can result in disease suppression.

Recently, commercial application of compost for plant disease suppression has been evaluated and promoted (Hoitink, 1990; McElroy, 1993).

Compost has been used for many years as a medium for biofiltration. More recently, it has been used as an air pollution control system for removal of volatile organic compounds (VOCs) from industrial applications. This chap-

417

ter will discuss biofiltration. The design and characteristics of compost biofilters are covered in the volume on technology.

Compost has also been used as a filter medium for storm runoff, and in Germany it has been used as a filler between panels for sound insulation.

The primary objective of this chapter is to provide an overview of the literature dealing with non-traditional uses of compost that have significant economic implications.

PLANT PATHOGEN DESTRUCTION DURING COMPOSTING

Very few references can be found to the effect of the composting process on plant pathogen destruction. With the increased use of green waste composting, it is conceivable that diseased plants are brought into a facility. Therefore, if the green waste is not composted properly, plant pathogens may survive the process, contaminate the compost and later infect plants treated with the compost.

Hoitink and Poole (1980) indicated that the tobacco mosaic virus may survive the composting process; therefore, compost prepared from tomato wastes may be contaminated.

Users of compost are concerned and interested in the effect of composting on the destruction of plant pathogens. Most of the literature on the destruction of pathogens in compost pertains to human pathogens. Regulations and specification for the destruction of human pathogens require that the composting process attains a temperature of 55°C for several days, the length of time depending on the methodology used. In many states this temperature requirement does not apply to green waste, however.

The practice of steam sterilization of planting media used for beds or potting in greenhouses or nurseries has been going on for many years. However, if the sterilized medium comes in contact with compost or other materials infested with plant disease organisms, considerably higher levels of disease organisms may be established. This has been shown to occur with human pathogens or indicator organisms. For example, the chapter on primary pathogens showed that a sterilized compost reinfected with salmonellae or coliforms could result in significantly higher levels of these organisms than found in the original material. This is the result of the lack of antagonistic or competitive organisms.

Hoitink et al. (1975) reported that root rot organisms were absent in composted bark, even if no fungicides were used. In a subsequent study (Hoitink et al., 1976), imbedded crowns and roots of *Rhododendron cataw-biensse* cv. 'Roseum Elegans' infested with *Phytophthora cinnamomi* Rands, *Pythium irregulare* Buism and an unidentified sphaerosporangial

Pythium sp. were placed in nylon bags into a hardwood bark compost pile and a non-compost pile. Similarly, field-grown sugar beets infected with *Rhizoctonia solani* Kuhn, a crown rot, and chrysanthemum cuttings inoculated with *Erwinia chrysanthemi* isolate EC12, and geranium stem and leaf tissue infected with *Botrytis cinerea* Pers ex. Fr. were placed in nylon bags and imbedded in compost and non-compost piles. Temperatures in the compost piles did not exceed 50°C, but did exceed 40°C for over 10 days. Temperatures in the non-compost piles decreased from 20°C to 5°C during the first 40 days.

Hoitink et al. (1976) reported that Grushevoi and Levykh (1940) found that pathogens of tobacco seedlings, including *Thielaviopsis basicola* (Berk.) Ferraris, *Rhizoctonia*, and tobacco mosaic virus, were destroyed during composting even though the temperature did not exceed 49°C.

In reviewing the literature on the effect of compost on suppressing soilborne plant diseases, Hoitink and Fahy (1986) indicated that for a variety of compost, heating to 60°C destroyed the suppressiveness to *Rhizoctonia*, *Pythium*, and *Fusarium* diseases. The addition of small amounts of compost that were not subject to the high temperatures restored the suppressiveness to specific diseases. Based on this finding, the authors concluded that the beneficial microflora that are antagonists to plant pathogens are able to survive the cooler zones of compost piles.

Composting of plant residues can be effective in the destruction of plant pathogens and, therefore, the compost can be used safely (Bollen, 1993). The exception was compost prepared from plant material infested with the tobacco mosaic virus.

Bollen (1993) recommended that compost prepared from plant residues containing this virus not be used on susceptible crops. However, he pointed out that other plant viruses need specific vectors for transmission to plants and that these vectors are destroyed during the composting process. According to Bollen (1993), several plant viruses become inactivated at temperatures ranging from 75°C to 93°C, which are higher than those attained during composting; however, their vectors succumb to temperatures attained during composting.

Bollen (1993) reviewed the literature on the effect of temperature on the survival of several fungal plant pathogens. In some cases, different researchers reported inconsistent data. In the case of *Plasmodiophora brassicae,* which infects cabbage, some investigators reported that the organism was destroyed during composting at temperatures of 54°C whereas others reported that the organism survived temperatures of 55°C to 63°C. The survival at high temperatures was suggested to be a result of low moisture. Temperature inactivation of fungal plant pathogens depends on moisture conditions as well as other factors.

PLANT DISEASE SUPPRESSION

Biological control can be achieved in three ways (1) antibiosis, (2) nutrient competition, and (3) parasitism. Antibiosis is the production of fungal inhibitors by saprophytic microbes or plant roots. Parasitism involves direct attack of pathogens by non-pathogens (Millner et al., 1982). Biological control depends on the establishment of a balanced soil microflora that exhibits antagonistic biological activities toward phytopathogens.

During the past 15 years considerable research has investigated plant disease suppression using composted products. Different species or varieties are affected by different organisms. Table 13.1 summarizes the literature.

In 1973 Hunt et al. (1973) reported that populations of *Helicotylenchus* spp., a plant parasitic nematode, were higher in fertilized and lowest in compost plots treated with 8, 16, 32 tonnes/ha.

Habicht (1975) evaluated the application of 4%, 8%, and 16% dry weight, biosolids compost to a soil infested with root-knot nematode, *Meloidogyne incognita acrita,* and planted with tomato. The effect of raw biosolids addition was also evaluated. Two types of infestation were investigated for each amendment. In one case, compost was applied to an infested soil and in a second the soil was sterilized and then reinfested (larval infested).

Biosolids compost significantly reduced root galling on tomato plants (Figure 13.1). Galling was rated by Daulton and Nusbaum's (1961) method. The authors believed that the effect was related to chemicals, such as ammonia, salts, or organic acids, released during 14 days of incubation rather than the action of antagonists. The larval and soil infestation treatments had a significant effect on galling within each amendment treatment. The soil-infested treatment had a higher gall rating. Plant growth increased significantly with increasing compost applications, but differences between the nematode treatments were not significant.

Mannion et al. (1994) evaluated the use of MSW and MSW with biosolids compost on nematode population dynamics for the production of tomato and squash. No consistent effects were obtained of the two MSW composts on nematode species. The plant-parasitic nematodes were not affected by the compost treatments. In the two years, 1991 and 1992, the final population of juvenile *Meloidogyne incognita* in squash was significantly greater in the control than in the compost treatments. This is one of the most important parasites of vegetables in the tropics and subtropics.

Hoitink et al. (1977) showed that concentrations of *Phytophthora cinnamomi* caused severe root rot to lupine seedlings in peat-sand potting mixtures, but little damage was found to similar seedlings grown in hardwood bark-sand media. (These data are shown in Table 13.2.) At the low inoculum level (10 *P. cinnamomi*-colonized millet seed mixed in 1 liter of

TABLE 13.1. Summary of Some of the Literature on Disease Suppression from 1970 to 1994.

Organism	Compost	Plant	Reference
Saprophytic nematode	MSW	Sorghum, oats	Hunt et al., 1973
Root-knot nematode	Biosolids	Tomato	Habicht, 1975
Phytophthora cinnamomi	Hardwood bark	Rhododendrons	Hoitink et al., 1977
Rhizoctonia, Fusarium, Pythium, Thielaviopsis	Hardwood bark	Miscellaneous	Hoitink, 1980
Bacillus megaterium	MSW	None	Kuster and Schmitten, 1981
Aphanomyces euteiches, Rhizoctonia solani, Sclerotinia minor, Phytophthora capsici, Pythium ultimum, Thielaviopsis basicola, Fusarium solani, Pythium aphanidermatum	Biosolids	Pea, bean, cotton, radish, lettuce, pepper	Millner et al., 1982
Miscellaneous	Several	Several	Hoitink and Fahy, 1986
Miscellaneous	Several	Several	Hoitink and Kuter, 1986
Pythium ultimum	Pine, bark, hardwood bark, biosolids	None	Chen et al., 1988
Rhizoctonia solani, Sclerotium rolfsii	Cattle manure, grape pomace	Radish, beans, chickpeas, pothos	Gorodecki and Hadar, 1990
Pythium aphanidermatum	Manure	Cucumber	Mandelbaum and Hadar, 1990

(continued)

421

TABLE 13.1. (continued).

Organism	Compost	Plant	Reference
Fusarium oxysporum, Gibberella zeae, Helminthsporium sigmoideum, Glomerella cingulata	Bark	None	Kai et al., 1990
Miscellaneous	Several	None	Hoitink et al., 1991
Rhizoctonia solani	Several	Turfgrass	Nelson and Craft, 1991
Rhizoctonia solani, Typhula spp., *Pythium aphanidermatum, Lanzia moellerodiscus* spp., *Laetisaria fuciformis, Magneporthe poae, Leptosphaeria korrae, Pythium graminicola*	Several	Creeping bent grass, annula bluegrass, perennial ryegrass, Kentucky bluegrass	Nelson, 1992
Pseudomonas solanacearum	Yard waste, biosolids mushroom	Tomato	Chellemi et al., 1992
Dollar spot	Various	Creeping bentgrass, annual bluegrass turf	Nelson and Craft, 1992
Rhizoctonia solani, Fusarium soloni, Pythium ultimum	Yard waste	Pea, radish, Douglas fir	McElroy, 1993
Meloidogyne javanica	Crab	Tomato	Rich and Hodge, 1993
Typhula blight	Various	Turf grass	Nelson et al., 1994
Nematodes	MSW	Tomato, squash	Mannion et al., 1994
Pythium, Rhizoctonia	Yard waste	Cucumber	Grebus et al., 1994

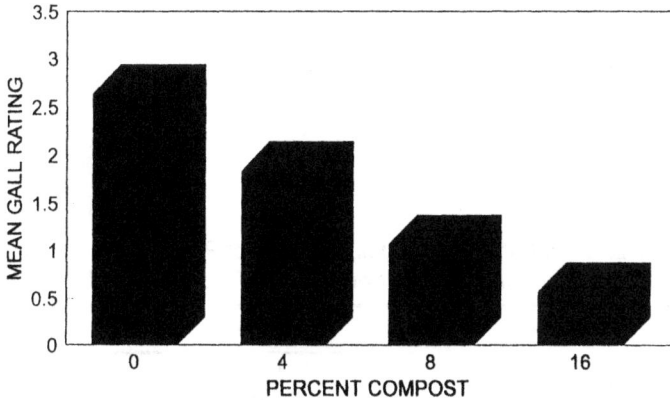

FIGURE 13.1. Effect of compost on root-knot nematode, *Meloidogyne incognita*. (Data from Habicht, 1975.)

potting medium), little effect on seedling survival was noted in the first 12–19 days. With the 100 and 1000 inoculum level treatments, lupine seedling survival was markedly depressed in the peat-sand and peat-perlite-sand media within 7 to 12 days. In the bark-sand compost, survival rate extended much longer and the percent survival was higher. The 1000 inoculum level in the peat-sand medium completely decimated lupine seedlings during the first seven days and within 12 days in the peat-perlite-sand medium. With the compost bark mixture, a 25% to 50% survival rate was observed in the 1000 inoculum level treatment during 12 to 32 days.

Studies were conducted on the leachate to determine sporangia and zoospore production by this organism. More sporangia were produced from the peat-sand leachates than from the compost medium. The authors noted that survival in compost appeared to depend on the formation of chlamydo-spores. Freshly prepared bark compost contained substances that lysed zoospores and cysts before germination. Based on these studies, they concluded that the effect of hardwood bark compost was due to antagonistic microorganisms, not to inhibitors.

Millner et al. (1982) found that biosolids compost showed a significant decrease in diseases caused by lettuce leaf-drop (*Sclerotinia minor*) root-rot and damping-off of bean, cotton, and radish caused by *Rhizoctonia solani*, and root-rot of peppers caused by *Phytophthora capsici* (Table 13.3). Disease caused by *Fusarium solani* and *Pythium aphanidermatum* was unaffected by compost application. *Pythium ultimum* and *Thielaviopsis basicola* were increased by compost application. This finding suggests that some antagonistic organism for *Pythium ultimum* and *Thielaviopsis basicola* was destroyed during composting, which allowed these two organisms to prevail and cause disease.

TABLE 13.2. Effect of Three Inoculum Levels of Phytophthora cinnamomi on the Survival of Lupine Seedlings in Three Potting Media.

Potting Medium	Inoculum Level[1]	Percent Survival				
		7 Days	12 Days	19 Days	25 Days	32 Days
Bark-sand compost	0	75	75	65	60	60
	10	70	70	65	65	65
	100	80	75	50	50	50
	1000	75	50	30	30	25
Peat-sand	0	75	75	65	60	50
	10	80	75	55	45	30
	100	60	40	20	10	5
	1000	0	25	5	5	0
Peat-perlite-sand	0	60	60	50	45	40
	10	65	65	50	40	35
	100	50	25	5	5	5
	1000	65	0	0	0	0

[1]Number of *P. cinnamomi*-colonized millet seed mixed in 1 L of potting medium.
Source: Hoitink et al., 1977.

424

TABLE 13.3. Effect of Biosolids Compost on Diseases Caused
by Several Soilborne Plant Pathogens.

Plant Pathogen	Host Plant	Disease[1] (percent)	
		Unamended	10% Compost Amended
Sclerotinia minor	lettuce—Paris White	84	33[2]
	lettuce—Boston	91	50[2]
Rhizoctonia solani	bean—Blue Lake	63	25[2]
	cotton—Stoneville	88	60[2]
	radish—Scarlet Globe	88	32[2]
Phytophthora capsici	pepper—California Wonder	97	55[2]
Fusarium solani	pea—Alaska pea—Perfected Freezer	33 65	49 55
Pythium aphanidermatum	bean—Blue Lake	14	12
Aphanomyces euteiches	pea—Alaska pea—Perfected Freezer	43 48	3 10
Thielaviopsis basicola	bean—Blue Lake bean—Cotton Stoneville	36[2] 49[2]	84 79
Pythium aphanidermatum	bean—Blue Lake	14	12
Pythium ultimum	pea—Alaska	39[2]	68

[1]Disease parameters based on root-rot severity (0–100%) for *Aphanomyces, Rhizoctonia,* and *Fusarium* and on disease plants (%) for *Sclerotinia, Phytophthora, Pythium,* and *Thielaviopsis.*
[2]Percent disease significantly less than the alternative treatment, $p = 0.05$.
Source: Millner et al., 1982.

Millner et al. (1982) also conducted field trials where the plots were artificially infected with *S. minor* and *R. solani.* In addition to fertilizer/lime or compost application, some of the plots were treated with two fungicides, thiram and PCNB. This allowed the researchers to determine if an addition of a fungicide to the compost provided better protection than when the compost was applied by itself. Three seasons were evaluated. For each crop season, lettuce leaf-drop was significantly lower in the compost plots than in the fertilizer plots. During the second and third crop seasons, the effect of compost was not as pronounced.

In the first season damping-off of peas by *Phythium* and *Rhizoctonia* in neither the compost alone nor with the fungicides treatments had any significant effect on disease reduction (Millner et al., 1982). However in the second season, damping-off of pea was significantly lower in the compost-fungicide treatment. In the third season, damping-off of pea in all compost treatments was significantly less than in all the fertilizer/lime treatments (Figure 13.2). For cotton, damping-off was lower in the fungicide-treated plots for both the fertilizer/lime and compost treatments. Compost alone did not differ significantly from the fertilizer treatment.

Two reviews were published in 1986 on the effect of compost on soilborne plant pathogens (Hoitink and Kuter, 1986; Hoitink and Fahy, 1986). Hoitink and Kuter (1986) indicated that compost type, stage of maturity, as well as physical and chemical properties affect plant growth and soilborne patho-

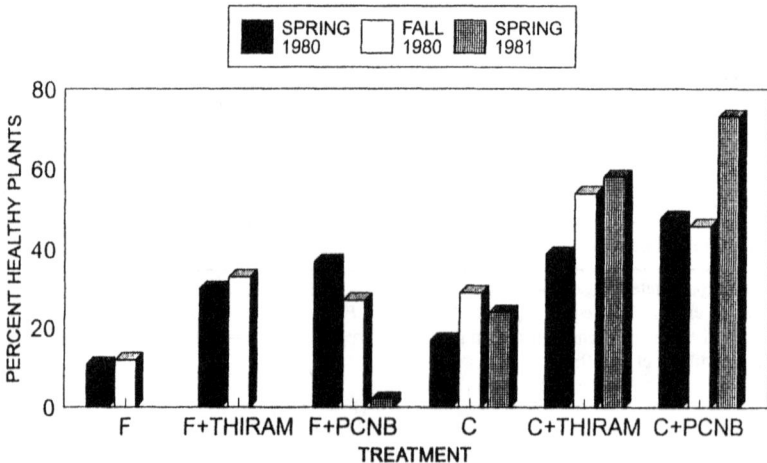

FIGURE 13.2. Effect of biosolids compost and fertilizer with thiram and PCNB on pea damping-off. (Reprinted with permission from Millner et al., 1982.)

TABLE 13.4. A Summary of Some of the Literature on the Effect of Compost Bark Extracts and Compost Maturity on Sporangia *Production* and Zoospore Release by Phytophthora cinnamomi.

Bark Species	Literature Reference	Compost Maturity	
		Fresh	Aged
Quercus (Oak)	Hoitink et al. (1977) Spencer and Benson (1982) Spring et al. (1980)	toxic	non-toxic
Pinus radiata	Gerritson-Cornell and Humphreys (1977)	toxic	not tested
Pinus radiata	Sivasithamparam, K. (1981)	not tested	non-toxic
Pinus pinaster	Sivasithamparam, K. (1981)	toxic	non-toxic
Pinus spp. (Pine)	Spencer and Benson (1982) Spring et al. (1980)	toxic	not tested

Source: Hoitink and Kuter, 1986.

gens and diseases caused by them. The following are some of their conclusions:

- Nitrogen fertility affects severity of soilborne diseases.
- Chemicals in tree bark that inhibit plant growth are destroyed during composting.
- The state of maturity of the compost affects the effect of compost inhibitors. (Table 13.4 summarizes the data obtained by several researchers.)
- The air capacity of the potting medium affects root-rot severity.
- Media moisture tension affects the release of sporangia and zoospores.
- Heat treatment negates the biological effects of disease suppression. Reinoculation with the disease-causing organism restores suppression.
- Incorporation of suppressive composts in media with appropriate chemical and physical properties can alleviate several soilborne plant diseases.

Other citations in the literature by Hoitink and Fahy (1986) include:

- Composted larch bark is used by vegetable growers in Japan to control the *Fusarium* brown rot of Chinese yam (Sekiguchi, 1977).

- Some types of compost are more effective than others for controlling Chinese cabbage (Tamura and Taketani, 1977). Compost prepared from tree leaves or rice hulls is more effective than that prepared from sawdust.
- Incidence of lettuce-drop caused by *Sclerotinia minor* is reduced by the application of biosolids compost over a four-year period with low applications of 7 mt/ha and 10 mt/ha (Lumsden et al., 1982).
- The application of biosolids compost (10% by weight) controls *Aphanomyces* root-rot of pea; *Rhizoctonia* root-rot of bean, cotton, and radish; *Fusarium* wilt of cucumber; and *Phytophthora* crown rot of pepper (Lumsden et al., 1983).
- Maturity of compost affects antagonistic activity. Nelson and Hoitink (1983) and Nelson et al. (1983) reported that immature composted hardwood bark high in cellulose was not suppressive to *Rhizoctonia* damping-off even though it contained high numbers of its antagonist *T. harzianum*. In mature compost, which was low in cellulose, the same antagonist was suppressive even at lower populations.

Chen et al. (1988) evaluated the relationship between *Pythium ultimum* disease severity of cucumber and microbial biomass and activity in an effort to be able to use microbial biomass or activity to predict the potential for disease. As shown in Figure 13.3, compost was the most effective treatment for *Pythium* damping-off disease suppression. A relationship was found between disease suppression and microbial biomass and microbial activity. Increase in microbial biomass or microbial activity resulted in a decrease in disease severity. The correlation coefficients were $r = -0.827$ and $r = -0.696$ for microbial biomass and microbial activity, respectively. Microbial biomass was better correlated with disease suppression than microbial activity. The data in this study indicated that the general microflora in the containers play a major role in the suppression of *Pythium ultimum* of cucumber. Other data showed similar relationships. For example, in a study involving 49 container media samples microbial activity was found to be a better predictor of disease severity than microbial biomass. The combination of microbial biomass and microbial activity could improve predictability of the disease. Further, the presence of thermophilic microorganisms decreased the validity of the correlation between microbial biomass and disease severity.

Hoitink et al. (1993) discussed the mechanisms for plant disease suppression as being either general or specific. General suppression is defined as suppression resulting from the activity of many different microorganisms. *Pythium* and *Phytophthora* spp. are examples of general suppression. Spe-

DISEASE SEVERITY

MICROBIAL BIOMASS AND ACTIVITY

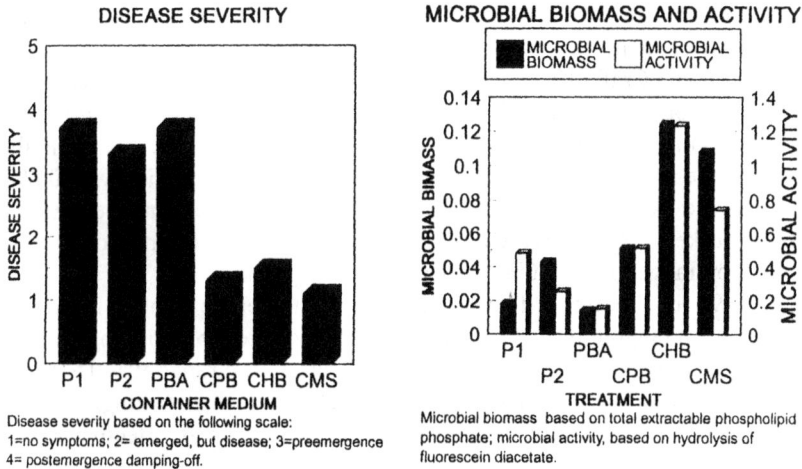

Disease severity based on the following scale:
1=no symptoms; 2= emerged, but disease; 3=preemergence
4= postemergence damping-off.

Microbial biomass based on total extractable phospholipid
phosphate; microbial activity, based on hydrolysis of
fluorescein diacetate.

FIGURE 13.3. The effect of peat (P1 and P2), processed bark ash (PBA), composted pine bark (CPB), composted hardwood bark (CHB), and composted biosolids on disease severity, microbial biomass, and microbial activity. (Data from Chen et al., 1988.)

cific suppression results when a single or a few organisms can explain the suppression of a pathogen or the disease. *Rhizoctonia solani* and *Sclerotium rolfsii* are cited as examples of specific suppression. The suppression may be the result of direct action of a microorganism as in the case of *Trichoderma* spp., a hyperparasite that colonizes sclerotia to reduce their inoculum potential, or the result of production of antibiotics or induce systemic resistance in plants.

As pointed out earlier, high temperatures during composting can result in a compost that does not have suppressive properties to soilborne diseases of plants (Hoitink et al., 1993). It is evident that the management of the composting process depends on the feedstock and the objectives for final product use. Biosolids, MSW and other feedstocks that contain human pathogens need to be disinfected and, therefore, require high temperatures. The compost derived from these wastes could be used in the traditional manner as soil amendments. Feedstocks containing green wastes or even industrial wastes (e.g., bark residue, pulp and paper, agricultural wastes) may not need to be composted at high temperatures and the resultant products could be effective as container media with disease-suppressive properties.

Chellemi et al. (1992) found that the effectiveness of compost to suppress bacterial wilt of tomato varied with the soil (Figure 13.4). The author cautioned that the results should not be extrapolated to other conditions since the disease was site specific. Soil pH had a significant effect on the incidence of the disease for one soil but not for the other. The reverse was true for organic C content.

Nelson and Craft (1992) found variation in the effectiveness of compost materials for suppression of dollar spot on creeping bentgrass and annual bluegrass turf. In 1989, biosolids compost from three facilities varied in the effectiveness of controlling dollar spot. The disease reduction was 0.0%, 12.6%, and 34.3%, respectively. Other composts studied were spent mushroom, turkey litter, brewery, and manure compost. The range of disease reduction ranged from 0.0% to 30.3%.

In 1990 top dressing with different compost materials resulted in disease suppression ranging from 33.3% to 99% 19 days after compost application. In a second evaluation 63 days after application, the range was 16.2% to 54.6%. Turkey litter compost was more effective than a fungicide and organic fertilizer. Finally, one biosolids compost was as effective as the fungicide.

Nelson et al. (1994) reviewed the biological control of turf grass diseases. Table 13.5 shows the plant pathogen and turf grass species that compost was effective in suppressing. The authors indicated that the microbiology of disease suppression has not been extensively studied. Furthermore, a given compost may not be consistent in suppressing disease from year to year, batch to batch, and site to site. The authors warned that the user needs to be assured that every batch of compost used specifically for disease control will be effective.

McElroy (1993) demonstrated that yard waste compost can be suppressive to both *Fusarium*- and *Pythium*-caused diseases of Douglas fir seedlings.

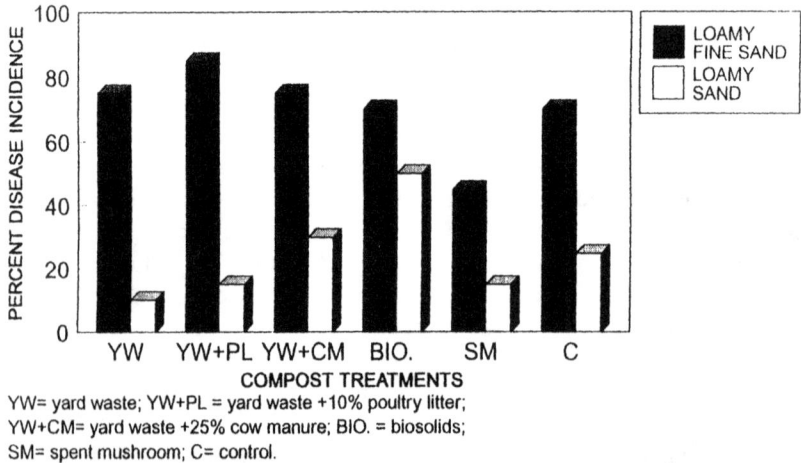

YW= yard waste; YW+PL = yard waste +10% poultry litter;
YW+CM= yard waste +25% cow manure; BIO. = biosolids;
SM= spent mushroom; C= control.

FIGURE 13.4. Effect of compost treatments on incidence of bacterial wilt of tomato. (Data from Chellemi et al., 1992.)

TABLE 13.5. Effect of Compost on Turfgrass Disease Suppression.

Disease and Pathogen	Turfgrass
Brown patch *Rhizoctonia solani*	Creeping bentgrass/annual bluegrass; tall fescue
Dollar spot *Sclerotinia homoeocarpa*	Creeping bentgrass/annual bluegrass
Necrotic ringspot *Leptosphaeria korrae*	Kentucky bluegrass
Pythium blight *Pythium aphanidermatum*	Perennial ryegrass
Pythium root rot *Pythium graminicola*	Creeping bentgrass/annual bluegrass
Red thread *Laetisaria fuciformis*	Perennial ryegrass
Typhula blight *Typhula* spp.	Creeping bentgrass/annual bluegrass

Source: Nelson et al., 1994.

The use of compost to suppress several soilborne pathogens is well documented. However, the effect of compost is not universally effective in the suppression of a disease for all plants affected. Therefore, we need to better understand the mechanism and factors affecting the suitability of a compost product to suppress soilborne plant pathogens.

BIOFILTRATION

This section discusses the scientific aspects of biofiltration. Design considerations are presented in *The Technology of Composting.*

Basic Concepts

Biofiltration refers to the use of a biologically active media to remove contaminants from a waste stream. Most of the literature refers to biofiltration as a process using biologically active media to remove and degrade odorous and volatile organic compounds. However, biofiltration has also been used to remove contaminants from the waste stream. Several types of media such as soil, peat, volcanic ash, shredded bark, or mixtures of these and other materials have been used (Bohn and Bohn, 1986; Bohn, 1991; Carlson and Leiser, 1966; Williams and Miller, 1993).

This discussion will cover only the use of compost. The use of biofilters in Europe has been quite intensive in biosolids treatment plants and for industrial application. For example, the Society of German Engineers (VDI, Dusseldorf, Germany) indicated that biofilters are used to scrub odors, aliphatics, aromatics, oxygen-containing organics, sulfur-containing organics, nitrogen-containing organics, halogenated organics, H_2S, NH_3, and aromatic oils (Fouhy, 1992). In Europe over 600 plants in the chemical industries were using biofilters and bioscrubbers in 1992.

An extensive literature review can be found in the Biofiltration Research Project Literature Review Report (County Sanitation Districts of Orange County, 1993). Carlson and Leiser (1966) investigated the use of several media in the laboratory, including compost, for the removal of sewage odors. During the research at Beltsville, Maryland, in 1975 biofilters were used to deodorize compost odors (Epstein et al., 1976). In the early 1980s E&A Environmental Consultants, Inc., assisted in the design of compost biofilters in Windsor, Ontario, for a dewatering building and in Springfield, Massachusetts, for sewer gases.

The European Community was much more active in the development and use of biofilters during the 1970s and 1980s (Ottengraf et al., 1974; Steinmuller et al., 1979, Koch et al., 1982). Major developments in compost biofiltration technology have been published in the European literature (Ottengraf and Van Den Oever, 1983; Bohnke and Eitner, 1983; Eitner, 1984). However, only in the past five years has there been much research and publication on compost biofilters.

Biofiltration involves two principal actions (1) physical adsorption on compost particles and water surfaces surrounding compost particles and (2) microbial degradation of the compound. Microbial action can involve metabolism, utilization or oxidation of organic compounds. The compost supports an indigenous population of bacteria, fungi and actinomycetes. It also demonstrates excellent adsorption characteristics with a fairly large surface area, high exchange capacity and nutrients for the microbial population. The compost environment in the biofilter must be maintained to support and allow for maximum growth of microorganisms.

Eitner (1984) described two scenarios involving the metabolism of an odorous gas molecule. In one scenario, the gas molecule diffuses to the surface of the compost medium where it is absorbed. It then enters the water phase where microbial action takes place. In the second scenario, the gas molecule enters directly into the microscopic water phase.

Eitner showed that the efficiency of the filter for several gases decreased at low moisture contents. The activity of the microorganisms as measured by respiration (oxygen uptake) decreased as the moisture content of the biofilter decreased from 60% to 35%. Van Langenhove et al. (1986) similarly pointed out that the most important parameter for a well-functioning biofilter is its moisture content, the optimum moisture content being approximately 65%.

Ottengraf (1986) reviewed some biofilter theoretical-design consideration studies as well as experimental studies. For a biological filter bed, he suggested that the constituent particles be surrounded by a wet biolayer. When gases flow around the particles there is a continuous transfer between the gas phase and the biolayer. Volatile organic compounds present in the waste stream are partially dissolved in the liquid phase of the biolayer and degraded or consumed by aerobic microorganisms. For easily biodegradable organic compounds, the addition of microorganisms from an activated biosolid suspension is useful. For more recalcitrant biodegradable organic compounds, or xenobiotics, inoculation with specially cultivated microorganisms would be useful. Ottengraf suggested that approximately 10 days are needed for acclimation of the biofilter bed. He concluded that different microbial populations are responsible for the degradation of different organic compounds.

Figure 13.5 depicts the theoretical mechanism of biofiltration as suggested by Eitner (1984) and Ottengraf (1986). Several important variables affect the performance of biofilters.

MOISTURE CONTENT

Microbial activity is affected by the moisture content of the medium. Odor removal occurs by biooxidation and is a function of microbial activity

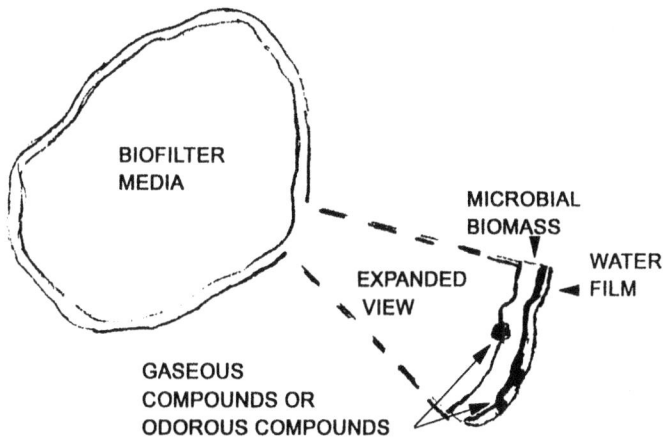

FIGURE 13.5. Conceptual model for adsorption and microbial degradation of gaseous or odorous compounds according to Eitner (1984).

(Williams and Miller, 1993). Ottengraf (1986) indicated that the optimum moisture content for a biofilter is in the range of 40% to 60% on a weight basis. Prokop and Bohn (1985) suggested that the required moisture content for a compost biofilter is 20% to 40%. Low moisture reduces microbial activity while high moisture content can result in anaerobic conditions and the emission of odorous volatile metabolic products (Ottengraf, 1986). In a laboratory study Yang and Allen (1994) evaluated the effect of compost water content on removal of H_2S through laboratory biofilter. They evaluated water contents from 0% (oven-dry) to 62% (maximum water-holding capacity). Biofilter inlet H_2S ranged from 80 to 110 ppm. Hydrogen sulfide removal efficiencies were over 99.9% when the compost water content was between 30% and 62% (Figure 13.6).

PH

The pH of the medium should be near neutral. Carlson and Leiser (1966) found that during the biodegradation of H_2S, the medium became more acid since sulfuric acid was formed. Similarly, degradation of nitrogenous compounds could lead to the production of nitric acid and lower the pH. Ebinger et al. (1987) found that propane removal by soil beds was lower at a pH of 5.3 than when the pH ranged from 6 to 8. Further, Hartenstein and Allen (1986) indicated that the filter bed should be kept in a slightly alkaline condition within a pH range of 7 to 8.5.

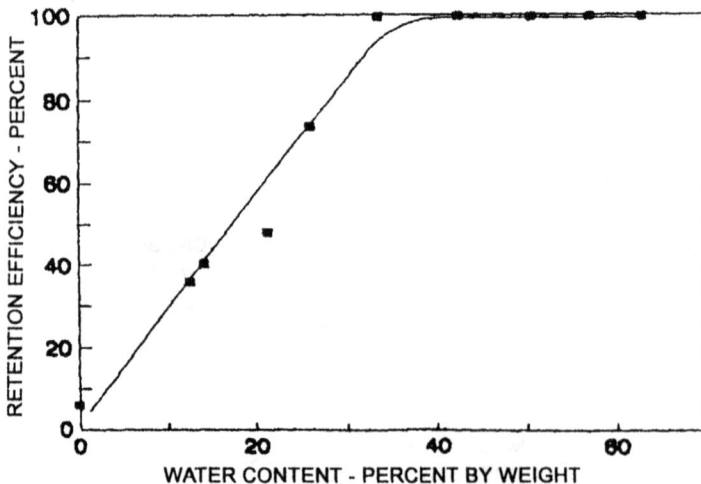

FIGURE 13.6. Effect of biofilter compost water content on hydrogen sulfide removal. (Data from Yang and Allen, 1994.)

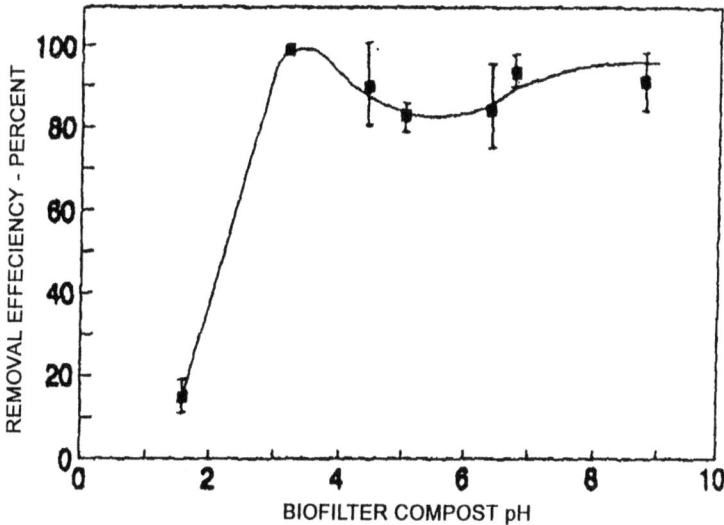

FIGURE 13.7. Effect of yard waste compost pH on hydrogen sulfide removal efficiency. (Data from Yang and Allen, 1994.)

Ottengraf (1986) noted that if acid conditions occurred during experiments, the addition of limestone, marl or other insoluble alkaline materials would correct the acidity. Finally, Yang and Allen (1994) treated compost with acid in a laboratory biofilter to study the effect of pH. pH values ranged from 1.57 to 8.76. The H_2S removal efficiency was highly dependent on pH below 3.2, but was independent at higher values (Figure 13.7).

NUTRIENTS

Generally, compost contains sufficient nutrients for microbial activity. Often the gases that are being scrubbed provide the nutrient requirement for microbial metabolism. Sulfur metabolizing bacteria can use sulfur as an energy source. Don (1985) found that addition of nutrients improved biofilter efficiency (Figure 13.8).

TEMPERATURE

Temperature affects the microbial population, and hence biofilter efficiency. Temperature also affects the water removal rate, which may cause excessive drying. Yang and Allen (1994) studied the effect of temperature on removal of H_2S in the range of $-1.5°C$ to $103°C$ (Figure 13.9). High removal efficiencies were obtained in the range of $25°C$ to $50°C$. When the

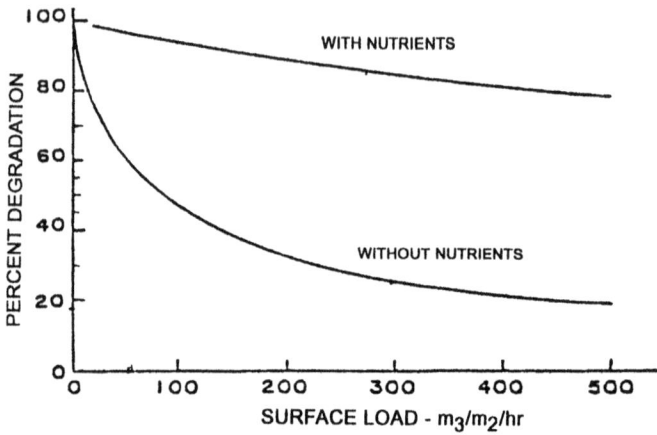

FIGURE 13.8. The effect of nutrient addition to a biofilter on the degradation of 100 mg/m³ of toluene. (Data from Don, 1985.)

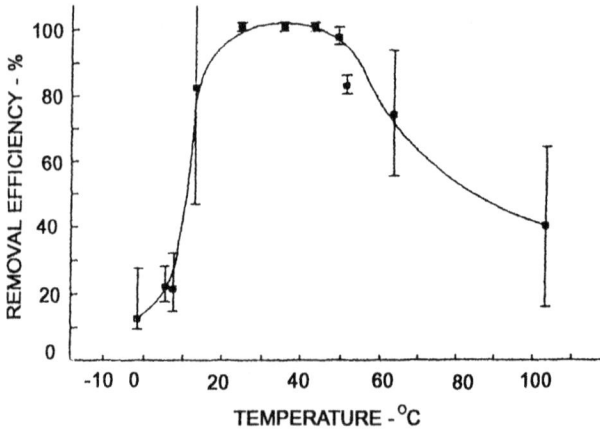

FIGURE 13.9. The effect of temperature on hydrogen sulfide removal efficiency in a biofilter. (After Yang and Allen, 1994.)

436

temperature of the biofilter bed was reduced from 25°C to 7.5°C, the H_2S was reduced by 80%. Similarly, increasing the temperature from 50°C to 110°C reduced biofilter efficiency for H_2S removal. This effect was not as dramatic as when the temperature was reduced below 25°C. However, the biofilter in Windsor, Ontario, Canada, which scrubs the odors from the biosolids dewatering building, operates very efficiently during the winter months.

MICROBIOLOGY

Although most bacteria use C as a source of energy, some can also utilize sulfur or nitrogen. Mineralization of C principally takes place by bacteria and fungi. Every known naturally occurring compound can be mineralized by bacteria and fungi (Krueger et al., 1973). For example, sulfur-metabolizing bacteria can use sulfur as a source of energy and are important in the removal of H_2S. Large quantities of sulfur must be oxidized for the organisms to receive sufficient energy (Hartenstein and Allen, 1986). The bacterial genus *Thiobacillus* is the primary oxidizer of sulfur compounds in the soil ecosystem (Tate, 1995). Under aerobic conditions, *Thiobacillus thiaparus* oxidizes H_2S to elemental sulfur. Heterotrophs such as bacillus and micrococcus oxidize sulfur (S) and/or thiosulfate (S_2O_3) to sulfuric acid and sulfates (Hartenstein and Allen, 1986). In biofilters the action of these organisms results in decreased pH and reduces its efficiency (Eitner, 1984).

Eitner (1984) indicated that *Pseudomonas,* an organoheterotroph, has been isolated from compost filters. This organism is common in soils and has been shown to degrade petroleum hydrocarbons. Eitner (1984) found that the total number of microorganisms was greatest at the bottom of the filter bed. Other organisms isolated were yeasts, fungi and actinomycetes.

Figure 13.10 shows the stratification of microorganisms in the filter bed. According to Eitner (1984), the microorganisms at the bottom of the biofilter consumed the more easily biodegradable compounds whereas the less degradable compounds were available for assimilation by microbes in the upper layers. Based on Eitner's data, Hartenstein and Allen (1986) indicated that a homogeneous mixing of the filter material does not appear to be beneficial since a stratified microbial distribution may be the most effective in removal of malodors.

Application

Bohn (1976) described the use of compost beds in Duisburg, Germany, to remove malodors. The Duisburg biofilter, operating for nine years, was 5 × 40 × 1 meters and treated an air flow of 3 m^3/sec (6300 scfm). Some

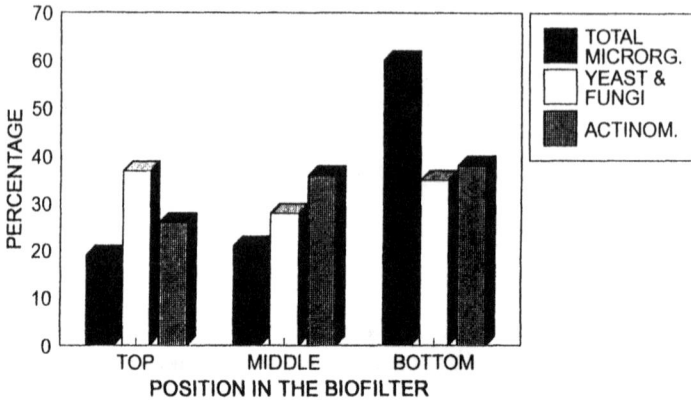

FIGURE 13.10. Microbial stratification, total microorganisms, yeasts and fungi, and actinomycetes in a biofilter. (Data from Eitner, 1984.)

of the limitations included (1) temperature should not exceed 60°C and (2) dust in the air being scrubbed should be low, presumably to avoid clogging of the media.

Bohn also noted that at that time compost was poorly adapted to control organic gases and that non-biodegradable organic gases had not been tested. Compost scrubbers were used at that time for organic gases whose microbial products are CO_2 and H_2O. He indicated that organosulfur and organonitrogen gases would also undergo microbial degradation but would leave a residue of sulfuric and nitric acids, which upon accumulation would inhibit microbial activity.

In a full-scale pilot study (Table 13.6) (Rands et al., 1981), compost was found to very effectively remove hydrogen sulfide (H_2S) from anaerobic digestion and rendering exhaust. Faint odors were detected above the biofilter when it dried out and channeling occurred.

Bohnke and Eitner (1983) evaluated three biofilter media for removal of six organic chemicals (Figure 13.11). The MSW with biosolids (ratio of 9:1) medium was most efficient for removal of five of the six compounds. All three media were efficient in the removal of terpene. Based on these data, specific media may be required to effectively remove different VOCs.

In 1984 and 1987, Eitner (1984; Eitner and Gethke, 1987) described the use of compost biofilters to scrub odors from wastewater treatment plants in Germany. Compost was a preferred medium for biofilters, not only for its properties but also because it was a way to recycle organic wastes. During a visit by the author to Duisburg, Germany, in 1986, he was told that 50% of the compost was being sold as biofilter material for industrial uses.

*TABLE 13.6. Effect of a Compost Biofilter on the Removal of
Hydrogen Sulfide from a Rendering Plant Exhaust.*

Date	Hydrogen Sulfide Concentration (ppm)				Average % Removal
	Influent	Influent	Effluent	Effluent	
1/79	696	63	0.17	0.02	99.80
2/79	498	221	0.64	0.00	99.95
3/79	448	111	0.15	0.23	99.95
4/79	276	54	0.01	0.00	100.00

Source: Rand et al., 1981.

Ostojic et al. (1991) evaluated biofilter efficiency for the removal of
ammonia (NH_4) and odors from a curing pile of biosolids compost at
Hamilton, Ohio. The average concentration of NH_4 into the scrubber was
176 ppm and 6.3 ppm for the outlet. The removal efficiency for ammonia
averaged 95%. Odor concentrations going into the scrubber averaged 1035
D/T; and outlet concentrations were 34.9. Biofilter removal efficiency was
96%.

Yang and Allen (1994a, 1994b) used a dual-tower laboratory experimental
biofilter to study design, operational kinetics, performance, and maintenance
for biofitration of H_2S. The effect of gas retention time on H_2S removal was

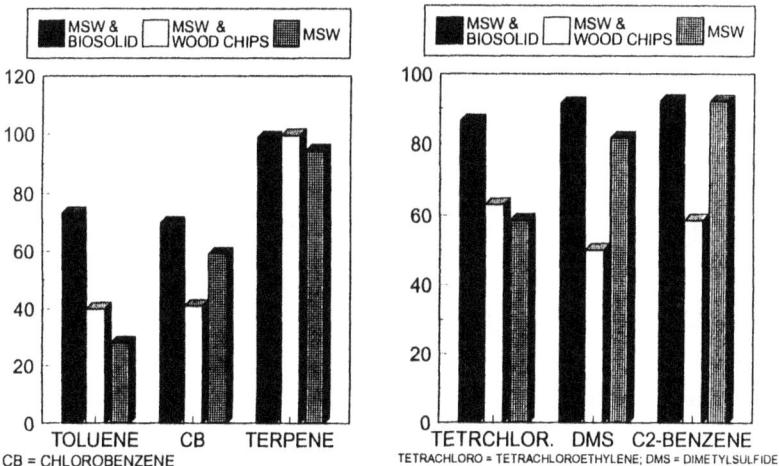

FIGURE 13.11. Removal efficiencies of three different media, MSW and biosolids, MSW and
wood chips, and MSW for six organic compounds. (Data from Bohnke and Eitner, 1983.)

studied by varying the gas flow rate through the biofilter tower. Hydrogen sulfide loading rates varied from 15 to 55 g-S/m³/hr. No effect on H₂S removal was found as long as the retention time was longer than 23 seconds (Figure 13.12).

Williams (1995) summarized test results for odor and VOC removal from several biofilters at composting facilities (Table 13.7). Five of the six facilities composted biosolids; the sixth composted MSW and biosolids. The facilities used four different composting systems. In all cases, the odor removal rate exceeded 90%. Removal of total reduced sulfur (TRS) ranged from 90% to 99%. Only three facilities reported VOC removal that ranged from 52% to 99%. Williams (1995) indicated that one of the key phenomena of biofilter performance was the consistent results of D/T values in the exhaust being below 25, regardless of inlet D/T values.

Ergas et al. (1995) evaluated VOC removal by a laboratory and pilot scale biofilter. The laboratory studies were not carried out for a long enough period to allow for firm conclusions. With a field pilot scale biofilter, removal of H₂S, benzene, toluene, and xylene ranged from 83% to 99.7% for one biofilter and 91% to 99.96% for a second biofilter. Odor removal was high.

Biofiltration has been shown to be an excellent method for removing malodors and VOCs from gas streams. The data by Yang and Allen (1994a, 1994b) indicated the importance of several parameters on biofilter efficiency. It must be realized that these studies, conducted on a small-scale

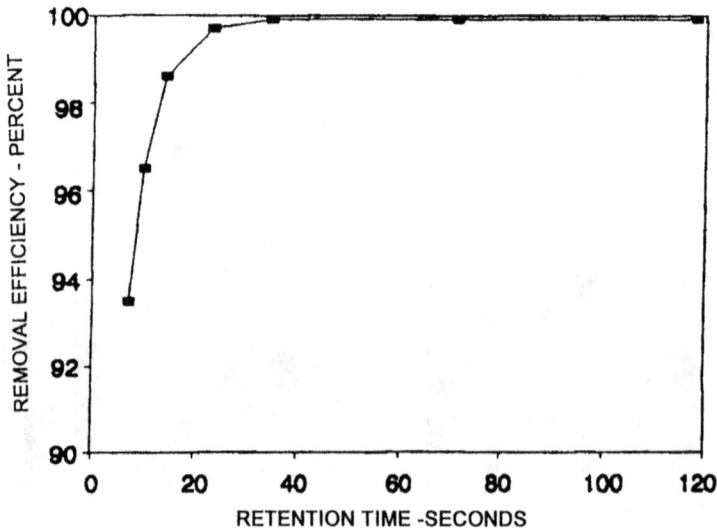

FIGURE 13.12. Effect of gas retention time on hydrogen sulfide removal efficiency using yard waste compost as the media. (Data from Yang and Allen, 1994.)

TABLE 13.7. Odor and VOC Removal from Several Biofilter Installations at Composting Facilities.

Facility	D/T Inlet		D/T Outlet		Odor Removal (%)		TRS[1] Removal	VOC[2] Removal	References
	Range	Ave.	Range	Ave.	Range	Ave.			
Hamilton, OH	180–1200	635	5–25	19	—	97	99	99	Wheeler, 1992
Hoosac, MA	—	—	—	—	—	95	99	52	E&A Environmental Consultants, Inc., 1993
Lewiston-Auburn, ME	71–158	115	7–11	8	90–94	93	—	—	Giggey et al., 1994; Ostojic and O'Brien, 1994
Plymouth, NH	170–318	227	<10–35	23	79–96	90	—	—	Kuter et al., 1993
Sevier, TN	—	1020	—	22	—	99	93	82	E&A Environmental Consultants, Inc., 1994
Yarmouth, MA	143–262	214	4–26	12	88–98	95	>90	—	Giggey et al., 1994

[1]TRS = total reduced sulfur.
[2]D/T = dilution to threshold or ED50.
Source: Williams, 1995.

laboratory biofilter, illustrate principals. The efficiency of biofilters will vary with the media and the gases to be removed.

SUMMARY

The outstanding feature of using compost is the return of organic matter to the soil. However, at the same time, the use of compost and other organic materials has been shown to control plant diseases and remove odors and VOCs from gas streams. The latter is primarily due to two characteristics of compost (1) high exchange capacity (i.e., adsorptive capacity) and (2) support of a large heterogeneous microbial population that could metabolize numerous natural and xenobiotic compounds.

Considerably more research is needed on disease suppression by compost to be able to draw any firm conclusions in this area. There is also a need to identify the mechanisms and the factors that produce an effective compost.

REFERENCES

Bohn, H. L. 1976. Compost scrubbers of malodorous air streams. *Compost Sci.* 17(5):15–17.

Bohn, H. L. 1991. Odor removal by biofiltration, pp. 135–147. In D. R. Derenzo and A. Gnyp (ed.). *Recent Developments and Current Practices in Odor Regulations, Controls and Technology.* Trans. Air & Waste Management Assoc.

Bohn, H. L. and R. K. Bohn. 1986. Soil bed scrubbing of fugitive gas releases. *Environ. Sci. and Health.* A21(6):561–569.

Bohnke, B. and D. Eitner. 1983. *Vergleichende Untersuchung verschiedener Kompost in einem mobilen Biofilter* (Investigation and comparison of different kinds of compost in a mobile biofilter). Gutachten I.A. der AG Kompostabsatz NM. Aachen, W. Germany.

Bollen, G. J. 1993. Factors involved in inactivation of plant pathogens during composting of crop residues, pp. 301–318. In H. A. J. Hoitink and H. M. Keener (eds.). *Science and Engineering of Composting: Design, Environmental, Microbiological and Utilization Aspects.* Renaissance Publ., Worthington, OH.

Carlson, D. A. and C. P. Leiser. 1966. Soil beds for the control of sewage odors. *J. Water Pollution Control Assoc.* 38(5):829–840.

Chellemi, D. O., D. J. Mitchell, and A. W. Barkdol. 1992. Effect of composted organic amendments on the incidence of bacterial wilt of tomato. *Proc. Fla. Hort. Soc.* Paper No. 122. Gainesville, FL.

Chen, W., H. A. J. Hoitink, and L. V. Madden. 1988. Microbial activity and biomass in container media for predicting suppressiveness to damping-off caused by *Pythium ultimum. Phytopathology.* 78:1447–1450.

County Sanitation Districts of Orange County. 1993. *Biofiltration Research Project Literature Review Report.* County Sanitation Districts of Orange County, Orange County, CA.

Daulton, R. A. C. and C. J. Nusbaum. 1961. The effect of soil temperature on the survival of the root-knot nematodes *Meloidogyne javanica* and *M. hapla. Nematologica.* 6:280–294.

Don, J. A. 1985. *The Rapid Development of Biofiltration for the Purification of Diversified Waste Gas Streams.* Paper presented at a Colloquium on Odorants, Baden-Baden, W. Germany.

Duddington, C. L. and C. M. G. Duthoit. 1960. Green manuring and cereal root eelworm. *Plant Pathol.* 9:7–9.

E&A Environmental Consultants, Inc. 1993. Report submitted to the Town of Hoosac, MA.

E&A Environmental Consultants, Inc. 1994. *Air Sampling Report for Sevier Compost Facility.* Report submitted to Bedminster Bioconversion Corp., Cherry Hill, NJ.

Ebinger, M. H., H. L. Bohn, and R. W. Puls. 1987. Propane removal from propane-air mixtures by soil beds. *J. Air Pollution Control Assoc.* 37(12):1486–1489.

Eitner, D. 1984. *Untersuchungen über Einsatz and Leistungsfahigkeit von Compostfilteranlagen zur biologischen Abluftreinigung im bereich Vonklaranlagen unter besonderer Berucksichtigung der Standzeit* (Investigations of the use and ability of compost filters for biological waste gas purification with special emphasis on operation time aspects). GWA, Band 71, RWTH, Aachen, W. Germany.

Eitner, D. and H. G. Gethke. 1987. Design, construction, and operation of biofilters for odour control in sewage treatment plants. *Proc. of the 80th Annual Meeting of the Air Pollution Control Assoc.,* New York, NY.

Epstein, E., G. B. Willson, W. D. Burge, D. C. Mullen, and N. K. Enkiri. 1976. A forced aeration system for composting wastewater sludge. *J. Water Poll. Control Fed.* 48(4): 688–694.

Ergas, S. J., E. D. Shroeder, D. P. Y. Chang, and R. L. Morton. 1995. Control of volatile organic compound emissions using a compost biofilter. *Water Environment Res.* 67(5):816–821.

Fouhy, K. 1992. Cleaning waste gas naturally. *Chemical Engineering.* December:41–46.

Gerritson-Cornell, L. and F. R. Humphreys. 1977. Results of an experiment on the effects of *Pinus radiata* bark on the formation of sporangia in *Phytophthora cinnamomi* Rands. *Phyton.* 36:15–17.

Giggey, M. D., C. A. Dwinal, J. R. Pinnette and M. A. O'Brien. 1994. Performance testing of biofilters in a cold climate, pp. 4/29–39. Presented at the *Water Environment Federation Specialty Conf. Series: Odor and Volatile Organic Compound Emission Control for Municipal and Industrial Facilities.* Jacksonville, FL.

Gorodecki, B. and Y. Hadar. 1990. Suppression of *Rhizoctonia solani* and *Sclerotium rolfsii* disease in container media containing composted separated cattle manure and grape marc. *Crop Protection.* 9:271–274.

Grebus, M. E., M. E. Watson, and H. A. J. Hoitink. 1994. Biological, chemical and physical properties of composted yard trimmings as indicators of maturity and plant disease suppression. *Compost Sci. & Util.* 2(1):57–71.

Grushevoi, S. E. and P. M. Levykh. 1940. Possibility of obtaining seed-bed soil free of infection in compost heaps. *Vses. Nauchn. Issledovatel. Inst. Tabach. Makhoroch. Prom. No.* 141:42–48.

Habicht, Jr., W. A. 1975. The nematicidal effects of varied rates of raw and composted sewage sludge as soil organic amendments on a root-knot nematode. *Plant Dis. Reptr.* 59:631–634.

Hartenstein, H. U. and E. R. Allen. 1986. *Biofiltration and Odor Control Technology for a*

Wastewater Treatment Facility. Report submitted to Dept. of Public Works, City of Jacksonville, FL.

Hoitink, H. A. J. 1980. Fungicidal properties of composted bark. *Compost Sci./Land Util.* 21(6):24–27.

Hoitink, H. A. J. 1990. *Production of Disease Suppressive Compost and Container Media, and Microorganism Culture for Use Therein.* United States Patent #4,900,348.

Hoitink, H. A. J., M. J. Boehm, and Y. Hadar. 1993. Mechanism of suppression of soilborne plant pathogens in compost-amended substrates, pp. 601–621. In H. A. J. Hoitink and H. M. Keener (eds.). *Science and Engineering of Composting: Design, Environmental, Microbiological and Utilization Aspects.* Renaissance Publ., Worthington, OH.

Hoitink, H. A. J. and P. C. Fahy. 1986. Basis for the control of soilborne plant pathogens with composts. *Ann. Rev. Phytopathol.* 24:93–114.

Hoitink, H. A. J., Y. Inbar, and M. J. Boehm. 1991. Status of compost-amended potting mixes naturally suppressive to soilborne diseases of floriculture crops. *Plant Disease.* 75(9):869–873.

Hoitink, H. A. J., L. J. Herr, and A. F. Schmitthenner. 1976. Survival of some plant pathogens during composting of hardwood tree bark. *Phytopathology.* 66:1369–1272.

Hoitink, H. A. J. and G. A. Kuter. 1986. Effects of composts in growth media on soilborne pathogens, pp. 289–306. In Y. Chen and Y. Avnimelech (eds.). *The Role of Organic Matter in Modern Agriculture.* Martinus Nijhoff Pub., Dordecht, Netherlands.

Hoitink, H. A. J. and H. A. Poole. 1980. Bark compost use in container media. *Compost Science/Land Utilization.* 21(3):38–40.

Hoitink, H. A. J., A. F. Schmitthenner, and L. J. Herr. 1975. Composted bark for control of root rot in ornamentals. *Ohio Rep.* 60:25–26.

Hoitink, H. A. J., D. M. VanDoren, Jr., and A. F. Schmitthenner. 1977. Suppression of *Phytophthora cinnamomi* in composted hardwood bark mix. *Phytopathology.* 67:561–565.

Hunt, P. G., C. C. Hortenstine, and G. C. Smart, Jr. 1973. Response of plant parasitic and saprophytic nematode populations to composted refuse. *J. Environ. Quality.* 2(2):264–266.

Kai, H., T. Ueda, and M. Sakaguchi. 1990. Antimicrobial activity of bark-compost extracts. *Soil Biol. Biochem.* 22(7):983–986.

Koch, W., H. G. Liebe, and B. Striefler. 1982. *Betriebserfahrungen mit Biofilters zur Reduzierung geruchsintensiever Emissionen* (Experiences with biofilters for the reduction of odorous air emissions). Staub-Reinhaltung der Luft, pp. 488–493. W. Germany.

Krueger, R. G., N. W. Gillham, and J. H. Coggin, Jr. 1973. *Introduction to Microbiology.* The Macmillan Co., New York.

Kuster, E. and I. in der Schmitten. 1981. On the occurrence of antimicrobial substances in the course of composting municipal garbage. *European J. Appl. Microbiol. Biotechnol.* 11:248–252.

Kuter, G. A., J. E. Harper, L. M. Naylor, and P. J. Gormsen. 1993. Design, construction and operation of biofilters for controlling odors at composting facilities. Presented at the *86th Annual Meeting and Exhibition of the Air and Waste Management Assoc.* Denver, CO.

Lear, B. 1959. Application of castor pomace and cropping of castor beans to soil to reduce nematode populations. *Plant Dis. Reptr.* 43:459–460.

Linford, M. B., F. Yap, and J. M. Olivera. 1938. Reduction of soil populations of the root-knot nematode during decomposition of organic matter. *Soil Sci.* 45:127–141.

Lumsden, R. D., A. J. Lewis, and P. D. Millner. 1982. Composted sludge as a soil amendment for control of soilborne plant diseases, pp. 274–277. In H. W. Kerr, Jr. and L. Knutson (eds.). *Research for Small Farms*. USDA, ARS, Misc. Publ. No. 1422, Washington, DC.

Lumsden, R. D., A. J. Lewis, and P. D. Millner. 1983. Effect of composted sewage sludge on several soilborne pathogens and diseases. *Phytopathology.* 73:1543–1548.

Mandelbaum, R. and Y. Hadar. 1990. Effects of available carbon source on microbial activity and suppression of *Pythium aphanidermatum* in compost and peat containers. *Phytopathology.* 80(9):794–804.

Mannion, C. M., B. Schaffer, M. Ozores-Hampton, H. H. Bryan, and R. McSorley. 1994. Nematode population dynamics in municipal solid waste–amended soil during tomato and squash cultivations. *Nematropica.* 24:17–24.

McElroy, F. D., 1993. *Commercial Development of Disease Suppressive Compost.* Final Report. Report No. 12. The Clean Washington Center, Seattle, WA.

Millner, P. D., R. D. Lumsden, and J. A. Lewis. 1982. Controlling plant disease with sludge compost. *BioCycle.* 23:50–52.

Nelson, E. B. 1992. *Biological Control of Turfgrass Diseases.* Information Bull. 220 Cornell Cooperative Extension. Cornell U., Ithaca, NY.

Nelson, E. B., L. L. Burpee, and M. B. Lawton. 1994. Biological control of turfgrass diseases, pp. 409–427. In A. Leslie (ed.). *Handbook of Integrated Pest Management for Turf and Ornamentals.* Chap. 38. CRC Press, Inc. Boca Raton, FL.

Nelson, E. B. and C. M. Craft. 1991. Suppression of brown patch with topdressing amended with composts and organic fertilizers, 1989. *Biol. and Cultural Tests Control Plant Dis.* 6:90–107.

Nelson, E. B. and C. L. Craft. 1992. Suppression of dollar spot on creeping bentgrass and annual bluegrass turf with compost amended topdressing. *Plant Disease.* 76(9):954–958.

Nelson, E. B. and H. A. J. Hoitink. 1983. The role of microorganisms in the suppression of *Rhizoctonia solani* in container media amended with composted hardwood bark. *Phytopathology.* 73:274–278.

Nelson, E. B., G. A. Kuter, and H. A. J. Hoitink. 1983. Effects of fungal antagonists and compost age on suppression of *Rhizoctonia* damping-off in container media amended with composted hardwood bark. *Phytopathology.* 73:1457–1462.

Ostojic, N., R. A. Duffee, and M. O'Brien. 1991. *Hamilton Wastewater Treatment and Compost Facility Odor Control Evaluation and Optimization. Report for the City of Hamilton, Ohio.* Odor Sci. & Eng., Inc., Hartford, CT.

Ostojic, N. and M. O'Brien. 1994. Control of odors from sludge composting using wet scrubbing biofiltration and activated sludge treatment, pp. 5/9–20. Presented at the *Water Environment Federation Specialty Conf. Series: Odor and Volatile Organic Compound Emission Control for Municipal and Industrial Wastewater Treatment Facilities.* Jacksonville, FL.

Ottengraf, S. P. P. 1986. Exhaust gas purification, pp. 425–452. In H. J. Rehm and G. Reed (eds.). *Biotechnology. Vol. 8.* VCH Verlagsgesellschaft, Weinheim, Germany.

Ottengraf, S. P. P. 1986. Exhaust gas purification. In H. J. Rehm and G. Reed (eds.). *Biotechnology. Vol. 8.* VCH Verlagsgesellschaft, Weinheim, Germany.

Ottengraf, S. P. P. and A. H. C. Van Den Oever. 1983. Kinetics of organic compound removal from waste gases with a biological filter. *Biotech. and Bioengin.* 25:3089–3102.

Ottengraf, S. P. P., A. H. C. Van Den Oever, and F. J. C. M. Kempenaars. 1974. *Waste Gas*

Purification in a Biological Filter Bed. Innovations in Biotechnology. Amsterdam, Netherlands.

Prokop, W. H. and H. L. Bohn. 1985. Soil bed systems for control of rendering plant odors. *J. of the Air Pollution Control Assoc.* 35(12):1332–1338.

Rands, M. B., D. E. Cooper, C. Woo, G. C. Fletcher, and K. A. Rolfe. 1981. Compost filters for H₂S removal from anaerobic digestion and rendering exhausts. *J. WPCF.* 53(2):185–189.

Rich, J. R. and C. H. Hodge. 1993. Utilization of blue crab scrap compost to suppress *Meloidogyne javanica* on tomato. *Nematropica.* 23:1–5.

Sekiguchi, A. 1977. Control of *Fusarium* wilt on Chinese yam. *Ann. Rep. Dep. Plant Pathol. Entomol. Veg. Floric. Exp. Stn., Nagano Japan.* 1:10–11.

Sivasithamparam, K. 1981. Some effects of extracts from tree barks and sawdust on *Phytophthora cinnamomi* Rands. *Aust. Plant Path.* 10:18–20.

Spencer, S. and D. M. Benson. 1982. Pine bark, hardwood bark compost, and peat amendment effects on development of *Phytophthora* spp. and lupine root rot. *Phytopathology.* 72:346–351.

Spring, D. E., M. A. Ellis, R. A. Spotts. H. A. J. Hoitink, and A. F. Schmitthenner. 1980. Suppression of the apple coller rot pathogen in composted hardwood bark. *Phytopathology.* 70:1209–1212.

Steinmuller, W., G. Claus, and H. J. Kutzer. 1979. Grundlagen der biologischen Abluftreinigung. Teil II: Microbiologischer Abbau von luftverunreinigenden Stoffen. (Fundamentals of biological waste gas purification. Part II. Microbiological degradation of air pollutants). *Staub-Reinhaltung der Luft.* pp. 149–152. W. Germany.

Tamura, M. and K. Taketani. 1977. Biology and control of clubroot of Chinese cabbage in the Ishikawa prefecture. *Ishikawa Pref. Agric. Exp. Stn. Bull.* 9:1–26.

Tate, R. L., III. 1995. *Soil Microbiology.* John Wiley & Sons, Inc. New York.

Van Langenhove, H., E. Wuyts, and N. Schamp. 1986. Elimination of hydrogen sulfide from odorous air by a wood bark biofilter. *Water Res.* 20(12):1471–1476.

Wheeler, M. L. 1992. Proactive odor management: The evolution of odor control strategies at the Hamilton, Ohio, wastewater treatment and sludge composting facility. *Presented at the Biocycle National Conf.* St. Louis, MO.

Williams, T. O. 1995. Odors and VOC emissions control methods. *BioCycle.* 36(5):49–56.

Williams, T. O. and F. C. Miller. 1993. Composting facility odor control using biofilters, pp. 262–281. In H. A. J. Hoitink and H. M. Keener (eds.). *Science and Engineering of Composting: Design, Environmental, Microbiological and Utilization Aspects.* Renaissance Publ., Worthington, OH.

Yang, Y. and E. R. Allen. 1994a. Biofiltration control of hydrogen sulfide 1. Design and operational parameters. *J. Air & Waste Manage. Assoc.* 44:863–868.

Yang, Y. and E. R. Allen. 1994b. Biofiltration control of hydrogen sulfide 2. Kinetics, biofilter performance, and maintenance. *J. Air & Waste Manage. Assoc.* 44:1315–1321.

CHAPTER 14

Regulations

INTRODUCTION

Regulations, guidelines and standards have been established for the composting process and for the use of compost by U.S. federal government, most states, Canada and Europe. In the United States the regulatory authority is the Environmental Protection Agency (USEPA). The USEPA regulates only biosolids compost under the 40CFR503 regulations. These regulations regard compost as only one form of biosolids and is part of the Land Application Regulations; process controls are only mandated in terms of biosolids quality and allowable use.

State regulations vary greatly. Some states such as New York have separate compost process and product regulations that cover all feedstocks. Others regulate compost under biosolid or solid waste regulations. Yet other states concentrate on product quality utilization, while others have detailed process requirements.

Several agencies in the United States, Canada, and Europe have chosen to issue guidelines rather than regulations. Guidelines leave the implementation or interpretation to local agencies, resulting in non-uniform application. For example, the state of Washington has issued guidelines that local health officials may implement. Consequently, two adjoining counties can have different criteria for compost utilization. Agencies providing guidelines usually do so for the following reasons:

- The staff does not wish to antagonize various political groups (e.g., agriculture or environmental) that may have different agendas.

447

- Guidelines are easy to issue and may not require public hearings or long periods to implement.
- Guidelines are easy to change and may not require legislative action.

Poor regulations are better than no regulations or guidelines. Poor regulations can be changed. With only guidelines or no regulations, it may be difficult to obtain permits or implement a program since any desired action can be stopped at the whim of staff or local authorities. Further, individual staff within agencies can impose their biases without considering the value of the project. Consequently, legislators should require regulators to issue regulations in lieu of guidelines.

The objective of this chapter is to present current regulations, guidelines, and standards prevailing in the United States, Canada, and several countries in Europe. The chapter will review the concepts and approaches to regulations and discuss the elements that should be regulated. Another objective is to provide the reader with current trends in compost regulations.

CONCEPTS AND APPROACHES TO REGULATIONS

Kennedy (1992) presented three basic approaches to developing regulations related to product use:

- no-net degradation
- risk-based approach
- best-achievable approach

The no-net degradation concept is based on the premise that application of compost should not increase the level of a heavy metal or other contaminant in the soil (Dillon, 1994). Several European countries and Canadian provinces have considered or are setting guidelines based on this concept. One problem with no net degradation is the question of what to use as a soil base level. Soil quality varies greatly within a small area; for example, urban soils may contain higher levels of lead from leaded gasolines than rural areas. As a result, regional standards have to be established based on fluctuations in soil quality.

If no-net degradation were used on a site-by-site basis, it would create excessive sampling requirements and allow the use of lower quality material on areas that are already contaminated.

Another problem with the no-net degradation concept is that soils are continuously amended with fertilizers, pesticides, herbicides, and other chemicals. This not only changes the baseline quality of the soil, but illustrates the illogic in singling out compost as the only regulated material.

The risk-based approach considers the potential risk to humans, animals, plants, soil biota, as well as environmental consequences. Thus, it evaluates the potential toxic effects of a chemical on the individual (human, animal or plant) or environmental entity. The risk-based approach also considers the risk in relation to other risks in the environment. The most comprehensive risk evaluation performed was for the USEPA 40CFR503 regulations for the disposal and use of biosolids.

Finally, the best-achievable approach ignores health and environmental aspects to primarily consider technology and economics. As a result, standards are based on what technology can achieve.

UNITED STATES

Federal Regulations

U.S. federal regulations dealing with waste materials are under the jurisdiction of the Environmental Protection Agency. Compost regulations promulgated by USEPA cover biosolids and any material containing biosolids. No regulations have been established for MSW or yard waste compost. The regulations governing compost were required by the Clean Water Act Amendments of 1987 [Sections 405(d) and (e) as amended (33 U.S.C.A. 1251, et seq.)]. The regulations were published in the *Federal Register* (58 FR 9248 to 9404) as the Standards for the Use or Disposal of Sewage Sludge, Title 40 of the Code of Federal Regulations Part 503. The 503 rule was published on February 19, 1993 and became effective on March 22, 1993. The Part 503 rule was amended on February 25, 1994 (59FR9095) for molybdenum. The pollutant concentration limits and annual pollutant loading rates for molybdenum were deleted. Only the ceiling concentration limit of 75 mg/kg was retained.

Two other pollutant (Cr and Se) limits were contested in the courts. Thus, law suits were filed by the leather industries of America and the Association of Metropolitan Sewerage Agencies, Milwaukee Metropolitan Sewerage District, and the city of Pueblo, Colorado. As a result of this action, the chromium pollutant limits were suspended and the selenium level was raised to 100 mg/kg. In addition to heavy metals, the 503 rule regulates pathogens and vector attraction.

Much of the discussion in this section is based on three USEPA documents.

(1) *Federal Register.* Friday February 19, 1993. Standards for the Use or Disposal of Sewage Sludge; Final Rules. Part II. Environmental Protection Agency. 40 CFR Part 257 et al.

(2) United States Environmental Protection Agency. Office of Wastewater Management (4204). *A Plain English Guide to the EPA Part 503 Biosolids Rule.* EPA/832/R-93/003. September, 1994.

(3) United States Environmental Protection Agency. Office of Wastewater Management (4204). *Guide to the Biosolids Risk Assessments for the Part 503 Rule.* EPA832-B-95-005.

The 503 rule was designed to protect public health and the environment from "any reasonably anticipated adverse effects of certain pollutants and contaminants that may be present in [biosolids]" (USEPA, 1994). The USEPA clearly stated that it promotes the beneficial use of biosolids. A very intensive risk assessment was conducted. The rule making took nine years and evaluated research from the past 25 years. In 1984, USEPA considered 200 pollutants identified in the "40 Cities Study." The selection of the 200 pollutants was based on the following criteria:

- human exposure and health effects
- plant uptake of pollutants
- phytotoxicity
- effects in domestic animals and wildlife
- effects in aquatic organisms
- frequency of pollutant occurrence in biosolids

This list of pollutants was submitted for review by four panels. The panels recommended that approximately 50 of the 200 pollutants be further studied.

In the final regulations, the USEPA addressed 25 pollutants using 14 exposure pathways (Ryan and Chaney, 1995). The 25 pollutants were:

Organics	Heavy Metals
Aldrin/dieldrin (total)	Arsenic
Benzene	Cadmium
Benzo(a)pyrene	Chromium
Bis(2-ethylhexyl)phthalate	Copper
Chlordane	Lead
DDT/DDE/DDD (total)	Mercury
Heptachlor	Molybdenum
Hexachlorobenzene	Nickel
Hexachlorobutadiene	Selenium
Lindane	Zinc
N-Nitrosdimethylamine	
Polychlorinated biphenyls	

Organics	Heavy Metals
Toxaphene	
Trichloroethylene	

The risk assessment (USEPA, 1995) followed four basic steps:

- Hazard identification: Can the identified pollutants harm human health or the environment?
- Exposure assessment: Who is exposed, how do they become exposed, and how much exposure occurs? Highly exposed individuals were identified and their exposure to pollutants in biosolids evaluated. Fourteen exposure pathways were identified for land application of biosolids (Table 14.1).
- Dose-response evaluation. The likelihood of an individual developing a particular disease as the dose and exposure increase. The two EPA toxicity factors used whenever available were: (1) risk reference doses (RfDs)—daily intake; (2) cancer potency values ($q_1 * s$)—conservative indication of the likelihood of a chemical inducing or causing cancer during the lifetime of a continuously exposed individual.
- Risk characterization: What is the likelihood of an adverse effect in the population exposed to a pollutant under the conditions studied? Risk is calculated as:

$$Risk = Hazard \times Exposure$$

Hazard refers to the toxicity of a substance determined during the hazard identification and dose-response evaluation; exposure is determined through the exposure assessment (USEPA, 1995). EPA made a policy decision to regulate risk at 1×10^{-4}.

The following is the general approach USEPA utilized in developing pollutant soil loading limits (Ryan and Chaney, 1995).

- delineation of pollutants of concern in biosolids
- identification of potential pathways for exposure and receptors (humans, soil biota, plants, and animals) to several pollutants through land application of biosolids
- identification of dose-response relationships for the receptors and pollutants of concern
- determination of the maximum level of each pollutant for each exposure pathway that would protect the highly exposed individual from adverse effects

Pathway	Highly Exposed Individual (HEI) or Receptor
1 Sludge-soil-plant-human	Protection of consumers who eat produce grown in soil using sewage sludge. 2.5% of a consumer's intake of grains, vegetables, potatoes, legumes, and garden fruit is assumed to be grown on sludge-enriched soil.
2 Sludge-soil-plant-home gardener	Home gardener who produces and consumes potatoes, leafy vegetables, legume vegetables, root vegetables, and garden fruit. 60% of the HEI's diet is assumed to be grown on sludge-amended soil.
3 Sludge-soil-child	Assessment of the hazard to a child ingesting undiluted sewage sludge. The sewage sludge ingestion was 0.2 grams dry weight per day for 5 years.
4 Sludge-soil-plant-animal-human	Human exposure from consumption of animal products. 40% of the HEI's diet of meat, dairy products, or eggs is assumed to come from animals consuming feed from soil to which sludge was applied. In a non-agricultural setting, a human consumes products from wild animals that consumed plants grown on sludge-amended soil. The HEI is also assumed to be exposed to a background intake of a pollutant.
5 Sludge-soil-animal-human	The direct ingestion of sewage sludge by animals and the consumption by humans of the contaminated tissue. Direct ingestion of sludge by animals, where it has been surface applied. When sewage sludge is injected into the soil or mixed into the plow layer, grazing animals ingest the soil containing sludge. The HEI is also assumed to be exposed to a background intake of a pollutant.
6 Sludge-soil-plant-animal-toxicity	Protection of the highly sensitive/exposed herbivorous livestock that consumes plants grown on sewage sludge–amended soil. It is assumed that the livestock diet consists of 100% forage grown on sewage sludge–amended land and that the animal is exposed to a background pollutant intake.

TABLE 14.1. (continued).

Pathway	Highly Exposed Individual (HEI) or Receptor
7 Sludge-soil-animal toxicity	Protection of the highly sensitive/highly exposed herbivorous livestock that incidently consumes sewage sludge adhering to forage crops and/or sewage sludge on the soil surface. The percent of sewage sludge in the live stock diet is assumed to be 1.5%, and the animal is exposed to a background pollutant intake.
8 Sludge-soil-plant toxicity	Evaluation of risk to plant growth (phytotoxicity) from pollutants in sludge. Probability of 50% reduction of plant growth associated with a low probability of 1×10^{-4}.
9 Sludge-soil-soil-biota toxicity	Protection of the highly exposed/highly sensitive soil biota. The criteria for this pathway have been set using the earthworm (*Eisenia foetida*) data.
10 Sludge-soil-soil biota-predator of soil biota toxicity	Protection of the highly sensitive/highly exposed soil biota predator. Sensitive wildlife that consume soil biota that have been feeding on sewage sludge–amended soil. Chronic exposure assumes that 33% of the sensitive species' diet is soil biota.
11 Sludge-soil-airborne dust-human	Tractor operator exposed to 10 mg/m³ total dust while tilling a field to which sewage sludge has been applied.
12 Sludge-soil-surface water-contaminated water-fish toxicity-human toxicity	Protection of human health and aquatic life. Risk to surface water associated with run-off of pollutants from soil on which sewage sludge has been applied. Water quality criteria are designed to protect human health assuming exposure through consumption of drinking water and resident fish, and to protect aquatic life.
13 Sludge-soil-air-human	Protection of members of farm households inhaling vapors of any volatile pollutant that may be in the sewage sludge when it is applied to the land. This pathway is not applicable to inorganic pollutants. It is assumed that the total amount of pollutant spread in each year would be vaporized during that year.
14 Sludge-soil-ground-water-human	Exposure of individuals drinking water from groundwater directly below a field to which sewage sludge has been applied.

- determination of maximum acceptable loading rates of biosolids to land for each pollutant based on the most limiting value for all evaluated pathways
- determination of the pollutant limits (cumulative soil pollutant application limit and maximum allowed biosolids pollutant concentration). This was obtained from maximum loading rates and biosolids concentration from the National Sewage Sludge Survey.

Several key aspects were used in determining the pollutant limits:

- The target organism was the highly exposed individual (HEI) rather than the most exposed individual (MEI). The HEI was a realistic individual, whereas the MEI was unrealistic and did not exist.
- EPA used the lifetime exposure criterion of 70 years. For home gardeners producing their own food, it was assumed that 59% of the food would be grown in a home garden amended with biosolids.
- Uptake slopes for pollutants by crops were assumed to be linear even though the data indicated a curvilinear slope. This was believed to be more conservative.
- Cancer risk for all biosolids use was set at 1×10^{-4}.
- Data for plant uptake was based on field data when available.
- Human dietary exposure to pollutants in biosolids was revised from the early assessment by apportioning food consumption among several age periods during the lifetime of the 70 years of the HEI.
- The final rule evaluated all organic pollutants proposed for the 503 rule. The levels found by the National Sewage Sludge Survey showed that organic pollutants were at low levels and in the evaluation did not pose significant risks to public health or the environment. EPA is currently considering a zero limit for PCB.

Examples of the risk assessment and the determination of the pollutant limits are shown below for As. The first analysis is for Pathway 1 where an adult consumes over a lifetime crops grown on biosolids-amended soil. The second example uses Pathway 3, child ingesting biosolids. Based on these analyses, it was determined that Pathway 3 was the limiting pathway. These analyses are based on USEPA (1995):

$$RIA = \frac{RfD * BW}{RE} - TBI * 10^3 \tag{1}$$

RIA = allowable dose of pollutant without adverse effects
RfD = reference dose in mg/kg-day; for As = 0.0008 mg/kg-day
BW = human body weight, 70 kg

RE = relative effectiveness of ingestion exposure, 1.0, no units
TBI = total pollutant intake from all background sources in water, food, and
air, 0.012 mg/day

For arsenic in biosolids as applied to Pathway 1:

$$RIA = \frac{0.0008 * 70}{1.0} - 0.012 * 10^3 = 44\,\mu g \qquad (2)$$

The RIA is used to determine the cumulative amount of a pollutant that can be land applied without adverse effects from biosolids for the selected pathway. In this case, Pathway 1 (an adult over a lifetime consumes crops grown on biosolids-amended soil) is used as an illustration.

$$RP_c = \frac{RIA}{\Sigma(UC \times DC \times FC)} \qquad (3)$$

RP_c = the cumulative amount of a pollutant that can be land applied without adverse effects from biosolids exposure through the pathway evaluated
UC = plant uptake slope for pollutant from soil amended with biosolids
DC = dietary consumption of different food groups grown in soils amended with biosolids
FC = fraction of different food groups assumed to be grown in soils amended with biosolids

The sum of UC × DC × FC is 0.00654. Therefore, for arsenic in biosolids as applied to Pathway 1, the cumulative amount that can be land applied without adverse effects is 6700 kg/ha.

$$RP_c = \frac{44}{\Sigma(UC \times DC \times FC)} = 6700\,kg/ha \qquad (4)$$

The most limiting pathway for As was Pathway 3, a child ingesting biosolids. This analysis is shown below:

$$RIA = \left(\frac{RfD * BW}{RE} - TBI\right) * 10^3 \qquad (5)$$

The principal difference in the calculation of Equation (5) versus Equation (2) is the body weight (BW) of a child (16 kg) compared to that of an adult (70 kg). Also, the total intake of As for a child is 0.0045 mg/day versus 0.012 mg/day for an adult.

$$RIA = \frac{0.0008 * 16}{1.0} - 0.0045 = 8.3\,\mu g \qquad (6)$$

The next step in calculating the concentration of a pollutant (RSC) in biosolids that can be expected not to produce adverse effects is as follows:

$$RSC = \frac{RIA}{I_s * DE}$$

where

RSC = the concentration of a pollutant in biosolids that can be ingested without expectation of adverse effects

RIA = the amount of pollutant ingested by humans without expectation of adverse effects

I_s = the rate of biosolids ingestion by children

DE = exposure-duration adjustment; an attempt to consider less than life-time exposure

The RSC for As concentration in biosolids ingested by children is calculated as follows:

$$RSC = \frac{8.3}{0.2 * 1} = 41\,\mu g$$

Similar assessments were conducted for other potential As toxicity pathways.

Phytotoxicity of inorganic elements (Pathway 8) was evaluated by two methods:

Method I

(1) A phytotoxicity threshold (PT_{50}) was established. This value is the concentration of a pollutant that can cause a 50% reduction in plant growth. This was based on short-term data.

(2) A calculation was made to determine the probability that the heavy

metal concentration in plants grown on biosolids-amended soil would exceed the PT_{50} at various metal loadings using field studies.

(3) An acceptable level of tolerable risk exceeding the PT_{50} was set at 0.01 (i.e., 1 out of 100 times).

(4) The allowable loading rate of biosolids (RP) was the rate that would have less than a 0.01 probability of causing the PT_{50} to be exceeded.

The example provided below is for zinc.

(1) The PT_{50} for Zn = 1975 μg Zn/g plant tissue dry weight.

(2) The probability that corn grown on biosolids-amended soils would exceed the PT_{50} was computed for 12 Zn loading ranges.

(3) The tolerable risk for exceeding PT_{50} was set at 0.01.

(4) None of the loading rates evaluated exceeded the probability of 0.01. Therefore, the highest loading rate evaluated (3500 kg Zn/ha) was chosen as the allowable loading rate (RP) for biosolids that would not cause a significant phytotoxic effect in corn. RP = 3500 kg Zn/ha.

Method II

This method evaluated the lowest-observed-adverse-effects-level (LOAEL). The reference cumulative application rate of a (RP) of Zn was calculated as follows:

$$RP = \frac{TPC - BC}{UC}$$

where

RP = the amount of a pollutant that can be applied to a hectare of land without expectation of adverse effects

TPC = the concentration of a pollutant in a sensitive plant tissue species (e.g., lettuce) as opposed to a less sensitive species such as corn (used in Method I)

BC = background concentration of pollutant in plant tissue

UC = plant uptake of pollutant from soil/biosolids

For Zn the following parameters were used:

TPC = 400 μg of Zn/g plant tissue in lettuce dry weight (μg/g DW)
BC = 47.0 μg of Zn/g plant tissue of lettuce DW
UC = 0.125 μg of Zn/g of lettuce plant tissue (kg of Zn per ha) (μg/g DW) (kg/ha)

The calculation of RP for Zn is as follows:

$$RP = \frac{400 - 47.0}{0} \cdot 125 = 2800 \text{ kg Zn/ha}$$

A comparison of the results of Methods I (3500 kg Zn/ha) and II (2800 kg Zn/ha) shows that the more restrictive result was an RP of 2800 kg Zn/ha. The limit set for Pathway 8 was the pollutant limit used in the Part 503 rule for Zn.

Compost regulations fall under Requirements for Land Application, Subpart B. Biosolids or biosolids products can be land applied in bulk or sold or given away in bags or other containers (buckets, boxes, or cartons or vehicles with a load capacity of less than one metric ton).

Compost applied to land must meet risk-based pollutant limits, pathogen reduction standards, and vector attraction criteria (USEPA, 1994). Prior to discussing the pollutant limits, the following definitions should prove helpful:

(1) Ceiling concentration—The maximum concentration in mg/kg of an inorganic pollutant (heavy metal) in biosolids compost allowed for land application. If compost contains pollutants above these levels, the product may not be applied to land. Below this limit, other criteria may restrict its use. States may issue regulations that have lower but not higher limits.

(2) Pollutant concentration (PC) limits—The maximum concentration in mg/kg of an inorganic pollutant applies to a Class B biosolids. It can only be applied in bulk to land. It normally does not apply to compost.

(3) Cumulative pollutant loading rate (CPLR)—The maximum amount of an inorganic pollutant that can be applied to an area of land. It applies to bulk distribution and must be tracked.

(4) Alternative pollutant limit (APL)—The highest level of a given heavy metal in biosolids compost permitted in materials to be marketed.

(5) Exceptional quality biosolids (EQ)—Although this term is not used specifically in the 503 regulations, it is used in USEPA documents explaining the 503 regulations (USEPA, 1994). It refers to the concentration of a low pollutant biosolids compost that meets the USEPA NOAEL (No Observed Adverse Effects Limits) criterion as well as the pathogen and vector attraction reduction requirements. The term applies to a biosolids compost in bulk or bagged or other containers without further federal regulations.

(6) Annual pollutant loading rate (APLR)— The highest annual (365 days)

rate of application of each pollutant to land in kg/ha. It applies to bagged biosolids compost sold or given away in a bag or other container.

Table 14.2 summarizes the pollutant limits for heavy metals in biosolids and biosolids products (USEPA, 1995). In addition to pollutant regulations, the 503 regulations require that before biosolids compost can be sold or given away it must meet Class A.

Table 14.3 summarizes the pathogen requirements for composted biosolids. USEPA designated six treatment alternatives. One alternative, biosolids treated in a process to further reduce pathogens (PFRP), applies to compost. Other alternatives deal with thermally treated biosolids, pH adjustment and other processes. Biosolids compost must also meet vector attraction reduction criteria.

The basis for the 503 pathogen requirements is found in the USEPA document "Technical Support Document for Reduction of Pathogens and Vector Attraction in Sewage Sludge" (USEPA, 1992). In the previous USEPA 257 regulations, the only requirements for composting were based on time-temperature relationships. In a 1988 study, Yanko (1988) demonstrated that regrowth of pathogens occurs in biosolids compost. In this study salmonellae were detected 165 times in 365 measurements. No salmonellae were detected in the 86 measurements for which the fecal coliform densities were less than 1000 MPN per gram. This indicated that the potential for finding salmonellae is low when the fecal coliform densities are less than 1000 MPN per gram. The correlation between fecal coliform densities and frequency of salmonellae detection is shown in Figure 14.1 (USEPA, 1992; Farrell, 1992). The reason for alternatively using either the fecal coliform test or the salmonellae test is that fecal coliform can regrow to levels exceeding 1000 MPN/g but salmonellae, once totally eliminated, can never grow (USEPA, 1992).

The vector reduction criterion that applies to composted biosolids containing partially decomposed organic bulking agents requires that the biosolids must be aerobically treated for 14 days or longer, during which time the temperature must always be over 40°C and the average temperature higher than 45°C. In addition to heavy metal and pathogen reduction criteria, Class B biosolids compost also has site restrictions, applying primarily to use on food crops, animal grazing, turf growing, and public access (USEPA, 1995).

The 503 regulations also provide sampling and analysis methodologies.

One of the most important aspects of the 503 regulations that impacts land application methodologies is liability. Whether by a public or private entity, direct land application is the legal responsibility of the producer of biosolids. If a municipality or its contractor violates the permit requirement for land application of biosolids, both the producer, its employees, and the contractor

TABLE 14.2. Pollutant Limits for Heavy Metals in Biosolids and Biosolids Products.

Pollutant	Ceiling Concentration Limits for All Biosolids Applied to Land (mg/kg)[1]	Pollutant Concentration Limits for EQ and PC Biosolids (mg/kg)[1]	Cumulative Pollutant Loading Rate Limits for CPLR Biosolids (kg/ha)	Annual Pollutant Loading Rate Limits for APLR Biosolids (kg/ha/365-day period)
Arsenic	75	41	41	2.0
Cadmium	85	39	39	1.9
Copper	4300	1500	1500	75
Lead	840	300	300	15
Mercury	57	17	17	0.85
Molybdenum[2]	75	—	—	—
Nickel	420	420	420	21
Selenium[4]	100	36	100	5.0
Zinc	7500	2800	2800	140
Applies to:	All biosolids that are land applied	Bulk biosolids and bagged biosolids[3]	Bulk biosolids	Bagged biosolids[3]
From Part 503	Table 1, Section 503.13	Table 3, Section 503.13	Table 2, Section 503.13	Table 4, Section 503.13

[1]Dry-weight basis.
[2]The limits for molybdenum were deleted from the 503 rule on February 25, 1994 (*Federal Register*, Vol. 39, No. 38, p. 9095).
[3]Bagged biosolids sold or given away in bag or other container.
[4]Chromium deleted from regulations and selenium modified in 1995.
Source: USEPA, 1994.

TABLE 14.3. Pathogen Requirements for Composting Alternatives.

1. Class A Requirements

The density of fecal coliform in the biosolids compost must be less than 1000 most probable number (MPN) per gram total solids (dry-weight basis).

or

The density of *Salmonella* sp. bacteria in the biosolids compost must be less than 3 MPN per 4 grams of total solids (dry-weight basis).

Either of these requirements must be met at one of the following times:

- when biosolids compost is used or disposed
- when biosolids compost is prepared for sale or give-away in a bag or other container for lane application
- when the biosolids compost or derived materials are prepared to meet the requirements for EQ biosolids

Pathogen reduction must take place before or at the same time as vector attraction reduction.

2. Processes to Further Reduce Pathogens (PFRP)

Using either the within-vessel composting method or the static aerated pile composting method, the temperature of the biosolids is maintained at 55°C or higher for three days.

Using the windrow composting method, the temperature of the biosolids is maintained at 55°C or higher for 15 days or longer. During the period when the compost is maintained at 55°C or higher, the windrow is turned a minimum of five times.

3. Processes to Significantly Reduce Pathogens (PSRP)

Using either the within-vessel, static pile, or windrow composting methods, the temperature of the biosolids is raised to 40°C or higher and maintained for five days. For four hours during the five-day period, the temperature of the compost pile exceeds 55°C.

Source: USEPA, 1995.

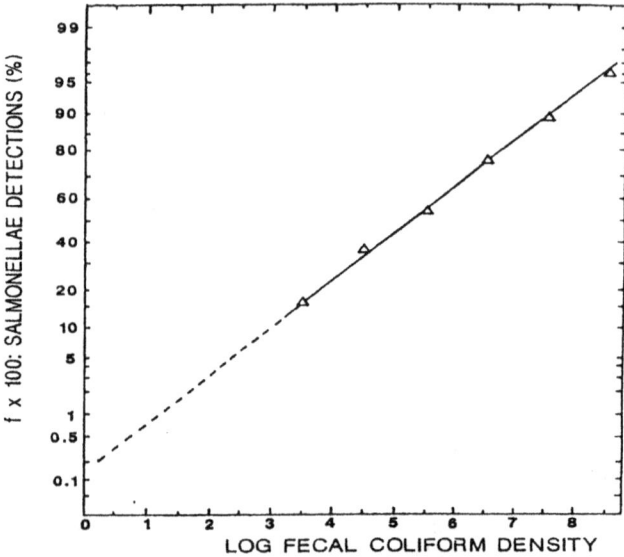

FIGURE 14.1. Relationship between log fecal coliform density and fraction of salmonellae detections. (USEPA, 1992.)

are subject to civil and criminal action. For example, if a contractor violates the municipality's permit to apply a specific quantity of biosolids containing the 503 heavy metal limitations, both the contractor, the municipality, and any knowledgeable individuals can be liable and sued for both criminal and civil damages.

Distribution and marketing of biosolids products such as compost does not entail similar liability. Thus, a contractor or the individual purchasing compost containing the limit of heavy metals and distributing or marketing the compost at excessive rates does not face criminal or civil charges. Only product liability litigation could result; for example, if the compost is provided to a user without adequate instruction on its use and it causes phytotoxicity.

This difference in responsibility and liability between direct application of biosolids to land and distribution and marketing of biosolids compost is an added reason for municipalities to consider composting.

State Regulations in the United States

States' regulations for compost vary. Most states regulate the product while some regulate both the product and the facility. The latter can include siting criteria, monitoring requirements, drainage and leachate control and treatment, and daily operating records.

Many states have elected to accept the USEPA Exceptional Quality heavy metal criteria. Other states have developed more stringent regulations. Table 14.4 shows some of the variations in contaminant levels for several states. Washington state, for example, has published interim guidelines for compost quality (Washington State Department of Ecology, 1994). To date, the promulgation of guidelines is unsatisfactory. It does not provide uniform criteria in a state and leaves decisions up to local authorities that often do not have the proper background. It also is subject to political pressures on a local level. For example, the state of Minnesota eliminates biosolids compost from being Class I regardless of its quality. Some rural areas in the state may produce a better-quality biosolids compost than yard waste or source-separated organic waste, yet they are under more restrictive criteria.

Several states place more restrictive regulations on mixed MSW compost than source-separated MSW or biosolids. This is indicative of regulators' lack of knowledge. Several studies (Epstein et al., 1992; Richard and Woodbury, 1992) have shown that there is relatively little difference in most of the heavy metals between source-separated MSW and mixed MSW compost [in some cases Pb and Zn are exceptions (see Chapter 6)].

The added restrictions on mixed MSW compost are designed to encourage source separation. Although this may be a noble idea, it is not backed by scientific evidence. Furthermore, placing more restrictions on mixed MSW compost removes the incentive by industry to develop separation technologies that could apply to communities that economically or otherwise cannot source separate. It would be much better to encourage composting of source-separated organics by showing that this is much more efficient and economical since less capital and O&M costs are involved (facilities are smaller and less equipment is used).

States should set product quality characteristics and let the system provider meet the criteria regardless of method of collection and feedstock characteristics. This is the concept in the federal 503 regulations that specifies the product leaving the facility regardless of the biosolids quality produced in the wastewater treatment system.

States should promulgate regulations for both product criteria and facility design. Soils in the United States vary considerably. As indicated earlier, pH is probably the most important parameter for heavy metal uptake by plants. In large areas in the United States, soils are alkaline, hence metal uptake criteria could be different than for acid soils. Further, salinity in agriculture is a greater problem in dry climates than in areas with rainfall. Therefore, states should tailor regulations to soils and crops rather than arbitrary numbers.

TABLE 14.4. *Element Levels (mg/kg dry weight in compost) as Regulated by Several States.*

Element	New York Maximum Conc.	North Carolina Class I	North Carolina	Oregon	Florida	California
Arsenic	41	NS	NS	NS	NS	41
Boron	NS	NS	NS	7	NS	NS
Cadmium	10	10	25	3	15	39
Chromium—total	1200	1000	2000	NS	NS	1200
Copper	1500	800	1200	750	450	1500
Lead	300	250	1000	150	500	300
Mercury	17	10	15	NS	NS	17
Nickel	290	200	1200	NS	50	420
Selenium	28	NS	NS	NS	NS	36
Zinc	2800	1000	2500	1400	900	2800
PCBs—total	1	2	10	NS	NS	NS

[1]NS = Not specified; New York (January 1995) also specifies 54 mg/kg DW for Mo.

The following examples indicate the extent to which some states are regulating compost facility criteria and product characteristics.

New York

New York State in 6 NYCRR Part 360, Solid Waste Management Facilities, regulates compost under subpart 360-5. The following illustrates some of the provisions for biosolids and mixed solid waste composting facilities:

(a) Permit application requirements:
 1. A regional map.
 2. A vicinity map.
 3. A site plan, minimum scale 1:2,400 with 1.5 m (5 feet) contours.
 4. Location and classification of wetlands and the location of floodplains within 305 m (1000 feet).
 5. A schedule of operations, daily traffic flow, procedure for unloading trucks, special precautions for operating under inclement weather, and a description of the fire and explosion prevention system.
 6. The source, quantity, and quality of all solid waste to be processed as well as amendment, bulking agent, admixture, or seed material.
 7. For biosolids or septage source of waste, the amount of waste to be processed and seasonal variations in quantity or quality during the year. Quality of biosolids or septage is to be provided for the following parameters: Total Kjeldahl N (TKN), NH_4, NO_3, total P, total K, pH, total solids, total volatile solids, As, Cd, Cr (total), Cu, Pb, Hg, Mo, Ni, Se, and Zn. Sampling and analytical requirements are indicated.
 8. Analysis of bulking agent, amendment, or admixture if deemed necessary by the department.
 9. For mixed MSW, a description of the recyclables separation program and household hazardous waste collection program.
 10. Detailed engineering plans and specifications including:
    (a) Type, size, detention time for all processes to demonstrate that equipment capacities are sufficient
    (b) Method of measuring, shredding, mixing, and proportioning input material
    (c) Storage capacity of facilities for all materials on-site
    (d) For solid waste, description of preprocessing and postprocessing equipment
    (e) Location of all temperature and other monitoring points and frequency of monitoring
    (f) Process flow diagram, including all major equipment. The flow

streams must indicate quantity of material on a wet weight, dry weight, and volume basis.

    (g) Aeration capacity of the composting system

    (h) Air emission collection and control equipment

    (i) Method of control of surface water runoff and leachate collection

11. Pathogen and vector attraction reduction. Class A, Alternative 1 (USEPA), for pathogens is required.

12. Maximum pollutant concentration for compost is specified.

13. The product must not contain more than 2% total gross contaminants by weight.

14. The particle size of the product must not exceed 10 mm (0.39 inch) except for wood particles used as a bulking agent.

15. The product must not contain sharp objects.

16. The compost product must be produced from a composting process with a minimum detention time (composting and curing) of 50 days.

17. The department may require certain process operational conditions or product restriction beyond those stated above.

18. Labeling requirements are specified.

19. The product may be distributed for use on food, feed or fiber crops, unless the product is derived from mixed solid waste. Products derived from mixed solid waste are not allowed for use on public contact areas, food and feed crops for at least the first year of operation, until a record of product quality consistent with the regulations is established.

20. The application rate of the product must be limited to agronomic rates.

21. A permittee may not operate a facility until certification has been obtained.

22. Records on pathogen and vector reduction requirements and certification that they were met are required.

23. Separate and similar regulations are provided for yard waste.

24. Regulations are provided for distribution and marketing of products generated outside the state. This includes requests for product distribution, compliance with pathogen and vector attraction methods, compliance with contaminant limits, and compliance with certain facility and management provisions required by the state, including record keeping and monitoring.

Tennessee

In January, 1995, the state of Tennessee Department of Environment and Conservation, Division of Solid Waste, amended Chapter 1200-1-7, Solid

Waste Processing and Disposal, to include a new rule, 1200-1-7.11 Composting. This rule covers both specifications for a composting facility and sale of compost in the state. The following are some of the major aspects of the rule.

COMPOSTING FACILITY GENERAL REQUIREMENTS

(1) New and existing facilities may not compost solid waste (includes biosolids, manure, yard waste) without a permit.

(2) The facility must be located, designed, constructed, maintained, and closed in a manner that controls disease vectors; release of solid waste constituents or other harmful material to the environment; the exposure of the public to potential health and safety hazards; and odors that constitute a public nuisance.

(3) The facility shall have trained personnel.

(4) Leachate collection and removal plans must be in place.

(5) The type and source of solid waste shall be determined and categorized for review.

(6) The type and source of any additives to be used for the production of compost shall be specified.

(7) All waste receiving areas, storage areas, processing, and curing areas except yard waste shall be paved.

(8) The operator shall take dust control measures to prevent a nuisance and safety hazard to adjacent landowners and the public.

(9) Runon/runoff control measures are specified.

(10) An odor control system shall be designed and implemented.

(11) Buffer zones are specified as follows: (a) 100 feet from property lines, (b) 500 feet from all residences unless the owners agree in writing to a shorter distance, and (c) 200 feet from water courses

FACILITY DESIGN AND CONSTRUCTION

(1) A master plan must be provided, which includes boundaries, drainage, runoff monitoring stations, primary access roads, wells, 100-year floodplain, and residences within a quarter of a mile of the site boundary.

(2) Design plans include structures, waste processing areas, waste storage areas, runoff/runon drainage, location of utilities and roads, and location of leachate collection and treatment facilities.

(3) A description of the facility and operation. This includes a description of how the facility would comply with all standards defined; waste

handling and processing equipment; the leachate management protocol; the odor control measures; and the final closure plan for the facility.

COMPOST STANDARDS

(1) The rule applies to compost produced outside the state used or sold for use in the state.

(2) Three types of wastes are specified: (a) yard waste only, (b) manure or yard waste and manure, and (c) solid waste including biosolids

(3) Three types of product maturity are specified, mature, semi-mature and fresh compost. Maturity is based on heating, color and reduction of organic matter.

(4) Three particle sizes, fine, medium, and coarse, are designated.

(5) Metal concentrations are the same as the USEPA 503 EQ pollutant concentrations.

(6) Labeling is required for individually packaged material or certified in writing on all bulk sales. Compost generated from municipal waste and/or biosolids shall be labeled, noting hazards for use on vegetables intended for human consumption.

(7) Testing is required for every 20,000 tons or every three months for moisture, total N, total P, total K, reduction in organic matter, and pH. Municipal solid waste shall be tested for all metals listed in the USEPA 503 document, foreign matter, and fecal coliform.

(8) The rule also specifies reporting requirements.

California

The California Integrated Waste Management Board (CIWMB) issued composting regulations on July 31, 1995. The regulations are under Title 14, California Code of Regulations, Division 7, and are to be used in conjunction with appropriate sections of 14CCR Division 7 and Public Resources Code (PRC). The regulations use a tier system requiring different levels of permitting depending on the feedstock and quantities of materials composted. Enforcement is up to the local enforcement agency (LEA). Three types of permits are specified: registration permit, standardized permit, and full solid waste facilities permit.

A registration permit is required for green material, animal material and biosolids facilities with less than 9040 m^3 (10,000 yd^3) of feedstock and compost on-site. The standardized permit applies to facilities with more than 9040 m^3 of feedstock and compost on-site. Finally, the full solid waste facilities permit applies to mixed solid waste composting facilities. Some of the key provisions of the regulations for standardized and or solid waste facilities permit are as follows:

(1) A description of the composting processes to be used, including estimated quantities of feedstocks, additives, and amendments

(2) A descriptive statement of the operations conducted at the facility

(3) A schematic drawing of the facility showing layout and general dimensions of all processes utilized in the production of compost including, but not limited to, unloading, storage, processing, parking, and loading areas

(4) A description of the proposed methods used to control litter, odors, dust, rodents, and insects

(5) A description of the proposed emergency provisions for equipment breakdown or power failure

(6) A description of the anticipated maximum and average length of time compost will be stored at the facilities

(7) A description of compost equipment used at the facility including type, capacity, and number of units

(8) Anticipated annual operation capacity in cubic yards for the facility

(9) A description of provisions to handle unusual peak loadings

(10) A description of the proposed method for storage and final disposal of non-recoverable or non-marketable residues

(11) A description of water supplies for process water required

(12) Identification of person(s) responsible for oversight of facility operations

(13) A description of the proposed site restoration activities, in accordance with specific requirements of the regulation

The regulations also require general design requirements, general operating standards, and facility siting on landfills. Environmental health standards include sampling requirements, product heavy metal content, and pathogen reduction (same as USEPA 503 regulations). Specific contaminant levels in green material are also indicated. The regulations specify record-keeping requirements, including that records must be kept at the facility for five years.

These state regulations representing three different regions illustrate the role state regulators have taken, far beyond "protecting human health and the environment." In many cases, regulators have far exceeded legislative mandates, resulting in excessive costs to taxpayers and communities. Cost of preparation of permit applications often exceeds several hundred thousand dollars. Regulators often see their role as protecting the user and the community from poor products and poor facilities. Product quality not related to human health or the environment would probably be best controlled by industry and the market.

CANADA

The Canadian government has established guidelines for compost quality. In addition, provinces can issue regulations or guidelines. National guidelines are published under the auspices of the Environmental Choice program as stated in Canadian Environmental Protection Act Paragraph 8(1)(b). The ministry hopes to encourage producers or importers to apply for the Environmental Choice Program for verification and subsequent authority to label qualifying products with the environmental choice Ecologo label. Presumably, this enhances marketing and acceptance of products.

Several requirements are imposed on the producer that go beyond the requirement for product quality. They tend to discourage the manufacture of compost products by eliminating many feedstocks. Consequently, municipalities and industries seek other non-beneficial use processes. The requirements are:

(1) Meet or exceed all applicable governmental and industrial safety and performance standards.
(2) Meet the requirements of all governmental acts and regulations, including the Canadian Environmental Protection Act (CEPA).
(3) Adhere to the policies and targets of the national packaging protocol and adopt the code of preferred packaging.

The following are the product criteria that must be met. The product must:

(1) Be uniformly exposed to temperatures exceeding 55°C for three consecutive days
(2) Have a pH between 5.0 and 8.0
(3) Not exceed the following amounts of heavy metals in mg/kg dry weight:

Arsenic (As)	13
Cadmium (Cd)	2.6
Chromium (Cr)	210
Cobalt (Co)	26
Copper (Cu)	128
Lead (Pb)	83

Mercury (Hg)	0.83
Molybdenum (Mo)	7
Nickel (Ni)	32
Selenium (Se)	2.6
Zinc (Zn)	315

(4) Have a minimum of 30% organic matter content on a dry-weight basis

(5) Have a maximum of 50% water content

(6) Have a sodium adsorption ratio (SAR) less than 5

(7) Have a maximum particle size of 13 mm

(8) The products must not contain plastic in excess of 0.4% dry weight and any combination of glass, rubber, and/or metal in excess of 1% by weight if such particles have a dimension in excess of 2 mm

(9) Have PCBs < 1.0 ppm

(10) Have a maximum electrical conductivity (Ec) of 3 ms/cm

(11) Be derived from source-separated municipal wastes

(12) Have undergone the entire composting process (compost must be stable; i.e., no longer decaying) under aerobic conditions

(13) If packaged in a plastic bag, be labeled in accordance with the Society of the Plastics Institute of Canada; plastic bottle and container material code system. It must contain over 15% recycled plastic by weight, and a minimum 5% of the total weight must be post-consumer plastic. It must not be formulated with inks, dyes, pigments or other additives that contain lead, cadmium, mercury or hexavalent chromium and must not have a combined contaminant concentration exceeding 250 ppm.

The objective of these guidelines is to minimize the negative impact of pollution generated by the use and disposal of goods and services available to Canadians. The aim is to encourage waste reduction and produce a valuable soil conditioner so as to reduce the volume of wastes disposed in landfills. "Based on a review of currently available life cycle information, the product category requirements will produce an environmental benefit through the reduction of waste entering the landfill and recycling of organic matter and plant nutrients." However, these guidelines could well result in more waste going into landfills as they are so restrictive and difficult to achieve for many feedstocks such as biosolids, MSW, food waste, industrial wastes and even yard waste from urban areas.

Another confusing aspect of the Canadian guidelines/regulations is the regulations promulgated by Agriculture and Agrifood Canada, Food Production and Inspection Branch, which differs from the Environmental Choice Program and Canadian Council of Ministries of the Environment guideline limits. These regulations are for composted manure, municipal waste tankage, garbage tankage, leather tankage, and industrial sewage and apply to the sale of dried processed sewage and other by-products as fertilizers or supplements.

Table 14.5 shows the maximum acceptable heavy metal concentrations and the maximum acceptable cumulative heavy metal additions to soil that would be reached in 45 years. The figures are based on the assumption that a product would provide 220 kg/ha of nitrogen annually.

Provinces vary in their standards, which may differ from the environmental choice criteria. Table 14.6 provides data from four provinces (Dillon, 1994). The restrictive limits that provide for unlimited use will be hard to comply with using separated organic wastes. Details of the Alberta guidelines as an example of provincial regulations are provided below.

Alberta issued revised draft guidelines for the production and use of compost in November 28, 1994 (Alberta Environmental Protection, 1994). Three classes of product are indicated (Tables 14.7a, 14.7b and 14.7c). The pathogen requirement calls for both fecal coliform and *Salmonella* testing.

TABLE 14.5. Maximum Acceptable Heavy Metal Concentrations and Maximum Acceptable Cumulative Heavy Metal Additions to Soil (Agriculture and AgriFood Canada, Trade Memorandum, T-4-93, July 1995).

Heavy Metal	Maximum Acceptable Metal Concentrations (mg/kg)[1]	Maximum Acceptable Cumulative Heavy Metal Additions to Soil (kg/ha)
Arsenic	75	15
Cadmium	20	4
Cobalt	150	30
Lead	500	100
Mercury	5	1
Molybdenum	20	4
Nickel	180	36
Selenium	14	2.8
Zinc	1850	370

[1]Maximum acceptable metal concentrations in compost or compost based products containing 2.5% N or less (as is) and 50% moisture or less, represented for sale as fertilizers or supplements.

TABLE 14.6. Canadian Provincial Guidelines for Heavy Metals.

Heavy Metal	Concentration (mg/kg dw)			
	Nova Scotia	Ontario	Alberta Class A	British Columbia
Cadmium	2.6	3	<2	2.6
Copper	100	60	<80	100
Chromium	210	50	<100	210
Lead	150	150	<50	150
Mercury	0.83	0.15	<0.2	0.83
Nickel	50	60	<32	50
Zinc	315	500	<120	315

Source: Dillon, 1994.

TABLE 14.7a. Quality Criteria, Heavy Metals, Organic Compounds, and Pathogens, for Compost Products in Alberta, Canada.

Heavy Metal	Class A (mg/kg dry weight)	Class B (mg/kg dry weight)	Class C (mg/kg dry weight)
Arsenic	<10	10–54	>54
Boron	<2	2–10	>10
Cadmium	<1	1–21	>21
Chromium	<100	100–1200	>1200
Copper	<80	80–1500	>1500
Lead	<50	50–300	>300
Mercury	<0.2	0.2–5	>5
Molybdenum	<4	4–20	>20
Nickel	<32	32–180	>180
Selenium	<2	2–14	>14
Zinc	<120	120–1850	>1800
Organic compounds	<1.9 ppm PCB and >90% germination rate for selected species	>90% rate for selected plant species	<90% germination rate for selected species
Pathogens	<1000 fecal coliform/g total solids (ts) and <3 MPN salmonella/4 g ts	<1000 fecal coliform/g total solids (ts) and <3 MPN salmonella/4 g ts	>1000 fecal coliform/g total solids (ts) and <3 MPN salmonella/4 g ts

Source: Alberta Environmental Protection, 1995.

TABLE 14.7b. Quality Criteria, Maturity, Soluble Salts, pH, and Inerts for Compost Products in Alberta, Canada.

Parameter	Class A	Class B	Class C
Maturity	<2.0 mg CO_2-C per gram of compost C, or equivalent in oxygen uptake and >90% germination rate for selected species.	2.0 to 5.0 mg CO_2-C per gram of compost C, or equivalent in oxygen uptake and >90% germination rate for selected species.	>5.0 mg CO_2-C per gram of compost C, or equivalent in oxygen uptake and >90% germination rate for selected species.
Soluble salts	<4.0 ds/m	<4.0 ds/m	>4.0 ds/m
pH	6.0 to 8.0	6.0 to 8.0	<6.0 to >8.0
Sharps— Classes A & B	Compost should not contain any sharps measuring more than 4 mm in any diameter.		
Total allowable foreign matter	Foreign matter (excluding sharps) exceeding 4 mm in any dimension shall not compromise more than 1.5% of class A or B composts on a dry-weight basis. No more than a third of this foreign matter may be plastic.		
Maximum particle size	Class A 13 mm	Class B 13 to 20 mm	

Source: Alberta Environmental Protection, 1994.

TABLE 14.7c. Classes of Compost for Use in Alberta.

Compost Categories	Relative Quality	Allowable End Uses
Class A	High	Unlimited use as either a soil amendment or soil replacement
Class B	Medium	Restricted to use as a soil amendment based on regulatory approval
Class C	Poor	Compost is unsuitable for application to Alberta soil. Must be disposed of based on the nature and extent of contaminants.

Source: Alberta Environmental Protection, 1995.

In addition to the criteria, a sampling protocol must be provided along with the specific methodologies to be used.

Two applications are required to establish and operate composting facilities. The "Short Form" is for (a) feedstocks limited to yard waste or source-separated fruit and vegetable waste or (b) facilities processing less than 5000 tonnes per year. The "Long Form" is for (a) facilities processing more than 5000 tonnes per year or (b) facilities that use separated municipal waste, municipal sludge, or industrial by-product(s) alone, or in combination, as a feedstock. The "Long Form" requires information on the following topics:

(1) General information (e.g., ownership, land use, zoning, proposed operation, facility size, project schedule, etc.)

(2) Geographic, bedrock characteristics, soil characteristics, and hydrological characteristics

(3) Environmental review, including a discussion of the potential environmental, social, cultural and economic impacts associated with the facility

(4) Design, operations and storage, including: (a) design capacity and process flow, (b) types of organic waste to be accepted, (c) personnel plan, (d) proposed dimensions of windrows or compost piles, (e) equipment and method to be used, (f) traffic pattern, (g) water sources, and (h) public information and education program

(5) Monitoring system

(6) Decommissioning

EUROPE

Most European countries have heavy metal limits either as guidelines, regulations or set as a voluntary quality standard by a variety of associations (Kulik, 1996). Heavy metal standards differ in the various countries (Table 14.8). Germany, Netherlands, and Belgium have the most stringent guidelines or regulations. Spain has the least stringent limits. The scientific basis for the metal limits is not indicated. Germany, Netherlands, and Belgium base the levels primarily on levels found in biowaste and a wish not to increase soil background levels (no net degradation approach).

Austria

The Austrian Institute of Standards produced standards defining compost quality requirements, analytical methods, quality control, and application

TABLE 14.8. *Heavy Metal Concentrations in Various European Countries.*

Country	Concentration (mg/kg)				
	Cadmium	Copper	Chromium	Lead	Mercury
Austria	4	400	150	500	4
Belgium	5	100	150	600	5
Denmark	1.2	1000	100	80	1.2
Germany	1	75	100	100	1
Italy (proposed)	3	200	150	200	2
Netherlands	0.7	25	50	65	0.2
Spain	40	1750	750	1200	25
Switzerland	3	150	150	150	2

Source: Dillon, 1994.

guidelines. These standards apply to compost produced from separately collected domestic wastes as well as organic agricultural and industrial wastes.

The Austrian standards are comprehensive. Only one classification for mature compost is defined. The standards define mature compost based on criteria for organic substances, nutrient content, physical criteria and foreign matter content. Maturity is defined based on germination tests using compost compared to a base substrate. The required testing includes heavy metals, biological comparability with vegetation, salmonellae, and enterococci levels (Dillon, 1994).

Denmark

Compost quality in Denmark is regulated by Statutory Order Number 736 issued on October 26, 1989. The statutory order allows some materials to be used for agricultural uses without permission. This applies to raw vegetable wastes, dairy and animal raw wastes, source-separated food wastes and biosolids.

The criteria used in defining compost are the feedstock, temperature attained during composting, acetic acid concentration, physical characteristics, total N, and heavy metal concentrations. Application of compost to land is limited by plant nutrient levels, dry matter content, and heavy metal concentrations. Further restrictions apply, depending on weather conditions and proximity to water supply plants, residential buildings, and food processing plants. There are two sets of limits for As and heavy metals. One set applies to compost to be used on lands growing food and fodder crops; the second set applies to parks and ornamental plants.

Germany

Germany has no governmental regulations. However, several associations or organizations have established standards for compost. For example, the German Quality Association for Compost has developed criteria for its "RAL" quality standard. Another quality standard for compost was established under the "Blue Angel" eco-label program. This standard applies to source-separated organic wastes.

Another group using the acronym LAGA has set high quality standards to encourage compost sale and use. Besides compost quality guidelines, LAGA makes recommendation for testing and use. The objective in all cases is to provide a high quality compost that is low in heavy metals and toxic organics (Dillon, 1994).

Italy

Compost guidelines have been published under Italian Directives for the First Application of Article 4 of the Decree of the President of the Republic (DPR) No. 915/1982. This decree defines compost, establishes guidelines for its manufacture and application, establishes allowable concentrations for various contaminants, and specifies analytical methods for testing compost quality. No administrative bodies current exist to impose the legislated standards (Dillon, 1994).

A new decree has been proposed that would establish two classes of compost. The highest quality compost produced from source-separated organic wastes would be class A. Use of this compost would not require authorization. Class B compost would be of medium quality and would require authorization, monitoring of heavy metals in soil, and monitoring of large application rates (Dillon, 1994).

The proposed criteria are very comprehensive. They include particle size, foreign matter content, moisture content, pH, nutrient concentrations, heavy metal concentrations, organic C, degree of humification, C/N ratio, salinity, salmonellae, fecal coliform, fecal streptococci, toxicity at germination, and respiration index (Dillon, 1994).

Netherlands

The basis of regulations is the Soil Protection Act (SPA) of 1987, which is administered by the General Administrative Orders (GAO). The Ministries of Housing, Planning and the Environment, and the Ministry of Agriculture have developed regulations governing the trade and use of compost. The main object of the SPA is to protect soil resources and prevent soil contamination (Dillon, 1994).

Three compost classes, very clean compost, clean compost, and compost, have been stipulated and heavy metal and As criteria have been established for each. The criteria for pollutants and application rates are based on the concept that additions to soil should not exceed that removed by crops and or leaching out of the soil profile.

CRITERIA FOR COMPOST QUALITY AND FACILITY DESIGN

Although most of the regulations center on compost product criteria designed to protect human health and the environment, other user and facility design criteria are also important. User criteria are needed to assure the market of good quality compost. This promotes the use of compost and composting as an important solid waste management practice. Further, facility criteria are needed to protect the community from potential health and environmental impacts.

Compost Quality Criteria

Two categories are indicated: quality criteria for protecting human health and the environment and product quality criteria to ensure the user of a high quality product. The following are the recommended criteria:

(1) Product quality criteria to protect human health and the environment: (a) heavy metals, (b) toxic organics, (c) pathogens, and (d) inerts
(2) Product quality criteria to provide a high quality product

In addition to all the criteria in item 1, this also includes the following: (a) stability, (b) maturity, (c) soluble salts, (d) foreign matter or inerts—amount and particle size, and (e) pH.

Product quality criteria not only protect the user but also the industry. By producing a high quality product, the industry builds consumer confidence in the product and the overall process. Many communities still question the availability of markets and are reluctant to select composting as a solid waste management system. High quality standards will help develop and enlarge the market while at the same time encouraging communities to select composting instead of other technologies.

Facility Design Criteria

Important facility criteria include:

- site aspects: buffers, runoff, protection of groundwater, boundary odor standards, and drainage
- odor control and mitigation measures
- facility design to protect worker health and the environment: building ventilation criteria, dust control and mitigation measures, bioaerosol control and mitigation measures, and worker safety criteria

CONCLUSION

Two principal methods are used to set regulations concerning the use of compost: no-net degradation and risk-based.

Several European countries use the no-net degradation approach, which tends to single out the compost product while not applying to other materials used on soils. United States Environmental Protection Agency has used the risk-based approach to regulate biosolids compost. Several states are using the same criteria for compost products produced from other feedstocks.

Several states in the United States have produced very intensive and specific regulations that go beyond the protection of human health and the environment. This often results in excessive permit costs. States in the United States or countries should not issue guidelines in lieu of regulations. Standards for product quality (except if related to health) should be a function of the compost industry. Such standards should be provided to promote the use of high quality compost products.

Several states in the United States are regulating the feedstock rather than setting standards for the product regardless of the feedstock source. For example, MSW is being regulated to a greater degree than biosolids, food waste, and yard waste. Since it results in excessive costs to communities, this discourages MSW composting and promotes incineration and landfilling instead. USEPA has set compost product quality as the criterion for land application, not the quality of the incoming biosolids feedstock.

Greater consideration should be given to facility design for control of odors, bioaerosols, and other aspects related to public and worker health.

REFERENCES

Alberta Environmental Protection. 1994. *Guidelines for the Production and Use of Compost in Alberta.* Draft prepared by Alberta Environmental Protection, Alberta Agriculture, Alberta Health, Edmonton Board of Health. Revised November 28, 1994.

Epstein, E., R. L. Chaney, C. Henry, and T. J. Logan. 1992. Trace elements in municipal solid waste compost. *Biomass and Bioenergy.* 3(3–4):227–238.

Farrell, J. B. 1992. Fecal pathogen control during composting, pp. 282–300. In H. A. J. Hoitink and H. M. Keener (eds.). *Science and Engineering of Composting: Design, Environmental, Microbiological and Utilization Aspects.* Renaissance Publ., Worthington, OH.

GCG Dillon, 1994. *Compost Quality Objectives Study.* Final Report to Alberta Environmental Protection, Alberta Agriculture, Alberta Health, and Edmonton Board of Health. Prepared by GCG Dillon, Canada and E&A Environmental Consultants, Inc. Canton, MA.

Kennedy, J. 1992. A review of composting criteria. *Proc. of the Composting Council of Canada.* Second Annual Meeting, Ottawa, ON.

Kulik, A. 1996. Europe cultivates organics treatment. *World Wastes.* 39(2):37–40.

Richard, T. L. and P. B. Woodbury. 1992. The impact of separation on heavy metal contaminants in municipal solid waste composts. *Biomass and Bioenergy.* 3(3–4):191–211.

Ryan, J. A. and R. L. Chaney. 1995. Issues of risk assessment and its utility in development of soil standards: The 503 methodology as an example. *Proc. Third International Symposium on Biogeochemistry of Trace Elements.* Paris, France.

USEPA. 1992. *Technical Support Document for Reduction of Pathogens and Vector Attraction in Sewage Sludge.* EPA 822/R-93-004.

USEPA. 1994. *A Plain English Guide to the EPA Part 503 Biosolids Rule.* U.S. Environ. Protection Agency. EPA/832/R-93/003.

USEPA. 1995. *A Guide to the Biosolids Risk Assessments for the EPA Part 503 Rule.* U.S. Environmental Protection Agency, Office of Wastewater Management. EPA832-B-93-005.

Washington State Department of Ecology. 1994. *Interim Guidelines for Compost Quality.* Pub. #94-38 Solid Waste Services Program. Olympia, WA.

Yanko, W. A. 1988. *Occurrence of Pathogens in Distribution and Marketing Municipal Sludges.* Rept. No. EPA-6/1-87-014 (NTIS PB 88-154273/AS). Cincinnati Health Effects Res. Laboratory, Cincinnati, OH.

Aeration
    effect of composting, 22–32
    methods of, 22
    rate, 35–36, 45
Aerobiology, 247
Agency for Toxic Substances and
        Disease Registry, 143
Air dispersion modeling, 330–340
    examples of, 334–340
    model parameters, 330–331
    regulatory models, 330
    reliability, 331–334
American Conference of Governmental
        Industrial Hygienists
        (ACGIH), 319, 327
Ammonia, 43, 62, 110, 115, 241, 304,
        305, 307, 308, 310, 420, 439
Anaerobic decomposition, 19
Antibiosis, 420
Arsenic (As), essentiality and toxicity,
        138–139, 147–151
Aspergillus fumigatus, 83, 250,
        253–283, 288, 290, 291
    in and around composting
        facilities, 263–271
    in environment, 261–263
    morphology, 253–254
    in MSW composting facility, 275
    pathogenicity, 254–261
    in yard waste facilities, 271–275

Best-achievable approach, *see*
        Regulations, approaches to
Bioaerosols, 247–291
Bioassays, 107
Biochemistry, 77–104
    humus formation, 101–104
    manifestations during compost
        application, 98–100
    manifestations during composting,
        83–98
    organic matter, 79–83
Biofiltration, *see* Compost utilization
Biological control, 420, 430
Biosolids, 2, 13, 15
Boron (B), essentiality and toxicity,
        139–140, 147–151, 385,
        386

Cadmium (Cd), essentiality and
        toxicity, 140–141, 147–151,
        160
Canadian Environmental Protection Act
        (CEPA), 470
Carbon (C), 39–42, 74, 83, 107, 109,
        118, 353, 359
Cation exchange capacity (CEC), 116,
        117, 157, 158, 343, 344; *see
        also* Soil
Century model, 99, 100
Clean Water Act, 13, 148, 164, 449
C/N ratio, 39, 41, 43, 44, 56, 82, 85, 90,

C/N ratio *(continued)*
91, 94, 96, 111, 112,
113–115, 118, 307
Compost stability and maturity
criteria for assessment, 133
definition, 109–112
methods to assess, 107, 110,
113–133
Compost utilization, 383–412
agricultural crops, 400–408
biofiltration, 431–442
demand, 384
horticulture, 389–400
plant disease suppression, 420–431
plant pathogen destruction,
418–419
silviculture, 408–411
Composting, xiii–xv, 1–17, 77, 79, 101,
107
advantages, 16
basic processes, 19–50
definition, 1, 19, 343
disadvantages, 16
effect on soil properties, 343–378
future trends, 13–16
history, 3–13
leachate characteristics, 152–155
philosophy of, 13–17
Composting Council, 9
Copper (Cu), essentiality and toxicity,
141–142, 147–151, 160

Desorption curves, 356
Dewar flask, 122, 123
Dilution-to-threshold (D/T) ratio, 302,
303, 308
Dilution-to-threshold (D/T) value, 335,
336, 337, 338, 339
Diseases associated with biosolids,
216–218
Diseases associated with MSW, 224

Electrical conductivity, *see* Soil
Enclosed system, *see* In-vessel system
Endotoxin, 250, 280, 283–290, 291
Enzyme activity, 127–130

Fermentation, 19
Food web, 77–79
Forced-air composting, 62

Free air space (FAS), 28, 29, 30
Fulvic fractions, *see* Humic substances

Gaussian dispersion, 331, 333
Germination index, 130

Health risks, 223–225, 241, 247, 248,
249, 250, 251–252, 253,
280–283, 285, 290–291, 374
Heavy metals, 14, 15, 110, 130, 137,
149, 152, 153, 155, 160, 161,
385, 448, 450, 459, 460, 463,
472, 473, 474, 476, 477
Humic acid, 94, 95, 102, 103, 120
Humic substances, 79, 91, 94, 96, 102,
103, 104, 120, 156, 164
Humification, 101
Humification index (HI), 89, 90, 94, 96,
119, 120
Humification ratio, 120
Humus, 77, 101; *see also* Biochemistry

Immobilization, 156
Indore composting process, 4
Industrial Source Complex (ISC), 330,
334
Inoculants, effect on composting, 73–74
International Commission on
Occupational Health (ICOH),
284
In-vessel system, 37, 461

Landfill, 2, 10, 201, 471
Leaching, 152–155, 199, 200, 201, 383
Lead (Pb), essentiality and toxicity,
142–143, 147–151
Liability, 459, 462

Marketing, 16, 383, 384, 408, 462, 466
Matric potential, 356
Mercury (Hg), essentiality and toxicity,
143–144, 147–151
Mesophilic temperature, 20, 21, 36, 50,
54, 58, 59, 63, 66, 70, 71, 72,
74, 235
Microbiological activity, 124–127
Microbiology, 53–74, 437
microbial populations, 56, 68–69,
74
moisture, 70, 74

nutrients, 70
temperature, 56–70, 74
Micronutrients, 137, 138, 383
Microorganisms
  role in composting, 53
  temperature tolerance, 53–55, 58
  types isolated from compost, 57
Mineralization, 21, 45, 437; *see also*
  Nitrogen
Moisture, 110, 433–434
  effect on composting, 32–36, 385;
  see also Microbiology
Molybdenum (Mo), essentiality and
  toxicity, 144–145, 147–151
Municipal solid waste (MSW), 2, 13

National Institute for Occupational
  Health (NIOSH), 288, 290
Nickel (Ni), essentiality and toxicity,
  145, 147–151, 160
Nitrogen, 43–48, 82, 101, 115
  mineralization, 367, 368, 369, 370,
  372, 373; *see also* Soil
No net degradation, *see* Regulations,
  approaches to
Non-point source pollution, 201, 360,
  383; *see also* Soil, runoff and
  erosion
Nutrients, 116, 378, 435, 383
  effects on composting, 39–48
  effects on microbial activity, 70–73

Occupational illnesses, 247, 249
Occupational Safety and Health
  Administration (OSHA), 215
Odors, 13, 15, 22, 28, 111, 112, 123,
  301–340, 431, 432, 433, 438,
  439, 440
  around composting facilities,
  304–316
  characterizations of, 302–304
  control, 301
  generation of, 301–302
  intensity, 302, 303
  management, 301, 314
  pervasiveness, 304
  scrubbing techniques, 308
Optical density, 121
Organic compounds, 53, 171–206

effect of composting method, 181
fate during composting, 194–198
pathways to food chain, 203–206
presence in compost materials and
  feedstocks, 173–194
reactions and movement in soil,
  198–203
Organic dust, 250, 251–252; *see also*
  Endotoxin
Organic dust toxic syndrome (ODTS),
  285, 288
Organic matter, 79–83
  changes during compost
  application, 98–100
  changes during composting, 83–98
  elements of, 79–80
  rate of microbial decomposition, 82
Oxygen, effect on composting, 19,
  22–32

Parasitism, 420
Pathogens, 36, 55, 213–241, 385, 418
  categories of, 213
  destruction by composting,
  225–235
  primary pathogens in wastes and
  compost, 215–223
  survival in soils and plants,
  235–241
Peak-to-mean conversion, 334
Penetrometer, 353, 355, 356, 357
pH, 48–49, 88, 115–116, 156, 157, 158,
  162, 163, 343, 344, 429,
  434–435, 463; *see also* Soil
Phytotoxity, 22, 109, 110, 123,
  130–133, 140, 144, 145, 383,
  385, 386, 450, 456–457
Plant assays, 123–124
Pollutants, 450–462
Proximate analysis of organic matter,
  80, 83, 109, 116

Recycling, 2, 15
Regulations, 14, 15, 16, 36
  approaches to, 448–449
  Canada, 470–476
  compost standards, 468, 479–480
  Europe, 476–479
  facility design and construction,
  467–470, 479–480

Regulations *(continued)*
    General Safety Order 29 CFR
        1910.1030, 215
    General Safety Order 40 CFR 503,
        2, 13, 37, 152, 194, 213, 225,
        447, 449, 450, 459, 468
    guidelines vs. regulations, 447–448
    land application, 13, 447, 458
    pathogen requirements for
        composting alternatives, 459,
        461
    states, 462–470
    U.S. federal, 449–462
Respiration, 20, 21, 24, 125–127, 128
Respiratory quotient (RQ), 25
Respirometry, 127
Retrofitting, 301, 334, 335
Risk assessment, 450–462
Risk-based approach, *see* Regulations,
    approaches to

Salmonella, 214, 216, 221, 222, 225,
    226, 228, 229, 231, 232, 233,
    234, 235, 237, 238, 239, 241,
    418, 459, 461, 462
Selenium (Se), essentiality and toxicity,
    145, 147–151
Self-heating, 122; *see also* Dewar flask
Sewage sludge, 2
Soil, 343–379
    acidity, 157
    biomass, 79
    bulk density, 351–353
    CEC, 363–365
    electrical conductivity (EC),
        365–367
    nitrogen, 367–378
    pH, 365
    porosity, 348–351
    runoff and erosion, 152, 360, 362
    strength, 353–356
    structure, 347–351
    temperature, 360–363
    water retention, 356–360
Soil conditioner, 343, 347, 379
Soil science, 101, 107
Soil Science Society of America, 363
Sorption, 199
Source reduction, 2

Source separation, 15, 150–151, 164,
    187, 463, 471
Static pile method, 23, 24, 27, 32, 35,
    37, 45, 115, 124, 187–193,
    230, 231, 232, 264, 266–267,
    269, 271, 301, 310, 311,
    314, 317, 318, 320, 321, 340,
    461
Steam sterilization, 418
Stokes' law, 48, 372

Temperature, 121–123, 418, 419, 429,
    435–437, 461
    effect on composting, 24, 29,
        36–39; *see also* Microbiology
Thermophilic temperature, 20, 21, 36,
    48, 50, 54, 56, 59, 63, 66, 67,
    71, 73, 74, 127, 417
Threshold limit value (TLV), 284, 288,
    306, 308, 317, 319, 326,
    327
Time-temperature relationship, 36, 214,
    225, 241
Trace elements, 15, 110, 137, 138, 379
    effect of compost on uptake,
        159–163
    environmental consequences of,
        152–155
    essentiality and toxicity, 138–146
    extraction procedures, 155
    occurrence in environment,
        146–152
    reversion, 158
    soil-plant interactions, 155–163;
        see also Heavy metals

U.S. Department of Agriculture, 8
U.S. Environmental Protection Agency
    (USEPA), 2, 8, 330, 334, 447,
    449
*U.S. EPA Guideline on Air Quality
    Models,* 330, 331
U.S. Public Health Service (now
    USEAP), 400

Van Maanen composting process, 4–5
Volatile organic compounds (VOC),
    180, 193, 304, 313, 316–329,
    417, 440

Volatilization, 43, 45, 156, 198, 199, 200, 316

Walkley-Black method, 107, 119
Waste management, 2
Water Environment Federation, 2
White rot fungus (*Phanerochaete chrysosporium*), 53, 171–173, 200

Windrow method, 4, 22, 23, 24, 37, 38, 58, 60, 61, 62, 117, 124, 152–153, 181, 187–193, 229, 230, 231, 232, 264, 301, 307, 309, 340, 461

Zinc (Zn), essentiality and toxicity, 146, 147–151, 160

For Product Safety Concerns and Information please contact our EU
representative GPSR@taylorandfrancis.com
Taylor & Francis Verlag GmbH, Kaufingerstraße 24, 80331 München, Germany